COSMETIC AND TOILETRY FORMULATIONS

Second Edition

Volume 4

by

Ernest W. Flick

np **NOYES PUBLICATIONS**
Park Ridge, New Jersey, U.S.A.

Library of Congress Catalog Card Number: 89-39099
ISBN: 0-8155-1383-6 (v. 4)

Published in the United States of America by
Noyes Publications
Mill Road, Park Ridge, New Jersey 07656

Transferred to Digital Printing, 2010
Printed and bound in the United Kingdom

Library of Congress Cataloging-in-Publication Data
(Revised for vol. 4)

Flick, Ernest, W.
    Cosmetic and toiletry formulations.

    1. Cosmetics. 2. Toilet preparations.
I. Title.
TP983.F55   1989   668'.55   89-39099
ISBN 0-8155-1218-X (v. 1)
ISBN 0-8155-1306-2 (v. 2)
ISBN 0-8155-1367-4 (v. 3)
ISBN 0-8155-1383-6 (v. 4)

Preface

This book contains 959 cosmetic and toiletry formulations, based on information received from numerous industrial companies and other organizations. This is Volume 4 of the Second Edition of this work; Volume 1 was published in 1989, Volume 2 in 1992, and Volume 3 in early 1995. There are no duplications in any of these volumes.

The data represent selections from manufacturers' descriptions made at no cost to, nor influence from, the makers or distributors of these materials. Only the most recent formulas have been included. It is believed that all of the trademarked raw materials listed are currently available, which will be of interest to readers concerned with raw material discontinuances. The 1995 market for cosmetic raw materials is estimated at $2 billion.

Each formulation in the book is identified by a description of end use. The formulations include the following as available, in the manufacturer's own words: a listing of each raw material contained; the percent by weight of each raw material; suggested formulation procedure; and the formula source, which is the company or organization that supplied the formula. The book is divided into the following 12 sections:

    I. Antiperspirants and Deodorants
   II. Baby Products
  III. Bath and Shower Products
  IV. Beauty Aids
   V. Creams
  VI. Hair Care Products
 VII. Lotions
VIII. Shampoos
  IX. Shaving Products
   X. Soaps and Hand Cleaners
  XI. Sun Care Products
 XII. Miscellaneous

Each formula is indexed in the section which is most applicable. The reader seeking a formula for a specific end use should check each section which could possibly apply.

In addition to the above, there are two other sections that will be helpful to the reader:

XIII. Trade–Named Raw Materials. Each raw material is listed with a brief chemical description and the name of the raw material supplier.

XIV. Suppliers' Addresses. Addresses of suppliers of trade–named raw materials and/or formulations, some of which are not available in the usual reference books.

It should be noted that some formulations in the book are translations. The manufacturer's exact wording has been used in these cases. Occasionally different companies have listed the same raw material differently; it is hoped that the reader will be able to identify the same of similar raw materials by consulting the Trade–Named Raw Materials section.

The table of contents of the book is organized in such a way as to serve as a subject index.

My fullest appreciation is expressed to the companies and organizations which supplied the information included in this book.

September 1995   Ernest W. Flick

NOTICE

Contents and Subject Index

Section I
Antiperspirants and Deodorants

Aluminum Chlorohydrate Antiperspirant Stick

Ingredients:	% W/W
A Dow Corning 345 fluid	45.00
Arlamol ISML	10.00
Stearyl alcohol	15.00
Aluminum chlorohydrate powder	25.00
Hydrogenated castor oil	5.00

B   *Fragrance

* q.s. these ingredients.

Procedure:
Heat all ingredients till they aɪ   11 liquid (app. 72C) and
mix thoroughly. Cool to just above sᴠ   ɔint and package in
antiperspirant stick containers.
Formula CP 1085

Aluminum Chlorohydrate Antiperspirant Stick

Ingredients:	% W/W
A Dow Corning 345 fluid	45.00
Stearyl alcohol	25.00
Aluminum chlorohydrate powder	25.00
Hydrogenated castor oil	5.00

B   *Fragrance

* q.s. these ingredients.

Procedure:
Heat all ingredients till they are all liquid (app. 72C) and
mix thoroughly. Cool to just above set point and package in
antiperspirant stick containers.
Formula CP 1086

Aluminum Chlorohydrate Antiper

Ingredients:	% W/W
A Dow Corning 344 fluid	45.00
Arlamol ISML	10.00
Stearyl alcohol	15.00
Reheis micro dry	25.00
Castorwax, hydrogenated castor oil	5.00

Procedure:
Heat all ingredients till they are all liquid (app. 72C) and
mix thoroughly. Cool to just above set point and package in
antiperspirant stick containers.
Formula PC 8220

SOURCE: ICI Surfactants: Suggested Formulations

Antiperspirant

Ingredients:	%W/W
Volatile silicone, D.C. 344 fluid	45.00
Arlamol ISML	10.00
Stearyl alcohol	15.00
Aluminum chlorohydrate powder, Micro-dry	25.00
Castor wax	5.00

Procedure:
   Heat all ingredients till they are all liquid (app. 72C)
and mix thoroughly. Cool to just above set point and package
in antiperspirant stick containers.

Antiperspirant

Ingredients:	%W/W
Volatile silicone, D.C. 344 fluid	48.50
Arlamol ISML	10.00
Al Zr Tetrachlorohydrex-gly, Rezal 36 GP, super ultrafine	20.00
Stearyl alcohol	16.00
Hydrogenated Castor Oil	5.00
Fragrance, Spicy lime #4851-AN, IFF	0.50

Procedure:
   Heat all ingredients till they are all liquid (app. 72C) and
mix thoroughly. Cool to just above set point and package in
antiperspirant stick containers.

Aerosol Antiperspirant

Ingredients:		%W/W
A	Quaternium-18 hectorite gel	8.00
	Arlamol E	3.00
	Cyclomethicone, Volatile silicone 7207	3.00
	Arlacel 80	0.50
B	Aluminum chlorohydrate, Macrospherical 95	6.00
C	Hydrocarbon propellant, propellant A-46	79.50

Procedure:
   Mix (A) thoroughly with medium sheer equipment. Add (B) at
a slow steady rate and mix thoroughly. Fill, vacuum crimp and
charge aerosol units with (C).

SOURCE: ICI Surfactants: Suggested Formulations

Anti-Perspirant Cream

Ingredients:	%W/W
A Arlacel 165, acid stable g.m.s.	5.00
Cetyl alcohol	5.00
B Water, deionized	50.00
Aluminum chlorhydroxide complex, 50% solution	40.00

Procedure:
Heat (A) to 70C. Heat (B) to 72C. Add (B) to (A) with agitation. Stir until set.

Anti-Perspirant Cream

Ingredients:	%W/W
A Arlacel 165	18.00
Spermaceti	5.00
B Glycerol	5.00
Water, deionized	53.00
C Titanium dioxide	1.00
D Aluminum chlorhydroxide complex	18.00
E *Perfume	

  * q.s. these ingredients.

Procedure:
Heat (A) and (B) to 80C. and add (B) to (A) with agitation. Stir for 15 minutes and add (C) slowly, continue to agitate and cool to 40C. Add (D) and continue to agitate until mixture is homogenous. Add (E) and stir.

Oil/Water Antiperspirant Cream

Ingredients:	%W/W
A Stearic acid, triple pressed	14.00
Beeswax	2.00
Mineral oil	1.00
Arlacel 60	5.00
Tween 60	5.00
B Water, deionized	33.00
C Aluminum Chlorhydrate, 50% sol.	40.00
*Perfume	

  *q.s. these ingredients.

Procedure:
Heat (A) to 85 deg. C and (B) to 85 deg. C. Add (B) to (A) while stirring. Mix while cooling to 40 deg. C. Add (C) and continue mixing until cool.

SOURCE: ICI Surfactants: Suggested Formulations

Antiperspirant Roll-On

Ingredients:	% W/W
A Arlamol E	3.00
Arlamol ISML	4.00
Brij 721	1.50
Brij 72	2.50
B Water	49.00
C Aluminum Zirconium Tetra Chlorohydrex-GLY (50% aqueous solution)	40.00
D *Preservative *Fragrance	

* q.s. these ingredients.

Procedure:
Heat (A) to 60C and (B) to 65C. Add (B) to (A) using propeller type agitation. Add (C) at about 50C and (D) at about 45C. Add water at 35C to compensate for loss due to evaporation. Package.
Formula AD-3

Roll-On Antiperspirant

Ingredients:	% W/W
A Arlamol ISML	3.00
Arlamol E	3.00
Brij 721	1.50
Brij 72	2.50
B Water, deionized	49.00
C Aluminum chlorohydrate (50% solution), Wickenol 303	40.00
D Germaben II	1.00

* q.s. these ingredients.

Procedure:
Heat (A) to 60C and (B) to 65C. Add (B) to (A) using propeller type agitation. Add (C) at about 50C and (D) at about 45C. Add water at 35C to compensate for loss due to evaporation. Package.

SOURCE: ICI Surfactants: Suggested Formulations

Antiperspirant Stick

Ingredients:	%W/W
A Brij 721	2.30
Arlamol E	11.50
Isopropyl myristate & stearalkonium hectorite	5.70
Ozokerite wax	22.40
Myristyl alcohol	17.20
B Aluminum-Zirconium trichlorohydrate	23.00
C Cyclomethicone	17.90

*q.s. these ingredients.

Procedure:
   Heat (A) to 65 deg. C and (B) to 68 deg. C. Add (B) to (A) with agitation and immediately add (C). Stir to 50 deg. C and pour into molds.

Oil-In-Water Antiperspirant Lotion

Ingredients:	%W/W
A Cetyl alcohol	0.50
Arlamol E	3.00
Brij 72	2.00
Brij 721	1.00
B Water	58.50
*Preservative	
C Aluminum chlorhydroxide, 50% aqueous solution	35.00

*q.s. these ingredients.

Procedure:
   Heat (A) to 85 deg. C and (B) to 85 deg. C. Add (B) to (A) while stirring. Mix while cooling to 40 deg. C. Add (C) and continue mixing until cool.
   Formula AD-5

Anti-Perspirant Lotion

Ingredients:	%W/W
A Arlacel 165, acid stable g.m.s.	15.00
Water, deionized	45.00
B Alumunum chlorhydroxide complex, 50% solution	40.00

Procedure:
   Heat (A) to 70C. Heat (B) to 72C. Add (B) to (A) with rapid agitation. Cool to room temperature with stirring.
SOURCE: ICI Surfactants: Suggested Formulations

Deodorant

	Wt%
A. Imwitor 960 flakes	6.0
Softisan 645	5.0
Hostaphat KL 340 N	5.0
Cetyl alcohol	2.0
Raluben TL	0.5
Aluminum acetate	0.5
B. Karion F	5.0
Hostacerin gel 1%*	12.0
Benzyl alcohol	1.0
Water	to 100.0
C. Perfume	qs

Preparation:
   The ingredients of A. are heated to 75-80C. Those of B. are brought to the same temperature and added slowly to A. with stirring. C. is added at approx. 30C.
         *Preparation of the Hostacerin gel: Hostacerin PN 73   1.0%
                                              Water   to 100.0%

   The Hostacerin is mixed with water until homogeneous and the mixture stirred until the gel is clear.

Deodorant Stick

	Wt%
A. Miglyol 812	30.0
Dynasan 114	20.0
Dynasan 110	18.5
Imwitor 960 flakes	10.0
Beeswax	20.0
B. Perfume	qs
Active ingredient	qs

Preparation:
   The ingredients of A. are melted at 75C and at about 40C perfume is added, the mix is homogenised and cast into moulds.

SOURCE: Huls America Inc.: Formulations for Cosmetics: Formulas

Dry Antiperspirant Stick

Materials:	% by Weight
1. Cyclomethicone	39.7
2. Stearyl Alcohol	22.0
3. Arlacel 165	2.0
4. Titanium Dioxide 3328	0.2
5. Bentone Gel VS-5 Rheological Additive	10.0
6. Aluminum Chlorohydrate (Micronized)	25.0
7. Talc	1.0
8. Fragrance	0.1

Manufacturing Directions
A. Heat the Cyclomethicone to 65C.
B. With stirring, slowly add ingredient 2, maintaining the temp-
   erature at 65C thru step E.
C. When all of ingredient 2 is melted, add ingredients 3 and
   4. Mix for 15 minutes.
D. Add Bentone Gel VS-5 Rheological Additive and mix for 30
   minutes.
E. Add ingredients 6 and 7, mix for 30 minutes.
F. Allow the batch to cool to 55C, add Fragrance, mix for 5
   minutes and pour into suitable containers.
Formula TS-289

Quick Drying Roll-On Antiperspirant

Materials:	% by Weight
1. Bentone Gel VS-5 Rheological Additive	15.0
2. Cyclomethicone	54.0
3. SDA Alcohol 40	3.0
4. Isopropyl Myristate	2.0
5. Aluminum Chlorohydrate	25.0
6. Fragrance	1.0

Manufacturing Directions:
A. Combine the Bentone Gel VS-5 with ingredients 2 and 3 using
   vigorous agitation. Mix until uniform.
B. Add ingredient 4, then ingredient 5, with constant agitation.
   Mix until the powder is uniformly distributed.
C. Add ingredient 6.
Formula TS-288

SOURCE: Rheox, Inc.: Suggested Formulations

Dry Roll-On Antiperspirant

Materials:	% by Weight
1. Bentone Gel IPM rheological additive	15.0
2. SDA 40	3.0
3. Isopropyl Myristate	2.0
4. Cyclomethicone	54.0
5. Aluminum Chlorohydrate	25.0
6. Fragrance	1.0

Manufacturing Directions:
1. Combine the Bentone Gel IPM additive with ingredients 2,3 and one-half of ingredient 4 in a mixing kettle. Mix until uniform. Add the remaining portion of ingredient 4, continue mixing.
2. Add ingredient 5 with agitation, stirring until the powder is uniformly distributed.
3. Add the fragrance, mix, fill units.
Formula TS-237

Talc Spray Antiperspirant

Materials:	% by Weight
1. Bentone Gel IPM rheological additive	8.0
2. SDA 40	2.0
3. Isopropyl Myristate	1.5
4. Cyclomethicone	5.0
5. Aluminum Chlorohydrate	6.0
6. Talc	2.0
7. Fragrance	0.5
8. Propellant A-46	75.00

Manufacturing Directions:
1. Combine ingredients 1 through 4 and thoroughly mix using medium shear equipment.
2. Add ingredients 5 and 6 at a slow rate and mix in thoroughly.
3. Add fragrance and mix.
4. Fill, vacuum crimp and gas aerosol units.
Formula TS-245

SOURCE: Rheox, Inc.: Suggested Formulations

General Purpose Stick Formulation-1

Ingredients:	Wt%
A-C Polyethylene 617A	8
Acetylated Lanolin Alcohol*	16
Ozokerite (170D)**	16
Mineral Oil (70 ss)***	55
Span 60	5

Cloud Point, C: 80

General Purpose Stick Formulation-2

Ingredients:	Wt%
A-C Polyethyelene 617A	7
Acetylated Lanolin Alcohol*	14
Ozokerite (170D)**	14
Mineral Oil (70 ss)***	60
Span 60	5

Cloud Point, C: 76

Procedure:
Heat all ingredients with slow agitation 2-5C above its cloud point until solution turns clear. Around 60-70C, add perfume, aluminum chlorohydrate and preservative. Deaerate and package.

Manufacturers:
*   * Malmstrom Chemical, Acetol, or Amerchol, Acetulan
*   ** Ross Wax or equivalent
*   *** Witco Chemical, Carnation Mineral Oil, or equivalent

Stick Antiperspirant

Ingredients:	Parts, Wt.
1. A-C 400	15
2. Isopropyl Palmitate	50
3. Dow Corning Silicone 344	15
4. Stearyl Alcohol	20
5. Aluminum Chlorohydrate	15
6. Perfume	Q.S.

Procedure:
Weigh 1-4 and heat until A-C 400 has dissolved or melted. Stir until wax blend is uniform. Add 5 and stir in. Cool mixture down to 55C and add Perfume.

SOURCE: Allied Signal Inc.: Suggested Formulations

Roll-On Antiperspirant

This low viscosity emulsion is stabilized with Veegum HV which also provides moderate thickening and excellent, dry after-feel. The Aluminum/Zirconium Complex has higher antiperspirant efficacy than aluminum chlorohydrate.

Ingredient:	% by Wt.
A Veegum HV, Magnesium Aluminum Silicate	1.00
Deionized Water	29.00
B Glyceryl Stearate (and) PEG-100 Stearate (2)	8.00
C Aluminum Zirconium Tetrachlorohydrex GLY, 30% Soln.(3)	33.00
Aluminum Chloride, 32 Baume Soln.	5.50
Aluminum Chlorhydrate, 50% Soln. (4)	16.50
D Cyclomethicone (5)	7.00
E Fragrance, Dye, Preservative	q.s.

(2) Arlacel 165
(3) Rezal 36G
(4) Chlorhydrol, 50% Solution
(5) Rhodorsil 700 45V2

Preparation:
    Add Veegum HV to water at 75C and mix with maximum available shear until smooth, uniform and completely free of undispersed particles. In another container, heat B to 75C. Heat C to 55C in a third container. Add B to A with slow mixing and cool to 50C. Add C to A&B. Mix at slow speed until the temperature reaches 25C. Add D and homogenize for 5 minutes. Add E. Mix until uniform and package.

Consistency:
    Viscosity measured after 30 days at room temperature is 600-800 cps.

Suggested Packaging: Roll-On Containers

Comments:
    This prototype formula is designed to serve as a guide for the development of new products or improvement of existing ones.

SOURCE: R.T. Vanderbilt Co., Inc.: Formula No. 443

Roll-On Antiperspirant

Ingredients:	%W/W
A Arlamol E	4.00
Brij 721	0.75
Brij 72	3.25
B Water, deionized	34.55
Allantion	0.20
C Aluminum Zirconium Tetrachlorohydrex Gly,	
Wickenol 386	57.25
D *Perfume and Preservative	

*q.s. these ingredients.

Procedure:
Heat (A) to 60C and (B) to 65C. Add (B) to (A) using prop-
eller type agitation. Add (C) at about 50C and (D) at about 45C.
Add water at 35C to compensate for loss due to evaporation.
Package.

Quick Dry Roll-On Lotion

Ingredients:	%W/W
A Cyclomethicone	10.00
Brij 72	2.00
Brij 721	0.70
B Water, deionized	52.30
C Aluminum chlorhydrate, 50% aqueous solution	35.00
D *Perfume	

*q.s. these ingredients.

Procedure:
Heat (A) to 70C and (B) to 72C. Add (B) to (A) with agitation.
Cool to 35-40C with stirring and slowly add the remainder of
ingredients.

Clear Roll-On A.P. Lotion

Ingredients:	%W/W
A Ethyl alcohol	69.00
Hydroxypropyl cellulose, Klucel GF	1.00
B Arlamol E	10.00
C Aluminum chlorhydroxide-propylene glycol complex	20.00

Procedure:
Mix (A) thoroughly using high shear if necessary. Add Arlamol
E and stir until homogeneous. Add (C) and stir until clear.

SOURCE: ICI Surfactants: Suggested Formulations

Spray Powder Anti-Perspirant

Ingredients:	% by Weight
(1) Ethanol (SD 40)	2.0
(2) Bentone Gel IPM	8.0
(3) Isopropyl Myristate	1.5
(4) Cyclomethicone	5.0
(5) Aluminum Chlorohydrate	6.0
(6) Kaopolite TLC	2.0
(7) Fragrance	0.5
(8) Propellant A46	75.0

Procedure:
   Add (1) to (2) and mix for 5 minutes. Add (5) and mix for 5 minutes. At high speed, add (3), (4), (6), and (7). Mix with (8) and fill.
   Follow recommended handling practices of the supplier of each product used.
   Good industrial practices should be used when handling flammable ingredients.

SOURCE: Kaopolite, Inc.: Suggested Formulation

Deodorant Stick

	% by Weight
Stearic Acid	8.00
Ethanol	74.13
Propylene Glycol	10.00
Isopropyl Palmitate	5.00
Phospholipid PTC	0.50
Sodium Hydroxide 50%	2.37

Procedure:
   Blend ingredients together except the sodium hydroxide. Heat and agitate to 65-70C, then slowly neutralize the stearic acid with the sodium hydroxide 50% and continue mixing until clear. Cool slightly to 70C. Add fragrance and color, and package.

Properties:
   Physical Appearance: Translucent Solid

SOURCE: Mona Industries, Inc.: Suggested Formulation

Stick Formulation

	Wt%
1. A-C Copolymer 400	15
2. Isopropyl Myristate	50
3. Dow 344 Fluid	15
4. Stearyl Alcohol	20

Procedure:
   Weigh all ingredients and heat with agitation. The cloud point of this blend is 72C. Above the cloud, the polyethylene will eventually dissolve in the blend. If a higher solvating temperature is used, solvation is much faster. Care must be taken, however, not to volatilize the Dow 344 Fluid. When all ingredients are dissolved, add 15% aluminum chlorhydrate and quick cool, with agitation, to 45C. Pour into container and allow to cool further.
Ref: 5189-12-2

Stick Formulation

	Wt%
A-C 400	20
2-Ethyl Hexyl Stearate	43
Myristyl Myristate	18
Dow Corning C344	16
Cetyl Alcohol	2.5
Span 65	0.5

Stick Formulation

	Wt%
A-C 400	20
2-Ethyl Hexyl Stearate	40
Myristyl Myristate	18
Dow Corning C344	16
Glycerol Mono Stearate	3
Cetyl Alcohol	2.5
Span 65	0.5

Stick Formulation

	Wt%
R762D	15
Isopropyl Palmitate	45
Dow Corning 344	15
Stearyl Alcohol (Alfol 18)	25

Procedure:
   The ingredients are mixed and heated until they form a clear solution. Cloud point of this stick is 76C. It can be poured at 65C.

Antiperspirant Stick:
   For antiperspirant stick to this base formulation add 25% aluminum chlorhydrate (particle size 12) and mix in hot.
Perfume is added just prior to pouring.

Deodorant Stick:
   For deodorant stick to this base formulation, add deodorant 1% just prior to pouring.

SOURCE: Allied Signal Inc.: **Suggested Formulations**

Section II
Baby Products

Baby Bath

		%
A.	Marlinat 242/28	22.0
	Marlinat CM 105	20.0
	Ampholyt JB 130	12.0
	Marlamid M 1218	2.0
	Water	to 100.0
B.	Camomile Special	0.5
	Avocado Special	1.0
	Perfume	qs
	Preservative	qs
	Magnesium sulphate x 7H2O	qs

Preparation:
   The components under A. are added together in sequence,
warmed and stirred until homogeneous. The components in B.
are added to A. at approx. 30C. The pH is adjusted to 5.5-6.6.

Baby Bath

	%
Marlinat 242/28	21.0
Ampholyt JB 130	17.0
Marlinat SL 3/40	8.0
Dionil OC/K	3.0
Lime Blossom Special	0.5
Camomile Special	0.5
Perfume	qs
Colour	qs
Preservative	qs
Water	to 100.0

Preparation:
   The components are mixed together in sequence and stirred
until homogeneous. The pH is adjusted to 5.5-6.5.

SOURCE: Huls America Inc.: Formulations for Cosmetics:
        Suggested Formulations

Baby Lotion Mousse

Phase A:	%(W/W)
Incroquat Behenyl TMS	1.17
Polawax	1.00
Crodacol S-95	0.33
Crodamol PMP	1.00
Crodamol PTC	0.67
Carnation Mineral Oil	1.00
Protopet 1S Petrolatum	2.00
Dimethicone 35 cs (200fl)	0.50

Phase B:	
Water, deionized	83.33
Glycerine	8.00
Germaben II	1.00

Procedure:
Combine "A" and heat to 75C. Combine "B" and heat to 75C.
Slowly add "B" to "A" with mixing. Cool. Fill into aerosol
containers and add propellant (pH 5.7).
Fill ratio: Concentrate 95%, Propellant A31 5%.

Baby Oil

Ingredients:	%W/W
1. Carnation, Light Mineral Oil NF	QS100.00
2. Ritalan	5.00
3. Ritacetyl	2.50
4. Propylparaben	0.10
5. Perfume	QS

Manufacturing Procedure:
Weigh and add all the ingredients into a container and begin
stirring while warming, until all ingredients which are solids at
ambient temperatures have melted (approximately 55C). Stir until
completely homogeneous and begin cooling. Stir continuously,
allow to remain undisturbed for 24 hours, and filter.

Baby Oil

Ingredients:	%W/W
1. Carnation, Light Mineral Oil NF	QS100.00
2. Ritalan "C"	1.00
3. Perfume	QS
4. Propylparaben	0.10

Manufacturing Procedure:
Weigh and add #1 into a container and begin stirring. Add #2,
3 and 4 and mix until a homogeneous dispersion occurs. Filter,
if necessary, and package fill into suitable containers.
**SOURCE: Witco Corp.: Petroleum Specialties Group: Suggested
Formulations**

Baby Shampoo

This baby shampoo features Crovol A-70 and Crosultaine C-50, natural derived surfactants which give this formula its mildness. Using a fragrance with a "baby-fresh" smell adds to its appeal.

Ingredients:	%
Part A:	
SLES (3 mole)	20.0
Crosultaine C-50 (Cocamidopropyl Hydroxysultaine)	12.0
Deionized Water	51.3
Part B:	
Crovol A-70 (PEG-60 Almond Glycerides)	15.0
Crothix* (PEG-150 Pentaerythrityl Tetrastearate)	0.5
Part C:	
Propylene Glycol (and) Diazolidinyl Urea (and) Methyl	
Paraben (and) Propyl Paraben**	1.0
Florasynth AB5697 (Powder Type)	0.2

Procedure:
   Combine the ingredients of Part A with mixing. Combine the ingredients of Part B with mixing and heat to 65C. Add Part B to Part A with mixing. Add the ingredients of Part C individually with mixing. Adjust pH if needed with a 10% HCl aqueous solution. Cool with mixing to desired fill temperature.
   pH: 7.0+-0.5   Viscosity: 5,000cps+-10% (@25C)
   **Germaben II   N.A.T.C. Approved
   *The use of Crothix in cosmetic and other formulations is
     covered under U.S. Patent #5,192,462.

SOURCE: Croda Inc.: Formulation SH-86

Baby Shampoo

	Weight,%
Mackadet BSC (Baby Shampoo Concentrate)	20.0
Mackstat DM (DMDM Hydantoin)	q.s.
Citric Acid to pH = 6.5-7.0	
Sodium Chloride qs to viscosity = 2000 cps	
Water, Dye, Fragrance qs to	100.0

Procedure:
1. Add Mackstat BSC to water and heat to 40 degrees C.
2. Add Mackstat DM.
3. Adjust pH with citric acid and viscosity with sodium chloride.
4. Add dye, fragrance and cool to room temperature.

SOURCE: McIntyre Group Ltd.: Suggested Formulation

Baby Shampoo

	Wt%
A. Marlinat CM 105	55.0
Marlinat 243/28	14.0
Ampholyt JA 140	7.0
Glucamate DOE 120	qs
Water	to 100.0
B. Lime-blossom Special	0.3
Camomile Special	0.2
Perfume	qs
Preservative	qs

Preparation:
    The constituents of A. are brought together in sequence and stirred while warm until homogeneous. The ingredients of B. are added to A. at approx. 30C. The pH is adjusted to 5.5-6.6.

Baby Shampoo

	Wt%
Marlinat 242/28	29.0
Ampholyt JA 140	7.5
Ampholyt JB 130	10.0
Softigen 767	3.0
Lamepon S	5.0
Camomile Special	1.0
Lime-blossom Special	1.0
Preservative	qs
Antil 141 liquid	qs
Water	to 100.0

Preparation:
    The components are mixed together in sequence and stirred until homogeneous. The pH is adjusted to 5.5-6.5.

SOURCE: Huls America Inc.: Formulations for Cosmetics: Formulas

Baby Shampoo

Ingredients: %W/W
A   Tween 20   6.00
    Cocoamphodiacetate   6.00
    Sodium lauryl sulfate   3.00
    Sodium laureth sulfate   3.00
    PEG-150 distearate   4.00
    Propylene glycol   3.00
    Water   75.00

B   *Citric acid

    *q.s. these ingredients.

Procedure:
    Mix (A) with gentle stirring and heat until homogeneous.
Adjust pH to 5.0 to 5.5 with (B).

Baby Shampoo

Ingredients: %W/W
A   G-4280   20.00
    Sodium trideceth sulfate   12.00
    Lauramphodiacetate   5.00
    Cocamidopropyl hydroxysultaine   2.50
    Sodium laureth-13 carboxylate   2.00
    Water   53.50
    *Preservative

B   PEG-150 distearate   5.00

C   *Citric acid

    *q.s. these ingredients.

Procedure:
    Mix (A) with gentle stirring and heat until homogeneous.
Heat to around 60C and add (B) and continuing stirring. When
clear, cool and adjust pH to 6.8 with (C).

SOURCE: ICI Surfactants: Suggested Formulations

Baby Shampoos

B 42/6:
Zetesol MS	25.0%
Amphotensid GB 2009	9.0%
Purton SFD	2.0%
Water, perfume, sodium chloride, preservative	q.s. to make 100.0%

B 42/7:
Zetesol MS	27.0%
Zetesol 2056	7.0%
Purton CFD	2.0%
Perfume	0.3%
Sodium chloride	2.0%
Water, preservative	q.s. to make 100.0%

Baby Foam Baths

B 70/139:
Zetesol MS	43.0%
Purton SFD	1.0%
Water, perfume, sodium chloride, preservative	q.s. to make 100.0%

B 70/129:
Extrakt 52	24.0%
Purton SFD	1.0%
Herbaliquid Kamille Spezial	5.0%
Water, perfume, preservative	q.s. to make 100.0%

Adjustment to approx. pH 7 by lactic acid or citric acid

B 70/101:
Amphotensid 9 M	40.0%
Setacin 103 Spezial	20.0%
Purton CFD	2.0%
Mulsifan RT 203/80	3.0%
Oxypon 2145	1.0%
Water, perfume, preservative	q.s. to make 100.0%

Adjustment to pH 6.5-7.0 by lactic or citric acid

SOURCE: Zschimmmer & Schwarz GmbH & Co.: Suggested Formulations

Tear Free Baby Bath

	Weight,%
Mackam 2C (Cocoamphodiacetate)	35.0
Mackol 70NS (Sodium Laureth Sulfate)	5.5
Mackam 35HP (Cocamidopropyl Betaine)	6.0
Mackanate DC-30 (Disodium Dimethicone Copolyol Sulfosuccinate)	4.0
Mackam CET (Cetyl Betaine)	1.5
Paragon (DMDM Hydantoin (and) Methyl Paraben)	0.7
Citric Acid q.s. to pH	5.0-5.5
Water q.s. to	100.0

Procedure:
1. Blend components and heat to 50C.
2. When product is clear adjust pH to 5.0-5.5 with citric acid.
   pH: 5.3
   Viscosity (cps, 25C): 500 cps
   Formula BN-127C

Baby Wipes

	Weight,%
Propylene Glycol	4.0
Mackam 2C (Cocoamphodiacetate)	2.0
Paragon (DMDM Hydantoin (and) Methyl Paraben)	q.s.
Citric Acid q.s. to	pH=6.0
Fragrance	q.s.
Water q.s. to	100.0

Procedure:
1. Blend components until clear.
2. Adjust pH with citric acid.
   This solution is combined with baby wipe tissues. It is very mild to skin and eyes.

Mild Children's Bubble Bath

	Weight,%
Mackanate EL (Disodium Laureth Sulfosuccinate)	10.0
Mackanate CP (Disodium Cocamido MIPA Sulfosuccinate)	10.0
Sodium Laureth Sulfate (30%)	9.0
Natrosol 250HHR	1.0
Mackstat DM (DMDM Hydantoin)	qs
Water, Fragrance, Dye qs to	100.0

Procedure:
1. Disperse Natrosol 250HHR in cold water.
2. Blend until completely dispesrsed.
3. Heat to 40 degrees C. and add remaining components.
4. Blend until clear.
5. Cool and fill.

SOURCE: McIntyre Group Ltd.: Personal Care Formulary

Section III
Bath and Shower Products

Aerosol Bath Oil

Ingredients:	%W/W
A Oleyl alcohol	15.00
Atlas G-1795	1.00
Alcohol, SDA No. 40	52.00
Perfume	2.00
B Propellant 12/114 (57/43)	30.00

Procedure:
Heat (A) and (B) to 90C. Add (B) to (A) with gentle stirring. Cool to 70C and add (C). Stir until uniform and pour while still fluid.

Dispersible Bath Oil

Ingredients:	%W/W
Mineral oil, Blandol	60.00
Active #4	20.00
Brij 93	13.00
Foam stabilizer, Ninol AA-62 Extra	2.00
Perfume	5.00

Procedure:
Mix all ingredients. Use slight heat if necessary and stir. Filter if necessary.

Bath Gel

Ingredients:	%W/W
G-9600	35.00
NaCl	0.80
*Color	
Fragrance	0.20
Preservative	0.10
Water	63.90

*q.s. these ingredients.

Procedure:
Mix well at room temperature.

SOURCE: ICI Surfactants: Suggested Formulations

Bath Gelee with Natural Lipid Protein

	Weight,%
Sodium Laureth Sulfate (60%)	20.0
Mackamide CS (Cocamide DEA)	20.0
Mackanate CP (Disodium Cocamido MIPA Sulfosuccinate)	20.0
Mackpro NLP (Quaternium-79 Hydrolyzed Animal Protein)	
(Natural Lipid Protein)	4.0
Mackstat DM (DMDM Hydantoin)	qs
Water, Dye, Fragrance qs to	100.0

Procedure:
1. Add Mackamide CS to sodium laureth sulfate and blend.
2. Add remaining components and heat to 45 degrees C.
3. Blend until homogenous and adjust pH to 6.5-7.0 with
   citric acid.
4. Cool and fill.

Bath Gelee

	Weight,%
Sodium Laureth Sulfate (60%)	34.6
Mackamide C (Cocamide DEA)	20.0
Mackanate EL (Disodium Laureth Sulfosuccinate)	45.0
Lactic Acid to pH = 6.0-6.5	
Mackstat DM (DMDM Hydantoin)	qs
Dye, Fragrance, qs to	100.0

Procedure:
1. Add components in order and heat to 45 degrees C.
2. Blend until homogeneous.
3. Adjust pH with lactic acid.
4. Add fragrance and cool to room temperature.

Emollient Bath Gelee

	Weight,%
Sodium Laureth Sulfate (60%)	20.0
Mackamide LLM (Lauramide DEA)	20.0
Mackanate EL (Disodium Laureth Sulfosuccinate)	20.0
Mackanate WGD (Disodium Wheatgermamido PEG-2	
Sulfosuccinate)	10.0
Mackstat DM (DMDM Hydantoin)	qs
Water, Dye, Fragrance qs to	100.0

Procedure:
1. Add Mackamide LLM to sodium laureth sulfate.
2. Add the remaining components and heat to 45 degrees C.
3. Blend until homogeneous.
4. Adjust pH to 6.5-7.0 with citric acid.
5. Cool and fill.

SOURCE: McIntyre Group Ltd.: Personal Care Formulary

Bath Milk

	%
A. Imwitor 960 flakes	5.0
Miglyol 812	15.0
Miglyol 840	10.0
Marlophor T10 Na Salt	8.0
B. Glycerol	3.0
Water	to 100.0
C. Extrapon Hamamelis Spec.	1.0
Perfume	qs
Preservative	qs

Preparation:
    The ingredients of A. are mixed and heated to 75-80C. B. is
heated to the same temperature and emulsified in A. C. is
added at 30C.

Washing Lotion

	%
Marlinat 242/28	22.0
Marlinat CM 105	6.0
Dionil OC/K	3.0
Marlamid PG 20	3.0
Perfume	qs
Colour	qs
Preservative	qs
NaCl	qs
Water	to 100.0

Preparation:
    The components are mixed together in sequence and stirred
until homogeneous. The pH is adjusted to 5.5-6.5.

SOURCE: Huls America Inc.: Formulations for Cosmetics:
        Suggested Formulations

Bath Oil, De Luxe with Plant Oils

	%
A. Miglyol 812	15.0
Miglyol 840	15.0
Softigen 767	10.0
Softisan 645	10.0
Soya bean oil	37.0
Hostaphat KL 340 N	13.0
Colour	qs
Antioxidants	qs
Perfume	qs

Viscosity: very low-viscous

Preparation:
All the components are mixed together and stirred at room temperature until homogeneous.

Foam Bath with Almond Oil

	%
A. Marlowet OA 4/1	30.0
Isopropyl myristate	22.0
Almond oil	8.0
B. Marlinat 242/70	25.0
Glycerol DAB 9	10.0
Marlamid DF 1218	3.0
Water	to 100.0
C. Perfume	qs
Preservative	qs

Preparation:
Heat A. and B. to 70-75C. Add B. slowly to A. with stirring. Add perfume and preservative at 30C.

Foam Bath Oil

	%
A. Marlowet R 11/K	30.0
Pine-needle oil	30.0
B. Marlinat 242/70	11.0
Lipoxol 600 MED	13.0
Water	to 100.0
C. Preservative	qs

Preparation:
Heat A. and B. to 70-75C. Add B. slowly to A. with stirring. Add perfume and preservative at 30C.

SOURCE: Huls America Inc.: Formulations for Cosmetics: Suggeted Formulations

Bubble Bath & Bath Oil

Ingredients:	%W/W
A Fragrance 2991 H, IFF Inc.	4.40
Tween 20	22.20
Foaming agent, Lanthanol LAL	10.00
B Water, deionized	62.20
C Foam stabilizer, Ninol AA-62 Extra	1.00
Preservative	0.20

Procedure:
   Add (B) to (A) with good agitation. Add (C) with stirring with slight heat to facilitate solution when necessary. Adjust the final product to a pH of less than 7.0 with an acid such as citric. Filter if necessary.

Bubble Bath

Ingredients:	%W/W
G-9600	20.00
NaCl	0.90
Sulfo succinate	7.00
Fragrance	0.20
Preservative	0.10
*Color	
Water	71.80

   *q.s. these ingredients.

Procedure:
   Mix well at room temperature.

Bubble Bath

Ingredients:	%W/W
A Triethanolamine lauryl sulfate, Maprpfox TLS-500	20.00
Coconut fatty acid diethanolamide, Super Amide L-9	7.50
B Perfume	5.00
C Water, deionized	57.50
Propylene glycol	5.00

Procedure:
   Warm (A) and mix until clear. Cool and add perfume, stirring until dissolved. Add (C) and stir.

SOURCE: ICI Surfactants: Suggested Formulations

Clear Gel

This clear gel is a classic formula that is based on Volpo and Crodafos surfactants. It exhibits the "ringing gel" phenomenon, characteristic of meny clear microemulsions. This effect is achieved by a system of water, oil and surfactant formulated in a ratio that forms a rigid micelle structure. The Volpos and Crodafoses are used as co-emulsifiers. (For laboratory preparation, a minimum batch size of 400 grams is recommended because they cool slower, allowing more time before the set point temperature is reached).

Ingredients: %

Part A:

Deionized water	54.0
Propylene Glycol	12.0

Part B:

Crodafos N3 Neutral (DEA Oleth-3 Phosphate)	2.0
Crodafos N10 Neutral (DEA Oleth-10 Phosphate)	4.0
Volpo 5 (Oleth-5)	4.0
Volpo 3 (Oleth-3)	7.0
Mineral Oil (70ssu)	17.0

Procedure:

Combine ingredients of Part A with mixing and heat to 90-95C. Combine ingredients of Part B with mixing and heat to 90-95C. Add Part A to Part B with mixing and cool to desired fill temperature (above set point). Set point approximately 85C.

N.A.T.C. Approved

Formula CG-8R

High Foaming Shower Gel

This formula uses an optimized combination of Crosultaine C-50, Incrodet TD-7C and SLES (3 mole) to produce a clear, high foaming, low color, and low odor bath and shower gel. Crothix is used to provide the body and viscosity seen in the formula, and Crovol A-70 is used to solubilize the fragrance and maintain the clarity of the product.

Ingredients: %

Part A:

SLES (3 mole)	20.0
Incrodet TD-7C (Trideceth-7 Carboxylic Acid)	7.0
Crosultaine C-50 (Cocamidopropyl Hydroxysultaine)	20.0
Disodium EDTA	0.1
Deionized water	48.3
Germaben II	1.0

Part B:

Crothix (PEG-150 Pentaerythityl Tetrastearate)	1.0
Crovol A-70 (PEG-60 Almond Glycerides)	2.0
BHT	0.1

Part C:

Perfume (BBA 860753)	0.5

Procedure:

Combine ingredients of Part A with mixing. Combine ingredients of Part B with mixing and heat to 65-70C. Continue mixing and cool to 50C. Add Part C to Part B with mixing. When clear, add Part B to Part A with mixing. Adjust pH with 10% NaOH solution.

Formula BP-41

SOURCE: Croda Inc.: Suggested Formulations

Foam Baths, Clear

B 70/136:
Zetesol NL	46.0%
Purton SFD	2.0%
Water, perfume, sodium chloride, preservative	q.s to make 100.0%

B 70/123:
Zetesol 2056	27.0%
Purton CFD	2.0%
Water, perfume, sodium chloride, preservative	q.s. to make 100.0%

B 70/117:
Zetesol 856T	35.0%
Setacin 103 Spezial	5.0%
Oxypon 2145	2.0%
Water, perfume, sodium chloride, preservative	q.s. to make 100.0%

B 70/148:
Zetesol 2056	40.0%
Oxypon 288	2.0%
Perfume	1.5%
Water, preservative	q.s. to make 100.0%

B 70/161:
Zetesol 856 T	35.0%
Oxypon 328	4.5%
Jojoba oil	0.5%
Purton CFD	2.0%
Perfume	1.0%
Water, sodium chloride, preservative	q.s. to make 100.0%

Foam Baths, Pearlescent

B 71/16:
Zetesol 2056	27.0%
Perlglanzmittel GM 4175	3.0%
Perfume	0.8%
Sodium chloride	2.5%
Water, preservative	q.s. to make 100.0%

B71/4:
Zetesol 856T	37.0%
Setacin 103 Spezial	6.0%
Perglanzmittel GM 4055	7.5%
Oxypon 2145	3.0%
Water, perfume, preservative	q.s. to make 100.0%

SOURCE: Zschimmer & Schwarz GmbH & Co.: Suggested Formulations

Herbal Foam Bath

		%
A.	Marlon PS 65	5.0
	Marlinat 242/28	35.0
	Dionil OC/K	3.0
	Ampholyt JB 130	10.0
	Marlinat CM 105	9.0
	Water	to 100.0
B.	Perfume	qs
	Extrapon 3-Special	2.0
	Preservative	qs
	Colour	qs
	NaCl	qs

Preparation:

The components under A. are mixed together in sequence and stirred while warmed until homogeneous. The components under B. are added to those of A. at about 30C. The pH is adjusted to 5.5-6.6.

Foam Bath, Transparent

	%
Marlinat 242/28	35.0
Marlinat SL 3/40	10.0
Marlamid DF 1218	2.0
Magnesium sulphate x 7H2O	0.5
Perfume	qs
Colour	qs
Preservative	qs
NaCl	qs
Water	to 100.0

Preparation:

The components are mixed together in sequence and stirred until homogeneous. The pH is adjusted to 5.5-6.5.

SOURCE: Huls America Inc.: Formulations for Cosmetics: Suggested Formulations

Mild Shower Gel

	%
Marlinat 242/28	22.0
Marlinat DFN 30	6.0
Ampholyt JB 130	16.0
Lamepon S	5.0
Camomile Special	2.0
Perfume	qs
Colour	qs
Preservative	qs
Antil 141 liquid	qs
Water	to 100.0

Preparation:
    The components are mixed together in sequence and stirred until homogeneous. The pH is adjusted to 5.5-6.5.

Shower Bath Gel with Pearl Lustre

	%
Marlinat 242/28	43.0
Marlinat CM 105	10.0
Marlinat SL 3/40	5.0
Dionil OC/K	3.0
Marlamid PG20	3.0
Panthenol	0.2
Perfume	qs
Colour	qs
Preservative	qs
NaCl	qs
Water	to 100.0

Procedure:
    The components are mixed together in sequence and stirred until homogeneous. The pH is adjusted to 5.5-6.5.

Clear, Mild to the Skin Shower Bath

	%
Ampholyt JB 130	18.0
Marlinat 242/28	14.0
Marlinat SL 3/40	11.0
Dionil OC/K	1.5
Perfume	qs
Colour	qs
Preservative	qs
NaCl	qs
Water	to 100.0

Preparation:
    The components are mixed together in sequence and stirred until homogeneous. The pH is adjusted to 5.6-6.5.

SOURCE: Huls America Inc.: Formulations for Cosmetics:
        Suggested Formulations

Refreshing Shower Gel

	%
Marlinat CM 105	32.0
Marlinat DFK 30	17.0
Ampholyt JB 130	11.5
Dionil OC/K	2.0
Rosemary Special	0.5
Stinging Nettle Special	0.5
Hamamelis Special	0.5
Marlowet R40/K	1.0
Menthol	0.5
Perfume	qs
Colour	qs
Preservative	qs
Antil 141 liquid	qs
Water	to 100.0

Preparation:
   The components are mixed together in sequence and stirred until homogeneous. The pH is adjusted to 5.5-6.5.

Shower Gel

	%
Marlinat 242/28	28.0
Marlinat SL 3/40	15.0
Ampholyt JA 140	5.0
Dionil OC/K	3.0
Perfume	qs
Colour	qs
Preservative	qs
NaCl	qs
Water	to 100.0

Preparation:
   The components are mixed together in sequence and stirred until homogeneous. The pH is adjusted to 5.5-6.5.

SOURCE: Huls America Inc.: Formulations for Cosmetics:
        Suggested Formulations

Shower Bath for Everyday Use

	%
Marlinat 242/28	35.0
Marlinat CM 105	16.0
Ampholyt JB 130	8.0
Dionil OC/K	2.0
Marlamid PG 20	1.0
Panthenol	0.1
Vitamin F	0.1
Perfume	qs
Colour	qs
Preservative	qs
NaCl	qs
Water	to 100.0

Preparation:
   The components are mixed together in sequence and stirred until homogeneous. The pH is adjusted to 5.5-6.5.

Mild Foam Bath

	%
Marlinat 242/28	30.0
Marlinat CM 105	18.0
Dionil OC/K	3.0
Ivy Special	0.3
Rosemary Special	0.1
Perfume	qs
Colour	qs
Preservative	qs
NaCl	qs
Water	to 100.0

Preparation:
   The components are mixed together in sequence and stirred until homogeneous. The pH is adjusted to 5.5-6.5.

SOURCE: Huls America Inc.: Formulations for Cosmetics:
        Suggested Formulations

Shower Baths, Clear

```
B 75/26:
Zetesol 2056                                          25.0%
Setacin 103 Spezial                                    9.0%
Purton CFD                                             1.0%
Perfume                                                1.0%
Sodium chloride                                        2.5%
Water, preservative                      q.s. to make 100.0%


B 75/5:
Zetesol 856 T                                         21.5%
Amphotensid GB 2009                                    8.0%
Purton SFD                                             2.0%
Oxypon 2145                                            1.0%
Perfume                                                1.0%
Water, preservative, sodium chloride     q.s. to make 100.0%
```

Shower Baths, Pearlescent

```
B 75/6:
Extrakt 52                                            42.0%
Perlglanzmittel GM 4055                                5.0%
Purton SFD                                             2.0%
Water, perfume, preservative             q.s. to make 100.0%


B 75/27:
Zetesol 2056                                          27.0%
Perlglanzmittel GM 4175                                4.0%
Purton CFD                                             2.0%
Oxypon 2145                                            1.0%
Perfume                                                1.0%
Sodium chloride, water, preservative     q.s. to make 100.0%


B 75/56:
Zetesol 856 T                                         22.0%
Perlglanzmittel GM 4055                                5.0%
Oxetal VD 20                                           2.0%
Oxypon 2145                                            1.0%
Perfume                                                1.0%
Water, preservative                      q.s. to make 100.0%
```

SOURCE: Zschimmer & Schwarz GmbH & Co.: Suggested Formulations

Shower Gel Cleanser

A gentle foaming gel cleanser designed to thoroughly deep cleanse and refresh the skin without disturbing the skin's natural moisture balance.

Ingredient/Trade Name:	% by Weight
Part A:	
Deionized Water	50.150
Polyquaternium-10	0.200
Citric Acid	0.100
Tetrasodium EDTA	0.100
Methylparaben	0.150
Sodium PCA/Ajidew N-50	0.500
Part B:	
Sodium Laureth Sulfate	30.000
TEA-Cocoyl Glutamate/Amisoft CT-12	10.000
Cocamidopropyl Betaine	5.000
PEG-150 Distearate	0.700
Lauramide DEA	2.000
Part C:	
Fragrance/#IY-67	0.250
Methylchloroisothiazolinone and Methylisothiazolinone	0.050
Part D:	
Sodium Chloride	0.800

Procedure:
    Disperse Polymer JR-125 in deionized water. Heat to 70C. Add remaining part A ingredients. Mix until uniform. Add part B ingredients in order. Mix at 70C until completely homogeneous. Cool to 40C. Add part C. Mix well. Add part D as needed to increase viscosity. Continue mixing and cooling to 35C.
Appearance: Clear liquid/pH: 5.20-5.70/Viscosity: 4,000-6,000 cps

Bath Crystals

A gentle milky bath crystal formulation that will relax your senses with the aroma of the hot baths of Japan. It will leave your skin clean and smooth while taking your mind to far and exotic places.

Ingredients:	% by Weight
Sodium Chloride	86.00
Cocoyl Glutamate/Amisoft CA	7.00
Fragrance/Takasago	1.50
Cabosil M-5	0.50

Blend ingredients 1,2,3. Premix ingredients 4 and 5. Add premix to pre-blend and mix until homogeneous and uniform.

SOURCE: Ajinomoto USA, Inc.: Suggested Formulations

Silky Bath Gel

Rhodigel is used as a thickener and foam stabilizer for this silky bath gel formula. Vanseal NALS-30 provides foam enhancement and skin conditioning properties. The cocoamidopropyl betaine and sodium laureth sulfate surfactants provide good cleaning and rinsing properties for this formula. The glycerin and PPG-3 myristyl ether are emollients which contribute to the elegant after feel.

Ingredient:	% by Wt.
A Rhodigel	0.50
Deionized Water	55.00
Glycerin	1.00
B Vanseal NALS-30	8.00
Cocoamidopropyl betaine (2)	16.00
Sodium laureth sulfate (3)	16.00
C PPG-3 myristyl ether (4)	3.50
Preservative, color, fragrance	q.s.

(2) Lexaine C
(3) Sipon ES-2
(4) Promyristyl PM-3

Preparation:
Add Rhodigel to the water slowly, while agitating at high shear. Rhodigel will become uniformly distributed throughout the water and thickening will begin. At this time, adjust to gentle mixing. Gentle, uniform mixing will solubilize the Rhodigel in time and will also minimize air entrapment. Add glycerin and mix until uniform. Add A to 50C and hold this temperature while adding remaining ingredients. Add B ingredients in order using gentle mixing to avoid air entrapment. Addition of C will thicken formula to a gel consistency.

Consistency: Flowable gel. (Viscosity: 8000-10,000 cps)

Suggested Packaging: Squeeze bottle or tube.

Comments:
This prototype formula is designed to serve as a guide for the development of new products or improvement of existing ones.

SOURCE: R.T. Vanderbilt Co., Inc.: Formula No. 429

Water White Shower Gel

Sandopan DTC acid, used as the secondary surfactant, contributes to the mildness of the surfactant system. Velsan P8-3 is an emollient ester that adds a velvety feel to the skin. This crystal clear, water white shower gel has good viscosity and foaming characteristics.

Ingredients:	%W/W
Standapol ES-3	41.25
Sandopan DTC Acid	10.00
Lauramide DEA	3.30
Velsan P8-3	2.00
Glucamate DOE-120	1.70
Germaben II	1.00
Versene NA	0.10
Fragrance	0.05
Deionized Water	qs

Procedure:
Phase A: Hydrate Glucamate DOE-120 and Versene NA in all of the deionized water.
Phase B: Add remaining ingredients in order. Combine Phase A&B, adjust pH to 6.0.
Properties:
Viscosity: 8500 cps   % Solids: 29.0%   pH: 6.0
Formulation CHS-51

Shower Gel

Sandobet SC and Sandopan DTC acid contribute to the mildness of the surfactant system and add to the foaming properties of the shower gel. Velsan P8-3 is an emollient ester that adds a velvety feel to the skin. This crystal clear shower gel has good viscosity and foaming characteristics.

Ingredients:	%W/W
Standapol ES-3	46.40
Sandobet SC	10.00
Sandopan DTC Acid	5.60
Lauramide DEA	3.30
Velsan P8-3	2.00
Glucamate DOE-120	1.70
Germaben II	1.00
Versene NA	0.10
Fragrance	0.05
Deionized Water	qs

Procedure:
Phase A: Hydrate Glucamate DOE-120 and Versene NA in all of the deionized water.
Phase B: Add remaining ingredients in order. Combine Phase A&B, adjust pH to 6.0
Formulation CHS-52

SOURCE: Sandoz Chemicals Corp.: Suggested Formulations

Section IV
Beauty Aids

Aloe Vera Skin Freshner

Ingredients:	Percent
A. Deionized water	49.50
Stearic acid triple pressed	18.00
Glycerol	5.00
B. Potassium hydroxide	1.20
C. Deionized water	5.00
Veragel 200 powder	0.25
D. Witch Hazel Extract	25.80

Procedure:
   Heat Part A to 70C. with stirring. Slowly add Part B to
Part A. Cool to 40C. with stirring. Combine Part C and add to
Part A and Part B. Add Part D with stirring, cool to 25C.
Package at room temperature.

Powder Mask Activator

Ingredients:	Percent
Deionized Water	95.3
Croquat M (Cocodimonium Hydrolyzed Animal Protein)	0.5
Incronam AL-30 (Almondamidopropyl Betaine)	0.5
Honey	1.2
Aloe Veragel Liquid	1.0
Lipofruit Cucumber	0.5
Germaben II	1.0

Procedure:
   Dissolve Germaben II in water. When uniform, add Incronam
AL-30. Next dissolve aloe, honey and cucumber extract. Mix well.
Add Croquat M. Adjust pH to pH 4.2+-0.1 with 10% TEA. Immediately
prior to application mix the two components to form a thick
paste. Recommended use 1:1 ratio.

SOURCE: Dr. Madis Laboratories Inc.: Suggested Formulations

Blemish/Scar Concealer

Protopet 1S Petrolatum	9.80%
Carnation Mineral Oil	16.00
Octyldodecanol	6.00
Ozokerite	2.50
Phenyldimethicone	1.60
Hydrogenated castor oil	2.00
Synthetic resin	0.50
Propylparaben	0.20
Antioxidant	0.10
Talc	8.00
Clay	8.00
Silica	0.30
Titanium dioxide	41.00
Iron oxides	4.00

Skin Cleanser with Porous Cellulose Particles

Beeswax	3.00%
Paraffin	5.00
Protopet 1S Petrolatum	15.00
Carnation Mineral Oil	41.00
Sorbitan Sesquioleate	4.20
Polyoxyethylene sorbitan monooleate	0.80
Water	25.00
Fragrance	1.00
Porous Cellulose particles	5.00

SOURCE: Witco Corp.: Petroleum Specialties Group: Suggested
Formulations

Body Scrub

Ingredients:	% by Weight
(1) Bentonite (NF grade)	3.0
(2) Deionized Water	60.2
(3) Beeswax	1.4
(4) Spermacetic, synthetic	1.4
(5) Sorbitol Monopalymatate	3.3
(6) Polysorbate 60	3.3
(7) Cetyl Alcohol	2.8
(8) Light Mineral Oil	18.6
(9) Siltex -50 +100 Mesh	6.0
Preservative, Color, & Perfume	q.s.

Procedure:
   Disperse (1) into (2) and heat to 70C. In separate container, mix and heat (3), (4), (5), (6), (7), and (8) to 80C. Add to water phase with high agitation. Slow mixer and add (9). Cool before filling containers.
   Follow recommended handling practices of the supplier of each product used.
   Good industrial practices should be used when handling flammable ingredients.

Body Powder

Ingredients:	% by Weight
(1) Zinc Stearate	5.0
(2) Zinc Oxide	5.0
(3) Magnesium Carbonate (light)	15.0
(4) Kaopolite TLC	75.0
(5) Fragrance, Pigments, & Preservative	q.s.

Procedure:
   Mix fragrance with (3) and allow to stand for approximately 16 hours. Dry mix remaining ingredients.
   Follow recommended practices of the supplier of each product used.

SOURCE: Kaopolite, Inc.: Suggested Formulations

Cheek Rouge

Materials:	% by Weight
1. Mineral Oil SUS 65/75	17.00
2. Ceraphyl 368	15.00
3. Castorwax	2.50
4. Beeswax	6.00
5. Super Corona Lanolin	10.00
6. Lexgard P	0.15
7. Bentone Gel MIO rheological additive	15.00
8. Pigment Concentrate*	3.00
9. Timiron Pearl Sheen MP-30	4.00
10.Talc 1623	18.00
11.328 Titanium Dioxide CTFA	6.00
12.Color No. 7055 Pur Oxy Yellow B.C.	1.50
13.Color No. 7153 Lo-Micron Pink B.C.	1.50
14.Perfume	0.35

*Pigment Concentrate:	
Color No. 3121 D & C Red. No. 21	30.00
Castor Oil	70.00

Manufacturing Directions:
1. Weigh items 1 through 5 into a stainless steel jacketed kettle. Heat to 85C until melted clear.
2. Bring temperature down to 80C. Add and mix in item 7 until it is homogeneous. Mix for 10 minutes.
3. Add and mix item 8 for 5 minutes or until dispersed.
4. Add items 9 through 13. Add each powder separately, making sure that each pigment is properly dispersed before the next addition.
5. Add item 14 and mix.

Note:
Items 10 through 13 should be micropulverized to achieve a smooth, homogeneous finished product.

SOURCE: Rheox, Inc.: Formula TS-129

Concealing Stick

This product covers fine facial lines and blemishes naturally, leaving a flexible barrier which also retains moisture and conditions the skin. Cera Bellina (Pg-3 Beeswax) is used as a pigment disperser which makes this formulation easier and thereby avoiding the expense of milling.

Phase A:

Castor Oil (Caschem)	27.3%
Petrolatum, White (Witco)	22.0%
Synthetic Candelilla (Koster Keunen)	10.0%
Isopropyl Palmitate (Unichema)	6.0%
Carnauba Wax (Koster Keunen)	5.0%
Orange Wax (Koster Keunen)	5.0%
Jojoba Wax (Flora Tech)	4.0%
Paraffin Wax 130/135 (Koster Keunen)	1.5%
Ozokerite 170 (Koster Keunen)	1.0%
Squalane (Barnet)	1.0%
Vitamin E (Roche)	0.5%

Phase B:

Cera Bellina (Pg-3 Beeswax, Koster Keunen)	9.0%
Titanium Dioxide (Whittaker C & D)	5.0%
Cosmetic Tan Iron Oxide (Sun Chemical)	1.5%
Brown Iron Oxide (Sun Chemical)	0.2%

Phase C:

Liquipar (Sutton)	1.0%

Procedure:
Heat and mix Phase B till pigments are evenly dispersed. Weigh Phase A and individually add to Phase B while mixing. Heat AB till homogeneous, add Phase C, cool and pour at 60C.

Adaption of formula and its influence on the product:
By reducing the wax concentration this product can be poured into a compact tray. Large variations can be made in wax and oil concentrations that are incorporated in stick products and still produce a similar finished product.

SOURCE: Koster Keunen, Inc.: Suggested Formulations

Cream Eye Shadow

Ingredients:	% by Weight
(1) Bentonite (NF grade)	2.0
(2) Carboxymethyl Cellulose (low viscosity)	0.2
(3) Deionized Water	49.3
(4) Sodium Lauryl Sulfate	0.3
(5) Propylene Glycol	5.0
(6) Triethanolamine	0.4
(7) Kaopolite TLC	20.0
(8) Titanium Dioxide (cosmetic grade)	5.5
(9) Iron Oxides (cosmetic grade, for desired color)	4.5
(10)Stearic Acid	0.8
(11)Glyceryl Monostearate	2.0
(12)Lanolin	4.0
(13)Sesame Oil	2.0
(14)Olive Oil	1.0
(15)Isopropyl Myristate	3.0
(16)Fragrance and Preservative	q.s.

Procedure:

Mix (1) into (3) with high shear until dispersed. Add (2) and continue to disperse. Slowly add (4), (5), (6), (7), (8), and (9). Continue mixing at high speed until well dispersed. Heat this mixture to 65C and deaerate. In a separate container mix (10), (11), (12), (13), (14), and (15), and heat to 70C until uniform. Add this mixture slowly to (1-9) while under agitation. Continue to mix until cool. (16) can be added to either (1-9) mixture or at the end depending on the recommendations of the supplier.

Follow recommended handling practices of the supplier of each product used.

Good industrial practices should be used when handling flammable ingredients.

Dry Powder Rouge Base

Ingredients:	% by Weight
(1) Lanolin Alcohol Acetate	4.0
(2) Zinc Stearate	6.0
(3) Kaopolite TLC	88.0
(4) Magnesium Stearate	2.0
(5) Fragrance, Preservatives, & Pigments	q.s.

Procedure:

Dry mix and press as required.

Follow recommended handling practices of the supplier of each product used.

SOURCE: Kaopolite, Inc.: Suggested Formulations

Cream Facial Scrub

	Wt%
Part A:	
Glyceryl Stearate SE*	5.0
Mineral Oil	5.0
Safflower Oil	1.0
Sesame Oil	1.0
Squalene	1.0
Dioctyl Adipate (and) Octyl Stearate (and) Octyl	
Palmitate**	1.0
Stearic Acid	2.5
Cetyl Alcohol	0.5
Hydroxypropyl Methylcellulose***	0.5
Part B:	
Water	68.7
Aloe Vera Gel	1.0
Germaben II	1.8
Part C:	
Triethanolamine	1.0
Part D:	
Acuscrub 44	10.0

Procedure:
   Separately combine Part A and Part B ingredients and heat
to 80C with agitation. Slowly add Part A to Part B with agit-
ation. Add Part C and begin to cool. At 50C, moderately add
Acuscrub 44 and stir until smooth and uniform.
   *   Henkel/Emery (Emerest 2407)
   **   Caschem (Wickenol 163)
   *** Dow Chemical Co. (Methocel 40-100)

Oil-In-Water Cream and Its Scrub Derivative

	%
1. A-C 617 Polyethylene	2.0
2. Stearic Acid	0.5
3. Lanogene (lanolin alcohol & mineral oil)	6.0
4. Isopropyl palmitate	12.5
5. Sorbitan Monostearate	1.3
6. Polyoxyethylene 20 Sorbitan Monostearate	1.8
7. Sorbo (sorbitol 70%, water 30%)	5.0
8. Carbopol 940	1.0
9. Germaben II	0.8
10.Water	68.35
11.Triethanolamine	0.75
12.Perfume	Q.S.

Procedure:
   Weigh 1-6, then weigh 7-10. Heat 7-10 with agitation using a
homomixer or colloid mill. When the aqueous solution reaches 85C,
heat oil phase to 90-95C with slow agitation until all the wax
has dissolved. Combine 1-10 and shear until mixture is homogeneous. Add 11 and shear well. Cool to 55C and add 12.

SOURCE: AlliedSignal Inc.: Suggested Formulations

Cream-to-Powder Eye Shadow with Dry-Floc PC

Dry Floc PC, a highly water resistant starch, provides a silky smooth elegant feel, and is talc free.

Dermacryl LT provides water resistance, increased moisture protection as well as improved coverage.

Ingredients:	%W/W
Phase A:	
Eutanol G	13.00
Dermacryl LT	1.00
Phase B:	
Multiwax W-835	0.30
Bentone Gel VS-5 PC	30.00
Phase C:	
Myritol 318	3.00
Trivent SS-20	0.50
Emerest 2452	4.00
Vitamin E Acetate	0.10
Isopar H	7.40
Cab-O-Sil TS-530	0.50
Spectrapearl BLG	10.00
Dry-Flo PC	30.00
Phenoxyethanol	0.20

Procedure:

Phase A, slowly sift Dermacryl LT into Octyl Dodecanol, heat to 80C, mix until complete. Add Phase B, mix thoroughly. Combine Phase C. Add Phase C to A/B mix thoroughly. Cool to room temperature.

Formula 8302-115

Liquid Talc

Tapioca Flour gives a soft powdery after feel and is talc free.

Ingredients:	%W/W
SD Alcohol 40 (190 Proof)	54.00
Crodacol S-95NF	0.75
Vitamin E	1.00
Dexpanthenol	1.00
Tween 80	0.75
Fancol ALA	1.00
Estalan 430	10.00
Dermol 89	6.00
Dermol 105	4.00
Ceraphyl 375	0.75
Ethomeen C-25	2.00
Carbopol 1382	0.75
Tapioca	18.00
Ethomeen C-25	qs pH 7.00

Combine all ingredients except Carbopol 1382 and Tapioca. Mix with good agitation until uniform. Slowly sift in Carbopol 1382. Mix until completely dispersed. Sift in Tapioca. Mix until uniform. Add Ethomeen C-25 to obtain pH of 7.0 package.

Formula 7915-68

SOURCE: National Starch and Chemical Co.: Suggested Formulas

Creamy Eye Shadow

Materials:	% by Weight
1. Isopropyl Lanolate-Distilled	2.0
2. Stearyl Alcohol	3.0
3. Beeswax	6.0
4. Ganex V-220	5.0
5. Cyclomethicone	3.0
6. Soltrol 100	5.0
7. Hi-Sil T-600	0.8
8. Bentone Gel SS71 rheological additive	45.0
9. Chroma-lite Dark Blue	6.0
10.50% 328 Titanium Dioxide	20.0
CTFA Extended with Talc 141	
11.Flamenco Velvet-100	4.0
12.Preservative	0.2

Manufacturing Directions:
1. Weigh items 1 to 4 into a stainless steel jacketed kettle. Heat to 70C until melted clear.
2. Bring temperature down to 65C. Add 5 and slowly while mixing to avoid sudden cooling. Mix for 5 minutes.
3. Maintain temperature to 65C. Add and mix items 7 and 8 using a homomixer at medium speed. Mix until homogeneous (lump free)
4. Add items 9 and 10. Mix thoroughly after each addition to disperse pigments uniformly.
5. Add items 11 and 12 and mix for 5 minutes.
6. Pour at 62-63C just before it starts congealing.
Formula TS-249

Waterproof Eye Liner

Materials:	% by Weight
1. Beeswax	16.5
2. Ganex V-220	5.0
3. Shell Sol 71	35.0
4. Bentone Gel SS71 rheological additive	33.5
5. Preservative	0.2
6. Chroma-lite Black	9.8

Manufacturing Directions:
1. Weigh items 1 and 2 into a stainless steel jacketed kettle. Heat to 70C and hold until melted clear.
2. Bring temperature down to 65C. Add item 3 slowly, while mixing, to avoid sudden cooling. Mix for 5 minutes.
3. Maintain temperature at 60C. Add and mix in item 4 using a homomixer at medium speed. Mix until homogeneous.
   NOTE: A clean chrome spatula dipped into the gel mass will show an even surface sheen, with no agglomerates when properly dispersed.
4. Add items 5 and 6 separately. Mix thoroughly after each addition.
5. Pour at 55-60C just before it starts congealing.
Formula TS-189
SOURCE: Rheox, Inc.: Suggested Formulations

Creamy Matte Make Up Base

Ingredients:	% by Weight
(1) Deionized Water	54.20
(2) Bentonite (NF grade)	3.00
(3) Carboxymethyl Cellulose (medium viscosity)	0.50
(4) Tamol N	0.30
(5) Propylene Glycol	5.00
(6) Kaopolite TLC	19.80
(7) Titanium Dioxide	3.70
(8) Iron Oxide (micronized)	1.50
(9) Isopropyl Myristate	5.00
(10)Sorbitol Monolaurate	0.75
(11)Polysorbate 20	2.25
(12)Stearyl Alcohol	2.00
(13)Amerchol-L-101	2.00
Perfume & Preservative	q.s.

Procedure:

Mix with high shear (1), (2), and (3) until well dispersed (approximately 20 minutes). Add (4), (5), (6), (7), and (8) and heat to approximately 60C. In a separate container, mix (9), (10), (11), (12), and (13) and heat to 70C while mixing. Add to the other mix and continue mixing until cool.

Follow recommended handling practices of the supplier of each product used.

Good industrial practices should be used when handling flammable ingredients.

Liquid Make-Up

Ingredients:	% by Weight
(1) Deionized Water	66.6
(2) Bentonite (NF grade)	1.5
(3) Carboxymethyl Cellulose (medium viscosity)	0.5
(4) Triethanolamine	1.0
(5) Tamol-N	0.4
(6) Lexanol PG 900	5.0
(7) Kaopolite TLC	7.0
(8) Titanium Dioxide (cosmetic grade)	2.0
(9) Iron Oxide (micronized)	1.0
(10)Lexmul P	0.5
(11)Isopropyl Myristate	5.0
(12)Lexate IL	4.5
(13)Stearic Acid	2.0
(14)Mineral Oil	3.0
Preservative & Fragrance	q.s.

Procedure:

Disperse (2) into (1) at high speed. Add (3) and continue to disperse until uniform. Reduce speed and add (4), (5), (6), (7), (8), and (9). Heat to 70C. In a separate container, heat and mix (10), (11), (12), (13) and (14) to 70C. Continue mixing until dissolved. Add this to the aqueous phase. Continue mixing and cool to 30C.

Follow recommended handling practices of the supplier of each product used.

Good industrial practices should be used when handling flammable ingredients.

SOURCE: Kaopolite, Inc.: Suggested Formulations

Creamy Roll-On Lip Gloss

Materials:	% by Weight
1. Indopol H-100	52.0
2. Isopropylan 50	24.5
3. Propylparaben	0.1
4. Tenox 4	0.1
5. Bentone Gel LOI Rheological Additive	20.0
6. Pigment Concentrate*	0.2
7. Timiron Super Red	3.0
8. Perfume	0.1

*Pigment Concentrate:	
Castor Oil	60.0
Color No. 3106 D&C Red No. 6	40.0

Manufacturing Directions:
1. Weigh items 1 through 4 into a stainless steam jacket kettle. Heat to 80C and hold until melted clear.
2. Bring temperature down to 75C; add and mix in item 5 using homomixer at medium speed. Mix for 10 minutes at 75C.
3. Add item 6 and mix for 5 minutes or until dispersed; then add item 7 and mix for another 5 minutes.
4. Continue mixing at medium speed and lower temperature to 60C. Add item 8. Remove heat and allow mass to come to room temperature.

Pigment Concentrate Preparation:
1. Weigh dry powder into the castor oil using a slow speed Hobart type mixer until uniform.
2. Give this concentrate two passes over a three roll mill at room temperature.

SOURCE: Rheox, Inc.: Suggested Formula TS-190

Deep Extra Body Conditioner

This formula utilizes the gelling properties of the Hexane-diol Behenyl Beeswax in a silicone oil and conditioning effects of Bee's Milk.

Oil Phase A:

Emulsifying Wax NF (Koster Keunen)	3.0%
Silicone Oil 245 (Dow)	2.5%
Silicone Oil 556 (Dow)	2.0%
Hexanediol Behenyl Beeswax (Koster Keunen)	1.0%
Lecithin (Am. Lecithin)	1.0%
Liquapar (Sutton)	0.5%
Propyl Paraben (Sutton)	0.2%

Water Phase B:

Carbopol 940 2% Solution (BF Goodrich)	10.0%
Sodium Lauryl Sulfate (DuPont)	0.5%
Triethanolamine (Dow)	0.1%
Cocamide (Croda)	0.1%
Silk Powder (Dasco)	0.1%
Marine Dew (Ajinomoto)	0.1%
Silk Soluble Liquid (Dasco)	0.1%
Water (Distilled)	68.3%
Methyl Paraben (Sutton)	0.3%

Phase C:

Bee's Milk (Koster Keunen)	10.0%
Fragrance (Aroma Tech)	0.2%

Procedure:

Mix and heat water phase to 70-75C. Melt oil phase and add to water phase at 70-75C under agitation. Continue mixing while cooling. At 40C add phase C and pour into container at 35C.

Adaptation of formula and its influence on the product:

Changes in the actives and the conditioners are straight forward, simple replacements. Gafquat's would be a class of conditioners which would fit this formula well.

SOURCE: Koster Keunen, Inc.: Suggested Formulation

Emollient Liquid Make-up

Conceals minor skin imperfections yet moisturizes, giving skin a soft dewy finish. This product has tremendous endurance through very vigorous activities.

Phase A:

Xanthan Gum (Kelco)	0.33%
Cellulose Gum (Aqualon)	0.23%
Water (Distilled)	50.55%
Triethanolamine (Dow)	0.70%
PEG 1450 (Union Carbide)	2.35%
Methyl Paraben (Sutton)	0.28%
Propylene Glycol (Dow)	1.88%
Polysorbate 60 (Gallard & Schlesinger)	0.47%

Phase B:

Titanium Dioxide (Whittaker C&D)	8.44%
Brown Iron Oxide (Warner-Jenkinson)	2.35%
Yellow BC Iron Oxides (Warner-Jenkinson)	0.94%
Orange Wax (Koster Keunen)	6.10%
Cera Bellina (Koster Keunen)	4.69%

Phase C:

Octyl Palmitate (Barnel)	6.57%
Silicone 245 (Dow)	2.00%
Isopropyl Palmitate (Unichema)	6.10%
Hydrogenated Castor Oil (Acme)	1.41%
Isostearic Acid (Unichema)	1.50%
Jojoba Oil (Jojoba Growers)	0.94%
Propylene Glycol Stearate (Inolex)	1.89%
Propylparaben (Sutton)	0.28%

Procedure:
Disperse pigments in the Orange Wax and Cera Bellina in a beaker with a glass stirring rod, allow to solidify. Repeat so-as-to break up visible applomerations. Mix and heat to 75C phase C until uniform, add to phase B under agitation, maintaining the temperature till homogeneous. Add the mixed and melted phase B and C to the mixed and melted phase A. Continue mixing while cooling, pour into containers at approximately 50C.

Adaption of formula and its influence on the product:
Sunscreens, vegetable oils and branched oils can be substituted in minor amounts since products such as these are sensitive to small changes that will affect stability.

SOURCE: Koster Keunen, Inc.: Suggested Formulation

Emulsion Makeup

Part A:
Amerchol L-101	4.5%
Amerlate	0.9
Stearic acid, XXX	2.7
Glyceryl monostearate, neut.	1.8
Carnation Mineral Oil	4.5

Part B:
Propylene glycol	4.5
Triethanolamine	0.9
Water	70.2
Titanium dioxide, talc & pigments	10.0
Perfume and preservative	q.s.

Procedure:
   Add Part B to 85 celsius to Part A at 95 celsius while
stirring. Continue mixing and cool to 30 celsius. Add to the
micronized powder blend in increments, mixing well after each
addition.

Water-in-Oil Lotion Makeup

	Percent
Magnesium aluminum silicate	1.2
Deionized water	37.9
Magnesium sulfate	0.4
Talc	1.5
Kaolin	1.5
Titanium dioxide	5.0
Iron Oxides	3.0
Carnation	15.0
Hydrogenated polyisobutene	8.0
Carnation (and) lanolin alcohol	8.0
Sorbitol 70%	5.0
Isopropyl lanolate (and) lecithin	7.0
Oleamide DEA	2.5
Preservative	q.s.

SOURCE: Witco Corp.: Petroleum Specialties Group: Suggested
        Farmulations

Eye Contour Balm

The skin around the eye is thinner, more sensitive and rapidly shows the signs of aging and fatigue such as; crow's feet, sagging skin, dark circles and puffiness. This is designed using nature as a starting point. Much of the product's act- ivity comes from mildly refined raw materials that keeps intact the natural properties, producing a non-oily, quick penetrating and fragrance free product. A low concentration of synthetic preservative is needed due to the minor components of a few of the raw materials. With regular use the appearance of aging will be reduced.

Oil Phase:

Sweet Almond Oil (Croda)	4.0%
Propolis Wax (Koster Keunen)	5.0%
Rice Bran Oil Filtered (Koster Keunen)	4.0%
Safflower Oil (Arista)	2.0%
Orange Wax (Koster Keunen)	1.0%
Hydrogenated Castor Oil (CasChem)	0.5%
Vitamin A Palmitate (BASF)	0.5%
Vitamin E (Roche)	0.5%
Isostearic Acid (Unichema)	2.5%
Cetylstearyl Alcohol (P&G)	1.5%
Squalene (Polyesther)	1.0%
Isopropyl Palmitate (Unichema)	2.0%
Glycerol Monostearate (Henkel)	2.0%

Water Phase:

Water (Distilled)	69.6%
Ginkgo Biloba (Vernin)	0.5%
Allantoin (Sutton)	0.5%
Triethanolamine (Dow)	1.0%
Xanthan Gum (Kelco)	1.0%
Magnabrite S (Whittaker C & D)	0.4%
Polysorbate 60 (Gellard & Schlesinger)	0.2%
Methyl Paraben (Sutton)	0.3%

Procedure:
Mix and heat oil phase to 80C. Add components of water phase under agitation and heat and homogenize to 75C. Add oil phase to water phase under agitation. Cool and pour into container at 45 to 50C.

Adaption of formula and its influence on the product:
With only slight changes this product can be all natural and free of synthetics preservatives.

SOURCE: Koster Keunen, Inc.: Suggested Formulation

Eye Mascara

Ingredients:	Wt%
1. A-C Polyethylene 617	12.0
2. A-C Copolymer 540	2.0
3. Mineral Spirits	68.0
4. Dihydroabietyl Alcohol	5.0
5. Candelilla Wax	2.4
6. Aluminum Stearate	0.5
7. Butyl Parahydroxy Benzoate	0.1
8. Iron Oxide	10.0

Procedure:
    Mix 1-5 and heat with agitation until all solid waxes have dissolved. Then sprinkle with stirring 6 and 7; when all is dissolved, add 8 and shear with homomixer or grind in with 3 roll mill.

Eye Mascara

Ingredients:	Wt%
1. Mineral Spirits	70
2. A-C 617A	12
3. Dihydroabietyl Alcohol	5
4. Candelilla Wax	2.4
5. Aluminum Stearate	0.5
6. Butyl Parahydroxy Benzoate	0.1
7. Iron Oxide	10

Procedure:
    Mix 1-4 and heat with agitation until all solid waxes have dissolved. Then sprinkle with stirring 5 and 6, when all is dissolved add 7 and shear with homomixer or grind in with a 3 roll Mill.

SOURCE: Allied Signal Inc.: Suggested Formulations

Eye Mascara

This formula demonstrates the stability, dispersive and rheological properties of Cera Bellina. The product is quick drying on the eye lash and uses a multi-phase emulsion system.

Oil Phase:

Cera Bellina (Pg-3 Beeswax, Koster Keunen)	15.0%
Glycerol Monostearate (Henkel)	3.5%
Orange Wax (Koster Keunen)	2.0%
Isostearic Acid (Unichema)	1.0%
Propyl Paraben (Sutton)	0.3%

Water Phase:

Water (Distilled)	63.6%
Sodium Borate (Borax)	1.5%
Propylene Glycol (Dow)	1.0%
Carboxy Methyl Cellulose (Hercules)	0.2%
Methyl Paraben (Sutton)	0.3%

Alcohol Phase:

SDA-30 (Quantum)	9.0%
Purified Black Oxide (Whittaker)	2.6%

Procedure:

Melt Cera Bellina to a maximum of 75C. Then add the rest of the oil phase while mixing. Make sure all ingredients are melted. In a separate vessel dissolve the components of the water phase. Add the water phase to the oil phase, maintaining the temperature of 75C for approximately 15 minutes. Allow to cool slowly. When the temperature is 35-40C, add the alcohol phase under high shear

Adaptation of formula and its influence on the product:

Changing the concentrations of water (increase) and solids (decrease) by small amounts will control the viscosity. A variety of pigments can be substituted to suit products needs.

Eyeliner Stick/Pencil

This product incorporates a large pigment load at the same time as having a smooth, even application to the skin. The texture is ridged, highly colored for long wear characteristics, will not bleed and is very stable.

1. Titanium Dioxide (Whittaker C & D)	25.0%
2. Ceresine Wax 140/150 (Koster Keunen)	15.0%
3. Red D&C #7 (Warner-Jenkinson)	12.5%
4. Purified Navy Blue (Whittaker C & D)	12.5%
5. Cera Bellina (Pg-3 Beeswax, Koster Keunen)	10.0%
6. Glycerol Trioctanoate (Barnet)	10.0%
7. Ethylene Glycoldistearate (Henkel)	5.0%
8. Glycerol Tribehenate (Croda)	5.0%
9. Hydrogenated Castor Oil (CasChem)	5.0%

Procedure:

Add 2,5,6,7,8 and 9 into a vessel. Melt, mix and maintain a temperature of 75C, add 1,3 and 4 under high shear. Maintain high shear mixing for 30-40 minutes. Cool and pour into molds.

Lower pigment load and higher wax and oil concentrations can be substituted so-as-to facilitate the greatest cost effectiveness and at the same time, produce a luxurious product.

SOURCE: Koster Keunen, Inc.: Suggested Formulations

Face and Body Milk

	Wt%
A. Miglyol 840	12.0
Imwitor 900	3.0
Marlowet TA 6	1.2
Marlowet TA 25	2.1
Jojoba oil	4.0
Beeswax	2.0
Propylene glycol monostearate	1.0
B. Keltrol gel 1%	15.0
d-Panthenol USP	3.0
Karion F	5.0
Water	to 100.0
C. Vitamin E	0.2
D. Perfume	qs
Preservative	qs

Preparation:
The constituents of A. are added together and heated to 75-80C. B. is stirred while heated to the same temperature. B. is emulsified in A. C. and D. are added at approx. 30C.
Preparation of the Keltrol gel: Keltrol F   1.0%
                                Water   to 100.0%
Both components are added together at room temperature and stirred until the homogeneous mix forms a clear gel.

Lip Care Stick

	Wt%
A. Softisan 100	20.0
Dynacerin 660	8.0
Miglyol 812	6.0
Softisan 649	5.0
Vaseline	30.0
Beeswax	20.0
Hard paraffin	5.0
Cetyl alcohol	5.0
Carnauba wax	1.0
B. Antioxidants	qs
Perfume	qs

Preparation:
All ingrediants are melted, stirred to a creamy consistency until cold, perfume incorporated before casting into moulds.

SOURCE: Huls America Inc.: Formulations for Cosmetics: Formulas

Face-Lotion

	%
Ethanol 96%	5.0
Softigen 767	3.0
Hamamelis Special	10.0
Marlowet R40/K	0.3
Glycerol	2.0
Ampholyt JB 130	0.2
Panthenol	0.2
Camomile Special	2.0
Perfume	qs
Preservative	qs
Water	to 100.0

Preparation:
   The components are mixed together in sequence and stirred
until homogeneous. The pH is adjusted to 5.5-6.5.

Skin Tonic Water without Alcohol

	%
Softigen 767	5.0
1,2-Propylene glycol	3.0
Extrapon 3-Special	4.0
Allantoin	0.5
Marlowet R40/K	0.5
Menthol	0.3
Vitamin F	0.2
Perfume	qs
Preservative	qs
Water	to 100.0

Preparation:
   The components are mixed together in sequence and stirred
until homogeneous. The pH is adjusted to 5.5-6.5.

SOURCE: Huls America Inc.: Formulations for Cosmetics:
        Suggested Formulations

Facial Cleanser

	Weight,%
Mackanate LO-Special (Disodium Lauryl Sulfosuccinate)	88.0
Mackol 16 (Cetyl Alcohol)	2.0
Brij 52	2.0
Mackstat DM (DMDM Hydantoin)	qs
Water, Fragrance qs to	100.0

Solids, %: 40.0
pH (as is): 5.5
Appearance: Pearly Cream

Procedure:
1. Add Mackol 16, Brij 52 and water to Mackanate LO-Special and heat to 70 degrees C.
2. Blend until homogeneous.
3. Adjust pH to 5.5 to 6.0 with sodium hydroxide.
4. Cool to 50 degrees C. and add Mackstat DM and fragrance.
5. Adjust solid to 40.0+-1.0 at this point.
6. Cool and fill.

Sting Free Facial Cleanser

	Weight,%
Mackam 2C	40.0
Sodium Laureth-1 Sulfate (25%)	15.0
Mackernium 007	1.5
Mackanate DC-30	4.0
Mackester SP	2.0
Mackstat DM	q.s.
Water, Dye, Fragrance q.s. to	100.0

Procedure:
1. Add Mackam 2C, Sodium Laureth-1 Sulfate, Mackanate DC-30 and Mackester SP to water.
2. Heat to 70C. and blend until homogeneous.
3. Cool to 50C. and slowly add Mackernium 007.
4. When completely dispersed, add the remaining components.

SOURCE: McIntyre Group Ltd.: Personal Care Formulary

Fluid Make-Up

	Wt%
A. Miglyol 812	7.0
Miglyol 840	5.0
Imwitor 960 flakes	3.0
Imwitor 900	2.0
Marlophor T10 Na salt	2.0
Paraffin oil	5.0
B. Keltrol gel 1%*	15.0
Water	43.0
Karion F	5.0
Glycerol	3.0
C. Perfume	qs
Preservative	qs
D. Titanium dioxide	3.0
Talc	3.0
Zinc oxide	3.0
Sicomet Brown 70 C.I. 77491	1.0

Preparation:

The ingredients of A. are melted and heated to 75-80C. The finely mixed pigments D. are homogenised together with A. The ingredients of B. are mixed, heated to 80C and slowly emulsified in A. + D. Perfume is added at around 35C.

*Preparation of the Keltrol gel: Keltrol F	1.0%
Water	to 100.0%

The Keltrol is mixed with the water until homogeneous and the mixture stirred until the gel is clear.

Compact Make-Up

	Wt%
A. Miglyol 812	18.0
Miglyol 840	10.0
Imwitor 900	8.0
Softisan 100	10.0
Softisan 649	4.0
Beeswax	7.0
Hard paraffin	10.0
Stearic acid	3.0
B. Sicomet Brown 70 C.I. 77491	0.5
Sicomet Brown 75 C.I. 77491	0.5
Talc	8.0
Zinc oxide	8.0
Titanium dioxide	8.0
C. Perfume	qs
D. Karion F	5.0

Preparation:

The ingredients of A. and D. are melted together and gradually added to the stirred-until-homogeneous phase B. The mix is then again heated to approx. 70C and stirred until cold. C is added and the whole mix is homogenised.

SOURCE: Huls America Inc.: Formulations for Cosmetics: Formulas

Fluid Make Up

Ingredients:	% w/w
A) Cutina GMS	4,00
Eumulgin B1	4,00
Eutanol G	10,00
Miglyol 812	6,00
Phenonip	0,30
B) Water demineralized	63,66
Euxyl K-200	0,20
Citric acid	0,04
Hyasol-BT	3,00
Bentone EW	0,60
Pigment paste white: Nr. 93975	5,00
Pigment paste yellow: Nr. 75577	1,50
Pigment paste red: Nr. 68775	1,00
Pigment paste black: Nr. 78375	0,40
C) Perfume Oil: Beauty 0/239870	0,30

Procedure:
Heat the ingredients of fatty phase A) to 70C.
Heat the ingredients of water phase B) to 75C.
Under stirring add phase B) to phase A), cool to 50C, homogenize and cool to 30C.
Then add phase C) and stir cold.
Application No. x 004.A/11.92

Regenerating Lipcare-Stick

Ingredients:	% w/w
1 Wax premix No. 1	72,25
2 Pigments	7,60
3 Arlacel 582	10,00
4 Imidazolidinyl Urea	0,15
5 Phytaluronate	5,00
6. Cephalipin	5,00

Procedure:
Melt Wax-Premix (1) at 80C in the main-mixer while mixing with paddles.
Incorporate item 2 and homogenize well.
Melt apart item 3 at 80C and slowly add the preheated (80C) items 4-6.
Add the obtained emulsion (4-6) to the pigmented wax-base while mixing at 80C.
Then pour into the moulds.
Application No. X 025.0/02.93

SOURCE: Pentapharm Ltd.: Suggested Formulations

Gelled Aloe Vera with Sunscreen

Ingredients:	Percent
A. DI water	81.18
Carbopol 940	0.92
Aloe Veragel Liquid 1:1	2.3
B. Triethanolamine	
C. DI water	10.0
Phenylbenzenimidazole Sulfonic Acid	2.0
Triethanolamine	to pH 7.0
D. DMDM Hydantoin	0.3
E. Disodium EDTA	0.1

Procedure:
Disperse Carbopol resin into water (under high agitation). Add Aloe Veragel extract. Neutralize with triethanolamine. Combine C, add triethanolamine to adjust pH to 7.0. Add C to neutralized A (moderate agitation). Add preservative and chelating agent.

Aloe Vera Jelly

Ingredients:	Percent
A. Water	91.94
Citric Acid (granular)	0.2
Glydant	0.3
Germall 115	0.25
Versene-220	0.05
Propyl Gallate	0.05
B. Propyl Glycol	4.5
Xanthan Gum	0.75
Aubygel x-125	1.5
C. Aloe vera 200 powder	0.46

Procedure:
Heat Phase A to 80C. Pre mix Phase B with good agitation and add Phase B to A. Mix slowly and cool to 55C. Add phase C, and mix until uniform. Package at between 35-45C.

SOURCE: Dr. Madis Laboratories Inc.: Suggested Formulations

Gentle Beauty Wash

This elegant beauty wash is based on mild surfactants that
result in a mild cleansing product. Its non-alkaline pH and
combination of ingredients are designed to leave the skin clean
and smooth without harsh soap effects.

Materials:	% by Weight
1. Deionized Water	46.20
2. Propylene Glycol	8.00
3. 50% Sodium Hydroxide	0.60
4. Bentone EW	0.40
5. Stearic Acid, triple pressed	8.00
6. Igepon AC78	11.00
7. Sipon ESY	11.00
8. Siponate DDB-40	5.00
9. Alconate SBR-3	2.50
10.Sodium Sethionate 55	6.50
11.Oxaban A	0.05
12.Fragrance KU 70	0.75

Manufacturing Directions:
Disperse the Bentone EW in the deionized water, sodium hydroxide and propylene glycol with rapid agitation. Start heating.
Decrease agitation and add the stearic acid and Igepon AC78.
Continue heating until the Igepon AC78 has dissolved (70 degrees
celsius). Add the Sipon ESY, Siponate DDB-40, Alconate SBR-3
and Sodium Isethionate 55. Start cooling. Add the preservative
and fragrance at 35C. The viscosity and pearl will develop upon
standing.

Ingredient Label:
Water, Sodium cocoyl isethionate, Stearic acid, Propylene
glycol, Sodium isethionate, Sodium laureth sulfate, Sodium
dodecyl benzene sulfonate, Disodium ricinolemido, MEA-sulfo-
succinate, Fragrance, Hectorite, Sodium hydroxide, 4,4-Dimethyl-
oxazolidine, 3,4,4-Trimethyloxazolidine.

Physical Properties:
pH: 5.8-6.1
Viscosity: Brookfield LV4 @ 25 degrees C (at rest 12 hours):
@ 3 RPMs 80,000-110,000 cp
@ 6 RPMs 50,000- 70,000 cp
@ 12 RPMs 30,000->50,000 cp
Appearance: White, pearled, pumpable cream

SOURCE: Rheox, Inc.: Rhone Poulenc Prototype Formulation 91-0401

Hair and Skin Moisturizing Mist

A refreshing spray containing special humectants to replenish much needed moisture to hair and skin. Contains Sodium PCA to help restore normal moisture balance and keep the skin young looking and fresh. Also contains Aloe Vera Gel and Panthenol to soothe, condition and protect the skin and hair from drying out. Recommended for use under dry climatic conditions, dry heat or during sun exposure.

Ingredient/Trade Name:	%
Part A:	
Deionized Water	91.13
Polyquaternium-4/Celquat H-100	0.05
Citric Acid	0.01
Propylene Glycol	2.00
Methylparaben	0.20
Part B:	
Sodium PCA/Ajidew N-50	2.00
Sorbitol & Sodium Lactate & Proline & Sodium PCA &	
Hydrolyzed Collagen/Prodew 100	2.00
Panthenol	0.10
Aloe Vera Gel/Aquasol 104	1.00
Dimethicone Copolyol/Dow Corning 193	0.30
Diazolidinyl Urea/Germall II	0.20
Part C:	
Tea-Cocoyl Glutamate/Amisoft CT-12	0.50
Polysorbate 20/Tween 20	0.50
Fragrance #X3110	0.01

Procedure:
Disperse Celquat H100 into rapidly agitating deionized water. Heat to 70 degrees Centigrade. Add remaining Part A ingredients. Mix until completely clear and uniform. Start cooling. At 50 degrees Centigrade, add Part B ingredients. Mix well after each addition. Premix Part C. At 40 degrees Centigrade add premixed Part C to Part A. Continue mixing and cooling to 35 degrees Centigrade.

Appearance: Clear, light yellow, water-thin liquid
pH: 5.00-5.40

SOURCE: Ajinomoto USA, Inc.: Suggested Formulation

Hair and Skin Moisturizing Spray

A refreshing spray containing special humectants to replenish much needed moisture to hair and skin. Contains Sodium PCA to help restore normal moisture balance and keep the skin young and fresh. Also contains Aloe Vera Gel to soothe dryness, and emollients to condition and protect the skin and hair from drying out. Recommended for use under dry climatic conditions, dry heat or during sun exposure.

Ingredient/Trade Name:	% by Weight
Part A:	
Deionized Water	88.95
Sodium PCA/Ajidew N-50	3.00
Aloe Vera Gel Decolorized 1x	5.00
Dimethicone Copolyol/Dow Corning 193 Surfactant	0.30
Diazolidinyl Urea/Germall II	0.10
Part B:	
Propylene Glycol	2.00
Methylparaben	0.15
Part C:	
TEA-Cocoyl Glutamate/Amisoft CT-12	0.50

Procedure:
Mix part A ingredients until everything is completely dissolved. Heat part B to 50 degrees Centigrade and mix until clear. Add part C. Mix well.

Appearance:
Clear, colorless, water-thin liquid.
pH: 5.30-5.60

Hair Liquid (Soft Type)

	Wt%
(O) Pyroter GPI-25	9.0
2-Hexyldecyl Alcohol	1.0
Ethyl Alcohol	40.0
Methylparaben	0.1
Perfume	q.s.
(W) Ajidew T-50	4.0
Water	45.9

Procedure:
1. Dissolve (O) and (W) separately at room temperature to be clear solutions.
2. Add (W) to (O) with stirring.
   pH: 7.0
   Viscosity: 6 cps

SOURCE: Ajinomoto USA, Inc.: Suggested Formulations

High Shine Lipstick

High gloss, firm lipstick with good moisturizing qualities. Liquefies instantly to an oil, slippery film while depositing very little sheer color and high pearlescence.

	%
Castor Oil	59.4
Candelilla Wax	8.0
Acetulan	7.5
Ross Wax 1275W	5.0
Propylene Glycol Monolaurate	5.0
Lanogene	5.0
Carnauba Wax	2.0
Propylparaben	0.1
Timiron MP-10	7.0
D & C Red #9 (31-3009)	0.6
D & C Red #7 Ca Lake (3107)	0.3
Pur. Navy Blue #7110	0.1
Fragrance	q.s.

Procedures:

Grind the pigments in part of the Castor Oil using either a 3-roll mill or mortar/pestle. Add all other ingredients (except for pearlescent pigment and fragrance and heat gently on steam bath to 80-85C. Add pearl, mix until homogeneous. Fragrance should be added at lowest possible temperature. Cast into molds.

Lipstick Formulation

	%
Ross Synthetic Candelilla Wax	10.9
Isopropyl Myristate	9.5
Lanolin N.F.	4.4
Ross White Beeswax N.F.	3.3
Ross Refined Paraffin Wax 130/35	2.0
Ross White Ozokerite Wax 77W	0.9
Castor Oil	54.3
Pigment	12.0
Teg. "P"	0.1

*Formulation developed by Precision Cosmetic of Mount Vernon, NY, in conjunction with the Frank B. Ross Company.

SOURCE: Frank B. Ross Co., Inc.: Suggested Formulations

Lip Balm White

	%
Rosswax 2639	85.0
Mineral Oil #7	13.5
Solar Chem O	1.5
Fragrance GP-58	q.s.

Procedure:
   Melt all ingredients to 190F in a stainless steel vessel.
Mix thoroughly with agitation, cool to 165F, fragrance and pour
into a container. Note: Capping may be necessary.

Lip Balm I

	%
Castor Oil Crystal O	46.0
Emery IPP	17.0
Emery 1723	10.4
Rosswax 2640	19.6
Acetulan	2.5
SDA Alcohol #40	2.0
Solar Chem O	1.5
Propylene Glycol	1.0
Fragrance GP-58	q.s.

Procedure:
   Melt all ingredients to 190F in a stainless steel vessel.
Mix thoroughly with agitation. Cool to 165F, fragrance and pour
into a container. Note: Capping may be necessary.

Lip Balm II

	%
Ross Base Oil 2539	55.4
Emery 1723	10.8
Rosswax 2641	29.3
SDA Alcohol #40	2.0
Solar Chem O	1.5
Propylene Glycol	1.0
Fragrance GP-58	q.s.

Procedure:
   Melt all ingredients to 190F in a stainless steel vessel. Mix
thoroughly with agitation, cool to 165F, fragrance and pour into
a container. Note: Capping may be necessary.

SOURCE: Frank B. Ross Co., Inc.: Suggested Formulations

Lip Gloss (Souffle)

Materials:	% by Weight
1. Beeswax	7.0
2. Ceraphyl 140A	12.7
3. Propylparaben	0.1
4. Tenox 4	0.1
5. Bentone Gel LOI rheological additive	70.0
6. Timeron Pearl Sheen MP-30	5.0
7. Pigment Concentrate**	0.4
8. Perfume	0.2
**Pigment Concentrate	
Castor Oil	60.0
Color No. 3106 D&C Red No. 6	40.0

1. Weigh item 1 through 4 into a stainless steel steam jacketed kettle. Heat to 85C until melted clear.
2. Bring temperature down to 78C. Add and mix in item 5 using a homomixer at medium speed. Mix for another ten minutes.
3. All item 6 and mix for five minutes or until dispersed; then add item 7 and mix for another five minutes.
4. Continue mixing at medium speed and lower temperature to 60C. Add item 8. Remove heat and stir slowly until congealed or pour into container.

Pigment Concentrate Preparation:
1. Weigh dry powder into castor oil using a slow speed Hobart type mixer until uniform.
2. Give this concentrate two to three passes over a three roll mill at room temperature.

Roll-On Lip Gloss

Materials:	% by Weight
1. Sucrose Acetate Isobutyrate	30.0
2. Isopropylan 33	44.5
3. Propylparaben	0.1
4. Tenox 4	0.1
5. Bentone Gel LOI rheological additive	22.5
6. Pigment concentrate*	0.2
7. Pearl White	2.5
8. Perfume	0.1
*Pigment Concentrate	
Castor Oil	60.00
Color No. 3106 D&C Red No. 6	40.00

1. Weigh items 1 through 7 into a stainless steam jacketed kettle. Heat to 80C until melted clear.
2. Bring temperature down to 75C; add and mix in item 5 using homomixer at medium speed. Mix for 10 minutes at 75C.
3. Add item 6 and mix for 5 minutes or until dispersed; then add item 7 and mix for another 5 minutes.
4. Continue mixing at medium speed and lower temperature to 60C. Add item 8. Remove heat and allow mass to come to room temp.

Pigment Concentrate Preparation:
1. Weigh dry powder into the castor oil using a slow speed Hobart type mixer until uniform.
2. Give this concentrate two passes over a 3 roll mill at room temperature.

SOURCE: Rheox, Inc.: Formulas TS-135 & TS-184

Lip Pencil
This product utilizes the dispersive properties of Cera
Bellina to produce a product which glides on easily, leaves
silky lasting colour that is water resistant.

Phase A:

Castor Oil (Caschem)	51.5%
Candelilla Wax (Koster Keunen)	8.0%
Carnauba Wax (Koster Keunen)	7.0%
Microcrystalline Wax (Koster Keunen)	5.0%
Ceresine 130/135 Wax (Koster Keunen)	3.0%
Mineral Oil (Witco)	3.0%
Cetyl Alcohol (P&G)	1.5%

Phase B:

Cera Bellina (Pg-3 Beeswax, Koster Keunen)	8.0%
Iron Oxide Brown (Warner-Jenkinson)	2.0%
Titanium Dioxide (Whittaker C&D)	8.0%
D&C Red #6 (Whittaker C&D)	3.0%

Heat and mix Phase B, dispersing pigments and breaking up
visible agglomeration. Add Phase A to Phase B, heat and mix
together under agitation. Allow to cool to 65C and pour into
molds. Changing the concentration of waxes and or oils will
soften and harden the product to maintain consistency when changing pigments and colours. Different pigments will affect the
gelling properties of the base and is compensated for with
changes in concentrations of waxes and oils used. Sunscreens
can also be incorporated without difficult formula changes.

New Look Lipstick
This red lipstick has a creamy texture and is long wearing.
The lipstick is very stable, although it has a high oil concentration. By using synthetic replacements of natural products
this formula is of very low cost but looks and performs like a
high end product.

1. NF Yellow Beeswax (Koster Keunen)	5.0%
2. Synthetic Candelilla (Koster Keunen)	9.9%
3. Ozokerite 170 (Koster Keunen)	10.6%
4. Synthetic Carnauba (Koster Keunen)	1.5%
5. Castor Oil (Caschem)	20.0%
6. Paraffin Oil (Kydol-Witco)	15.3%
7. Edible Coconut Oil (Cocochem)	32.7%
8. D&C Red #30 (Warner-Jenkinson)	4.5%
9. Titanium Dioxide (Whittaker C & D)	0.5%

Add 1, 5, 8 and 9 into a vessel large enough that will allow
the other components to be added after the beeswax, castor oil
and pigments have been melted (75C.) and mixed together with the
intention of breaking down as many of the visible agglomerations.
After the visible agglomerations have been separated, add 2, 3
and 4 to the dispersion while mixing. Increase the temperature
slightly if needed to melt the synthetic carnauba. Then the
remainder of the components can be added and mixed at a temperature of 70C. After the product is homogeneous use a shear
mixer at low speed for 45 to 60 minutes, maintaining a similar
temperature.

Slight changes in oils and waxes will not alter the products
appearance or performance. Sunscreens are easily incorporated.

SOURCE: Koster Keunen, Inc.: Suggested Formulations

Lipstick

Dermacryl LT provides increased moisture protection, rub-off resistance and water resistance.

Ingredients:	%W/W
Phase A:	
Eutanol G	6.00
Prisorine 3515	20.00
Ceraphyl ICA	2.70
Ceralan	5.00
Dermacryl LT	2.00
Phase B:	
Fluilan	10.00
Candelilla Wax	12.00
Ozokerite Wax	5.00
Cutina LM	10.00
Drakeol 35	5.00
Tenox BHA	0.10
Propylparaben	0.20
Phase C:	
DC Red #6 (Dispersion in Castor Oil)	15.00
Flamenco Superpearl	5.00
Phase D:	
Vitamin E Acetate	1.00
Vitamin E Palmitate	1.00

Procedure:
Combine Phase A, except Dermacryl LT. Heat to 80C. With good agitation, slowly sift in Dermacryl LT, mix until complete. Combine Phase B, heat to 80C. Add Phase B to Phase A at 80C, mix thoroughly. Phase C: Mix pigments together, add to A and B at 80C. Cool to 60C. Add Phase D, mix thoroughly. Pour into molds and cool to room temperature.

SOURCE: National Starch and Chemical Co.: Formula 7661-144B

Lipstick

	Wt%
A. Softisan 645	10.0
Dynacerin 660	8.0
Softisan 100	7.0
Miglyol 812	7.0
Beeswax	11.0
Eutanol G	9.0
Prosolal S9	5.0
PCL liquid	5.0
Protegin X	4.0
B. Carnauba wax	9.0
Castor oil	6.0
Hexylene glycol	3.0
Stearic acid	2.0
Timiron Starlustre MP-115	10.0
Perfume oil Tendresse	1.0
D. Colours:	qs
Pink: Sicomet Rot P 15630 CA	0.5
Red: Sicomet Rot P 15630 CA	3.0
Violet: Sicomet Violett P 77007 G	0.5

Preparation:
All ingredients are heated to 75C and homogenised with a high speed stirrer. The mix is poured into moulds before it has got cold.

Lip-Gloss

	Wt%
A. Softisan 645	44.5
Softisan 649	10.0
Rewopal PIB 1000	30.0
Lanfrax	10.0
Candelilla wax	2.5
B. Pearl lustre pigment*	3.0
Colour	qs
Perfume	qs

*Timiron Starlustre MP 115

Preparation:
A. is melted at 60C and stirred until cooled down to 40C. The ingredients of B. are mixed together and added to A. Finally, the mix is homogenised.

SOURCE: Huls America Inc.: Formulations for Cosmetics: Formulas

Lipstick

	Wt%
A. Miglyol Gel B	10.0
Miglyol 812	7.0
Miglyol 840	5.0
Softigen 701	7.0
Paraffin oil	7.0
Isopropyl myristate	5.0
Castor oil	9.0
Eutanol HD	5.0
B. Carnauba wax	8.0
Wool wax alcohol	5.0
Beeswax	10.0
Imwitor 900	4.0
Oxynex 2004	0.20
C. Perfume oil GC 10776	0.28
D. Talc	3.0
Titanium dioxide	3.0
Zinc oxide	3.0
Timiron Starlustre MP-115	5.0
Flame Orange 5305 No. 5 C.11994	1.7
Lo Micron Tan 3088 C. 11997	3.0

Preparation:
A. is mixed, B. is added and heated to 75-80C. D is finely
ground with A, the being stirred into D. in small portions.
Perfume is added to the mix and at 60C. it is poured into the
mould, which is not cooled. The mix is set by storing the mould
one hour in a refrigerator.

Lipstick, Glossy

	Wt%
A. Miglyol Gel B	20.0
Softisan 645	15.0
Imwitor 780K	6.0
Miglyol 829	6.0
Lanfrax	10.0
Beeswax	5.0
Candelilla wax	5.0
Castor oil	4.5
Sodium stearate	1.0
B. Rewopal PIB 1000	15.0
Antioxidants	qs
C. Perfume	qs
D. Sicomet Erythrosinlack E 127	0.5
Timiron Starlustre MP-115	10.0

Preparation:
A. and B. are mixed together and heated to approx. 80C. D. is
added and the mix is homogenised. Perfume is incorporated at
approx. 30C.

SOURCE: Huls America Inc.: Formulations for Cosmetics: Formulas

Lipstick

This red lipstick has a creamy texture and is long wearing. The lipstick is very stable, although it has a high oil concentration. Cera Bellina allows for high coloration without large pigment load and imparts a satin like feel to the lips.

Formula:

1. Cera Bellina (Pg-3 Beeswax, Koster Keunen)	16.70%
2. Candelilla (Koster Keunen)	3.10%
3. Ozokerite 160/164 (Koster Keunen)	4.20%
4. Carnauba #1 Yellow (Koster Keunen)	1.20%
5. Castor Oil (Caschem)	43.45%
6. Paraffin Oil (Kydol-Witco)	16.30%
7. Octyl Palmitate (Inolex)	6.20%
8. Jojoba Oil (Jojoba Growers)	1.00%
9. Vitamin E (Freeman)	0.50%
10.D&C Red #7 (Warner-Jenkinson)	4.00%
11.Orange Wax (Koster Keunen)	3.00%
12.Titanium Dioxide (Whittaker C&D)	0.25%
13.Propyl Paraben (Sutton)	0.10%

Procedure:

Add 1, 10, 11, and 12 into a vessel large enough that will allow the other components to be added after the Cera Bellina (Pg-3 Beeswax), Orange Wax and pigments have been melted (75C) and mixed together with the intention of breaking down as many of the visible agglomerations. After the visible agglomerations have been separated add 2, 3 and 4 to the dispersion while mixing. Increase the temperature slightly if needed to melt the carnauba. Then the remainder of the components can be added and mixed at a temperature of 70C. After the product is homogeneous, use a shear mixer at low speed for 45 to 60 minutes, maintaining a similar temperature.

Adaptation of formula and its influence on the product:

Other oils natural or synthetic can easily be substituted to meet the formulators needs, cost, availability, etc. without significantly changing the characteristics. Higher melting points can easily be achieved with only slight alterations in wax concentration. Biologically active compounds are easliy incorporated into the formulas. Depending on the desired color and the blend of pigments used, there will be a need for small ingredient concentration changes in the formula due to the pigment's effect on the gelling properties of the mixture.

SOURCE: Koster Keunen, Inc.: Suggested Formulations

Lipstick

	Wt%
A. Carnauba Wax	2.0
Candelilla Wax	6.5
Beeswax	5.5
Hydrogenated Castor Oil	2.0
Liquid Lanolin	16.8
Microcrystalline Wax	3.0
2-Octyldodecanol	15.0
2-Octyldodecyl Myristate	10.0
Castor Oil	23.2
B. Pearl Pigment	4.5
Titanium Dioxide	4.5
Amihope LL	3.5
Red Oxide of Iron	1.3
Organic Pigment (Red)	2.2

Procedure:
1. Mix (A) and (B), and dissolve at 80C.
2. Knead them with a roll.
3. Dissolve at 80C.
4. Remove small bubbles in a vacuum, and then press.

Note: This lipstick has smooth touch, and spreads well.

Lipstick Formulae

The addition of Eldew CL-301 results in a preparation with remarkable spreadability, improved luster and excellent color without reducing the strength.

	Wt%
A Castor Oil	55.0
Candelilla	4.8
Carnauba	2.0
Ozokerite	4.0
Caprylic/capric triglyceride	15.0
Eldew CL-301	5.0
B Mica	7.4
Yellow iron oxide	2.2
D&C Red #7	3.0
Amihope LL	1.6

SOURCE: Ajinomoto USA, Inc.: Suggested Formulations

Liquid Foundation

Part A:	Weight%
Hydroxypropyl Methyl Cellulose	0.4
Magnabrite	0.4
Water	56.26
Talc	3.00

Part B:	
Ceteareth 15	5.00
Propylene Glycol	5.00
Glyceryl Stearate	3.00
Isopropyl Palmitate	5.00
Cetyl Alcohol	4.00
Isostearic Acid	4.00
Carnation Mineral Oil	6.00

Part C:	
Titanium Dioxide	6.00
Iron Oxides	0.94

Part D:	
Germaben II	1.00

Procedure:
Blend together Magnabrite and hydroxypropyl methyl cellulose and sift into vortex of water with maximum shear. When dispersed, add talc. Heat to 80C. Add oil phase, stir until uniform. Add titanium dioxide and iron oxides. Mill to disperse pigments. Add D. Cool with mixing to 32C.

Powder Foundation

Talc	17.0%
Titanium dioxide	10.0
Sericite	40.0
Spherical silica	10.0
Polytetrafluoroethylene powder	10.0
Red iron oxide	0.5
Yellow iron oxide	1.0
Black iron oxide	0.1
Dimethylpolysiloxane	1.0
Carnation Mineral Oil	9.0
Sorbitan sesquioleate	1.0
Preservative	0.3
Fragrance	0.1

SOURCE: Witco Corp.: Petroleum Specialties Group: Suggested Formulations

Liquid Make-Up

This formula demonstrates the stability, dispersive qualities
and rheological properties of Cera Bellina in a high pigment
load, vegetable oil and hydrocarbon oil free product. The
product is quick drying on the skin leaving a soft velvet like
feel, for a flawless, long wearing natural look.

Oil Phase:

Isopropyl Palmitate (Unichema)	14.50%
Cera Bellina (Pg-3 Beeswax, Koster Keunen)	6.00%
Silicone Oil 556 (Dow)	4.50%
Ceresine Wax 130/135 (Koster Keunen)	1.50%
Glycerol Monostearate (Witco)	0.80%
Emulsifying Wax NF (Koster Keunen)	0.70%
PEG-100 Stearate (Lipo)	0.50%
Orange Wax (Koster Keunen)	0.30%
Propyl Paraben (Sutton)	0.20%

Water Phase:

Water (Distilled)	51.05%
Sodium Hydroxide (Aldrich)	0.60%
Citric Acid (Aldrich)	0.50%
Bermocall E230 (Whittaker C&D)	0.20%
Methyl Paraben (Sutton)	0.20%

Pigment Phase:

Titanium Dioxide (Whittaker C&D)	13.50%
Iron Oxide Yellow 7055 (Warner-Jenkinson)	2.70%
Iron Oxide Brown 7058 (Warner-Jenkinson)	1.20%
Mica (Rona)	1.00%
Iron Oxide Red 7067 (Warner-Jenkinson)	0.50%

Procedure:
   Melt oil phase to a maximum of 75C and mix, making sure all
ingredients are melted. In a separate vessel dissolve the components of the water phase. Add the oil phase to the water phase,
maintaining the temperature of 75C for approximately 15 minutes.
Allow to cool slowly. When the temperature is 55 to 60C, add the
pigment phase under high shear.

Adaptation of formula and its influence on the product:
   Changing the concentrations of water (increase) and solids
(decrease) by small amounts will control the viscosity. A variety
of pigments can be substituted to suit products needs.

SOURCE: Koster Keunen, Inc.: Suggested Formulation

Liquid Makeup

Materials:	% by Weight
1. Talc 1623	5.00
2. 328 Titanium Dioxide CTFA	5.00
3. Color No. 3170 Pur Oxy Yellow B.C.	0.40
4. Color No. 3315 Pur Oxy Umber B.C.	0.20
5. Color No. 2511 Lo-Micron Pink B.C.	0.40
6. Water (deionized)	79.95
7. Methylparaben	0.25
8. Propylene Glycol	5.00
9. Cerasynt 840	2.00
10.Bentone LT rheological additive	1.50
11.Perfume	0.30

Manufacturing Directions:
1. Part 1-In a powder blender, add items 1 through 5 and mix together for 20 minutes. Remove powder from blender and put through a micropulverizer.
2. Part 2-Add items 6 through 9 to a stainless steel vessel and heat to 65C. Using a homomixer at medium speed, add and mix in item 10. Continue mixing for 20 to 30 minutes until completely dispersed.
3. Cool Part 2 to 40C and add and mix in item 11.
4. With continuous mixing at medium speed, add Part 1 to Part 2 and mix together for 15 minutes.
5. Put entire mixture through a colloid mill for uniform pigment grinding.

Formula TS-116

Silky Facial Mask

Materials:	% by Weight
Part A:	
1. D.I. Water	76.3
2. Propylene Glycol	5.0
3. Miranol 2MHT Modified	3.0
4. Methyl Paraben	0.1
Part B:	
5. Paricin 9	2.0
6. Isopropylan 50	4.0
7. Propyl Paraben	0.1
Part C:	
8. Bentone MA Rheological Additive	4.0
9. Zinc Oxide	3.0
10.328 Titanium Dioxide	2.5

Manufacturing Directions:
1. Weigh items 1 through 4 into a stainleess steel steam jacketed kettle. Heat Part A to 65C.
2. In a separate vessel, add items 5 through 7 and heat to 65C. Mix until completely melted and homogeneous.
3. Mix Part B slowly into Part A while mixing. Mix for 15 to 20 min., and bring temperature down to 55C while mixing.
4. Mix Part C separately using a homomixer and add to above mixture while stirring. Mix it for 15 minutes. Maintain temperature and pass through Colloid Mill.

Formula TS-252

SOURCE: Rhone Poulenc Inc.: Suggested Formulations

Lotion Scrub

Part A:	Wt%
Water	36.2
Aloe Vera Gel	1.0
Germaben II	1.8
Magnesium Aluminum Silicate (1)	1.0
Part B:	
Glycol Stearate (2)	5.0
Safflower Oil	0.5
Sesame Oil	0.5
Hydroxypropyl Methylcellulose (3)	0.5
Part C:	
Sodium Lauryl Ether Sulfate (4)	20.0
Sodium Lauryl Sulfate (4)	18.0
Cocamide DEA (5)	0.5
Cocamidopropyl Betaine (6)	5.0
Part D:	
Acuscrub 44	10.0

Procedure:
    Slowly add thickener (2) with agitation to Part A ingredients until dissolved. Heat to 70C. Separately combine Part B and Part C ingredients, then add to Part A, mixing well between additions. At 50C, moderately add Acuscrub 44 and stir until smooth & uniform.
    1. Veegum
    2. Emerest 2350
    3. Methocel 40-100
    4. Sipon ES, Sipon LCP
    5. Ninol 49-CE
    6. Lexaine C

Lotion Hand Scrub

Part A:	Wt%
Sodium C14-16 Olefin Sulfonate (1)	30.0
Cocamidopropyl Betaine (2)	6.7
Cocamide DEA (3)	2.0
Glycol Stearate	1.0
Ammonium Chloride	2.5
Germaben II	0.8
Water	50.0
Part B:	
Hydrolyzed Animal Protein (4)	2.0
Part C:	
Acuscrub 44	5.0

Procedure:
    Heat water to 70C, then slowly add remaining Part A ingredients. Cool to 55C, add Part B with agitation. Add Part C with agitation and mix until homogeneous.
    (1) Witconate AOS   (3) Ninol 49-CE
    (2) Lexaine C   (4) Crotein SPA

SOURCE: AlliedSignal Inc.: Suggested Formulations

Low Cost, Extended Wear Lipstick

Resulting lipstick has exceptional shine, adherence and wear attributes. When applied to lips, the formula produces a comfortable non-waxy film, which resists migration and feathering. The shade of this formula is a clean, true, high chroma red with moderate opacity.

	%w/w
A. Castor Oil NF (Crystal O)	61.66
Candelilla Wax (S&P 75)	8.30
Carnauba Wax (S&P #1 Yellow 73)	2.05
Microcrystalline Wax (S&P 18)	4.20
Beeswax (S&P White NF 422P)	1.87
Tocopherol (Copherol F-1300)	0.12
Propylparaben (Lexgard P)	0.08
Butylparaben (Lexgard B)	0.05
B. D&C Red No. 6 Barium Lake, CI #15850:2 (C19-012)	1.38
D&C Red No. 7 Calcium Lake, CI #15850:1 (C19-011)	0.97
FD&C Yellow No. 5 Aluminum Lake, CI #19140:1	
(C65-4429)	2.74
Castor Oil (Crystal O)	11.88
C. Glycerin/Diethylene Glycol/Adipate Crosspolymer	
(Lexorez 100)	4.70

Procedure:
1. Combine section "A" heating to 90C. Mix slowly (be sure not to entrap air) until the phase is clear and homogeneous, then reduce temperature to 80 to 85C. Continue slow mixing.
2. Disperse pigments in section "B" in the castor oil contained in Phase "B" by adding the dry pigments to the castor oil and blending with a spatula, then disperse using a Cowles dissolver and finally pass the blend through a three roller mill three times or until the particle size of the pigments are less than number 7 on a Haggeman gauge.
3. When section "B" is adequately dispersed, add section "B" to section "A". Continue slow agitation and maintain a temperature of 80 to 85C.
4. When sections "AB" is homogenous add section "C" to "AB". Maintain a temperature of 80 to 85C and mix slowly until homogeneous.
5. Mold sticks at 80-85C.

SOURCE: Inolex Chemical CO.: Formulation 398-152-1

Low Cost Lipstick

Resulting lipstick is a low cost formulation that exhibits good application characteristics and is comfortable on lips. Wearability of the lipstick is limited. Feathering and migration of the lipstick film would be expected. The shade of this formula is a clean, true, high chroma red with moderate opacity.

	%w/w
A. Castor Oil NF (Crystal O)	66.36
Candelilla Wax (S&P 75)	8.30
Carnauba Wax (S&P #1 Yellow 73)	2.05
Microcrystalline Wax (S&P 18)	4.20
Beeswax (S&P White NF 422P)	1.87
Tocopherol (Copherol F-1300)	0.12
Propylparaben (Lexgard P)	0.08
Butylparaben (Lexgard B)	0.05
B. D&C Red No. 6 Barium Lake, CI #15850:2 (C19-012)	1.38
D&C Red No. 7 Calcium Lake, CI #15850:1 (C19-011)	0.97
FD&C Yellow No. 5 Aluminum Lake, CI #19140:1	
(C65-4429)	2.74
Castor Oil (Crystal O)	11.88

Procedure:
1. Combine section "A" heating to 90C. Mix slowly (be sure not to entrap air) until the phase is clear and homogeneous, than reduce temperature to 80 to 85C. Continue slow mixing.
2. Disperse pigments in section "B" in the castor oil contained in phase "B" by adding the dry pigments to the castor oil and blending with a spatula, then disperse using a Cowles dissolver and finally pass the blend through a three roller mill three times or until the particle size of the pigments are less than number 7 on a Haggeman gauge.
3. When section "B" is adequately dispersed, add section "B" to section "A". Continue slow agitation and maintain a tempera- ture of 80 to 85C.
4. Mold sticks at 80-85C.
   Formula 398-152-2

Lip Balm

Resulting product is a protective lip balm that exhibits extended wear characteristics due to Lexorez 100. The product could be positioned as an over the counter (OTC) pharmaceutical skin protectant with petrolatum as the active ingredient, and as a sunscreen, with octyl methoxycinnamate as the active ingredient

	%w/w
Octyl Methoxycinnamate (Escalol 557)	10.00
Petrolatum USP (Super White)	20.00
Glycerin/Diethylene Glycol/Adipate Crosspolymer	
(Lexorez 100)	10.00
Petrolatum (and) Lanolin Alcohol (Lexate PX)	49.00
Candelilla Wax (S&P 75)	8.00
Carnauba Wax (S&P #1 Yellow 73)	3.00

Procedure:
1. Combine all ingredients and heat to 80C.
2. Mix until homogeneous.
3. Cool to 65C and fill into propel-repel containers.
   Formula 383-186
SOURCE: Inolex Chemical Co.: Suggested Formulations

Makeup Foundation

This rich, elegant foundation incorporates Geahlene 750 for added moisturization and smooth application.

Ingredient/Trade Name:	Weight%
A Deionized Water	46.50
Magnesium Aluminum Silicate/Veegum Ultra	0.80
Xanthan Gum/Keltrol	0.30
Nylon-12/Orgasol 2002D Natural Extra Cos	1.20
Sodium PCA/Ajidew N-50	1.00
B Deionized Water	10.00
Propylene Glycol	7.00
Iron Oxides (and) Talc/Lo Micron Pink	0.40
Iron Oxides (and) Talc/Lo Micron Yellow	2.00
Iron Oxides (and) Talc/Lo Micron Black	0.15
Titanium Dioxide/3328 Titanium Dioxide	5.60
C Glyceryl Stearate/Lexemul 55G	2.00
Stearic Acid/Emersol 132	2.00
DEA-Cetyl Phosphate/Amphisol	2.00
Methylparaben (and) Butylparaben (and) Ethylparaben (and) Propylparaben/Nipastat	0.25
Isostearyl Neopentanoate/Dermol 185	3.00
Mineral Oil (and) Hydrogenated Butylene/Ethylene/ Styrene Copolymer (and) Hydrogenated Ethylene/ Propylene/Styrene Copolymer//Geahlene 750	15.00
Phenoxyethanol/Emeressence 1160	0.70
Tocopherol Acetate/Vitamin E Acetate	0.10

Procedure:
Disperse Veegum Ultra into rapidly agitated deionized water. Mix well. Add the remaining part A ingredients, mixing well after each addition. In a separate container, homogenize part B until smooth. Add part B to part A. Heat to 80C. Mix until uniform. Heat part C to 80C and mix until the solids are dissolved. Add part C to part A. Mix for 30 minutes until homogeneous. Continue mixing and cooling to 35C.

SOURCE: Penreco: Suggested Formulation

Mascara

	Wt%
A. Veegum pharm	2.0
Tylose CB 30000	0.1
1,2-Propylene glycol	1.5
Water	63.0
B. Miglyol Gel B	10.4
Imwitor 780 K	3.0
Beeswax	5.0
Miglyol 812	2.0
Sicomet Black 80 C.I. 77499	4.0
C. Morpholine	0.4
Colophonium	1.5
Luviskol VA 64	2.0
Ethanol 96%	5.0
Preservative	qs

Preparation:
    The ingredients of A. are brought together and heated to
75-80C. Those of B. are brought to the same temperature and
homogenised. A. is emulsified in B.
    C. is dissolved and emulsified in A. and B. at approx. 30C.

Eyebrow Pencil, Heat Stable

	Wt%
A. Softisan 100	16.0
Softisan 378	8.0
Miglyol 812	8.0
Softisan 645	3.0
Softisan 154	2.0
Beeswax	3.0
B. Talc	20.0
Titanium dioxide	18.0
Zinc oxide	18.0
Sicomet Black 80 C.I. 77499	5.0
C. Perfume	qs

Preparation:
    The ingredients of B. are very finely ground. The fats A.
are melted and stirred slowly into the pigments. Perfume is
incorporated at around 30C. This is followed by pouring into
moulds or extrusion.

SOURCE: Huls America Inc.: Formulations for Cosmetics: Formulas

Moisture Stick Base

	%
Mineral Oil 80/90 Visc.	47.0
Ross Wax 26-1152	28.0
Ross Wax 15-1182	2.0
Ross Wax 1824	10.0
Jojoba Oil	2.0
Amerlate P	10.0
Vitamin E	1.0

Procedure:
Melt all ingredients together in a kettle to 170F under agitation. When mixed thoroughly pour into molds. Capping may be necessary.

Cream Rouge Formula

Part A:	%
Ross Refined #1 Yellow Carnauba Wax	6.0
Ross Ozokerite Wax 77W	10.0
Mineral Oil	24.0
Isopropyl Palmitate	27.0

Part B:	
Talc	10.0
Titanium Dioxide	20.0
Color	3.0

Procedure:
Melt Part A to 70C. When cooled run together with Part B on a three roll mill.

Pot Eyeshadow

Ingredients:	%
Mineral Oil 70/80	40.0
Petrolatum	15.0
Ross Ozokerite Wax 77W	20.0
Ross Refined Candelilla Light Flakes	4.0
Pigment Paste	20.0
Preservative	1.0

Procedure:
Grind color with oil and petrolatum in roller mill. Heat waxes until melted and add pigment paste. Maintain 85C for 30 minutes with agitation. Pour into molds.

SOURCE: Frank B. Ross Co., Inc.: Suggested Formulations

Moisturizing Liposomal Gel

Ingredients:		% w/w
1	A) Cremophor RH 410	5,00
2	Fragrance: PCV 1454	0,10
3	B) Water demineralized	81,50
4	Glycerin	2,00
5	Nipagin	0,15
6	Euxyl K-200	0,30
7	Phytaluronate	5,00
8	Fitobroside	5,00
9	Allantoin	0,30
10	Carbopol 941	0,30
11	Triethanolamine	0,35

Procedure:
    Dissolve item 2 in item 1.
    Dissolve items 4-9 in water (3).
    Under stirring, slowly add phase B) to phase A). For thicken-
ing, incorporate and dissolve item 10. Adjust the pH with item
11 to 7.0. A turbid, fluid solution is obtained.
Application No. D 018.A/02.93

Skin Tightening Gel

Ingredients:		% w/w
1	A) Water demineralized	87,95
2	Phenonip	0,30
3	Euxyl K-200	0,20
4	Glucam E-10	5,00
5	B) Carbopol 940	0,75
6	Triethanolamine	0,80
7	C) Pentacare HP	5,00

Procedure:
    Dissolve items 2-4 in water (1).
    Stir fast and incorporate item 5.
    Neutralize and thicken item 6.
    Slowly add item 7.
Application No. D 032.0/06.93

SOURCE: Pentapharm Ltd.: Suggested Formulations

Moisturizing Make-Up Foundation

		% By Weight
(A)	Phospholipid SV	3.0
	0.5% Kelzan AR in 1% NaCl	64.5
	Propylene Glycol	8.0
	Pigment (9/12 White/Brown)	15.0
	Steareth-20	1.6
	Methyl Paraben	0.25
(B)	Isopropyl Palmitate	2.0
	Hexyl Laurate	2.0
	Steareth-2	2.4
	Dow Fluid 200/100 cs.	1.0
	Propyl Paraben	0.25

Procedure:
   Combine ingredients in phases A and B as shown and heat to 55C. Blend phase B into phase A with sufficient homogenization to ensure good emulsification. Stir cool to 40C, add fragrance, and package.

Comments:
   This liquid foundation is enhanced by the presence of Phospholipid EFA which aids in binding the pigment to skin for a longer lasting application. Further benefits obtained by using Phospholipid EFA include a smooth, non-chalky application and a non-drying afterfeel on the skin.

Skin Cleanser

	% By Weight
Sodium Laureth Sulfate (1 mole 25%)	28.00
Monamate LNT-40	12.50
Monamid 716	3.00
Phospholipid PTC	2.50
Water	52.75
Sodium Chloride	1.25

Procedure:
   Blend ingredients in order listed at room temperature. Adjust pH to 5.5-6.0 with citric acid. Add color and fragrance as required. Package.

Typical Properties:
   Appearance: Clear liquid
   Viscosity: Approximately 3,000 cps

SOURCE: Mona Industries, Inc.: Suggested Formulations

Moisturizing Milk with High Electrolyte Content

Ingredients:	%W/W
A Arlatone 985	4.00
Brij 721	2.00
Mineral oil	5.00
Caprylic/capric triglycerides	5.00
B Propylene glycol	3.80
Urea	10.00
Sodium chloride	5.00
Water	65.20
C *Preservative	
*Perfume	

*q.s. these ingredients.

Procedure:
Heat (A) and (B) to 70C separately. Slowly add (B) to (A) stirring thoroughly. Homogenize mixture. Remove from the homogenizer and allow to cool to 35C while stirring. Add (C) and cool to 30C. Package.

W/O Hydrating Body Milk

Ingredients:	%W/W
A Arlacel 1689	3.50
Arlamol HD	8.00
Arlamol M812	4.00
Arlamol DOA	4.00
Paraffin oil	5.50
B Glycerol	4.00
MgSO4.7H2O	0.50
*Preservative	
Water	70.35
C Perfume	0.15

*q.s. these ingredients.

Procedure:
Heat (A) and (B) to 70C separately. Slowly add (B) to (A) stirring thoroughly. Homogenize mixture for one minute. Allow to cool to 30C while stirring. Add (C) and while stirring. Package.

SOURCE: ICI Surfactants: Suggested Formulations

Nail Polish Remover

	Wt%
Ethyl acetate	25.0
Butyl acetate	25.0
Driveron S	23.0
Dionil OC/K	10.0
Solvent APV	10.0
Butyltriglycol	7.0

Preparation:
The ingredients are added to each other in sequence and stirred until homogeneous.

Nail Polish Remover

	Wt%
Ethyl acetate	50.0
Vestinol C	28.0
Ethyltriglycol	10.0
Castor oil	10.0
Paraffin oil	2.0

Preparation:
The ingredients are added together in sequence and stirred until homogeneous.

SOURCE: Huls America Inc.: Formulations for Cosmetics: Formulas

Non-Feathering Extended Wear Liptsick

Resulting product is a moldable lipstick with exceptional shine, application and extended wear attributes. When applied to lips, the formula produces a comfortable non-waxy film, which resists migration and feathering. Droop point of the stick is above 43C.

	%w/w
A. Castor Oil (Crystal O)	QS to 100
Candelilla Wax (S&P 75)	8.23
Carnauba Wax (S&P #1 Yellow 73)	2.07
Microcrystalline Wax (S&P 18)	4.19
Beeswax (S&P White NF 422P)	1.08
Glyceryl Stearate (and) PEG-100 Stearate (Lexemul 561)	1.09
Petrolatum (and) Lanolin Alcohol (Lexate PX)	2.01
Caprylic/Capric Triglyceride (Lexol GT-865)	1.08
Octyl Stearate (Lexol EHS)	2.06
Propylene Glycol Dicraprylate/Dicaprate (Lexol PG-865)	1.06
Silica (Cab-O-Sil M-5)	0.13
Tocopherol (Copherol F-1300)	0.12
Propylparaben (Lexgard P)	0.08
Butylparaben (Lexgard B)	0.05
Glycerin/Diethylene Glycol/Adipate Crosspolymer (Lexorez 100)	4.77
B. Pigments	5.00-10.00

Procedure:
1. Combine sufficient castor oil from section "A" with pigments in section "B" in a ratio of approximately 2 parts castor oil to 1 part pigments.
2. Disperse pigments in step number 1, first with a Cowles dissolver, then reduce the pigment particle size to less than number 7 on a Haggeman gauge by passing the dispersion through a three roller mill.
3. Combine the balance of ingredients in section "A", heating to 85-90C.
4. Begin agitation as soon as waxes begin to melt. Agitate slowly. Do not entrap air.
5. When section "A" is homogeneous and at 85-90C, add section "B" to section "A".
6. Pour into molds at 80-85C.

SOURCE: Inolex Chemical Co.: Formula 398-62-2

Oil/Water Matte Cream Makeup

Ingredients:		%W/W
A	Magnesium aluminum silicate	2.60
	Sodium carboxymethylcellulose	0.40
	Water, deionized	42.40
B	Dispersing agent	0.30
	Propylene glycol	5.00
	Water	12.30
C	Talc	18.50
	Kaolin	1.30
	Titanium dioxide	3.70
D	Iron oxides	1.50
	Isopropyl myristate	5.00
E	Arlacel 20 and Tween 20	3.00
F	Stearyl alcohol and Americhol	4.00

Procedure:
Blend dry ingredients of (A) and add to water slowly agitating until smooth. Micropulverize (C) and add to (B). Mill to obtain a smooth paste. Add (B-C) to (A) and heat to 65C. Heat (D,E,F) to 70C and add to (A,B,C) blend. Mix until cool.

Lip Balm

Ingredients:		%W/W
A	Synthetic Spermaceti	10.00
	Petrolatum	40.00
	Glycerol monostearate	26.00
	Caprylic/Capric Triglyceride, Neobee M-5	13.60
	Squalane, Robane	10.00
B	Allantoin	0.20
	Lemon oil, USP	0.10
	Menthol	0.10

Procedure:
Melt (A) until uniform. Add (B) and stir until homogeneous. Pour in molds.

SOURCE: ICI Surfactants: Suggested Formulations

Oil-In-Water Milk Emulsion

Ingredients:	%W/W
A Arlatone 985	4.00
Brij 721	2.00
Caprylic/capric triglycerides	5.00
Arlamol HD	5.00
B Propylene glycol	1.25
G-2330	1.25
Water	81.50
C *Preservative	
*Perfume	

*q.s. these ingredients.

Procedure:
Heat (A) and (B) to 70C separately. Slowly add (B) to (A)
stirring thoroughly. Homogenize mixture when emulsion reaches
approximately 40C. Add (C) below 35C.
Formulation F41-3-2

Oil-In-Water Milk Emulsion

Ingredients:	%W/W
A Arlatone 985	4.00
Brij 721	2.00
Arlamol S7	6.00
Mineral oil	5.00
B G-2330	4.00
Water	79.00
C *Preservative	
*Perfume	

*q.s. these ingredients.

Procedure:
Heat (A) and (B) to 70C separately. Slowly add (B) to (A)
stirring thoroughly. Homogenize mixture. Remove from the
homogenizer and allow to cool to 35C while stirring. Add (C)
and cool to 30C. Package.
Formula F44-6-4

SOURCE: ICI Surfactants: Suggested Formulations

One Coat Mascara

This is a long wearing, non-flaking mascara which coats the lashes evenly in one coat. The formulation separates lashes easily during application and does not clump. The formulation does not flake during wear and has the additional benefit of washing off easily without the need for the use of special eye make-up removers or harsh scrubbing.

	%w/w
A. Deionized Water	57.70
Colloidal Magnesium Aluminum Silicate (Veegum HV)	1.80
Hydroxyethylcellulose (QP-15,000-H)	2.70
Xanthan Gum (Keltrol T)	4.45
Pigment (Pur Oxy Yellow BC and Pur Oxy Black BC)	8.95
Methylparaben (Lexgard M)	0.18
Propylparaben (Lexgard P)	0.09
B. PVP/VA Copolymer (Luviskol 73W)	10.00
Imidazolidinyl Urea (Germall 115)	0.20
Tetrasodium EDTA (Hamp-ene Na4)	0.08
C. Glyceryl Stearate (Lexemul 515)	2.00
Beeswax NF (S&P White NF 422P)	4.75
Glyceryl Stearate (and) PEG-100 Stearate (Lexemul 561)	3.50
Glycerin/Diethylene Glycol/Adipate Crosspolymer (Lexorez 100)	3.50
Sorbitan Stearate NF (Arlacel 60)	0.10

Procedure:
1. Combine "A" except for the pigments and heat to 80-85C. When temperature is reached, homogenize mixture slowly adding pigments.
2. Combine "C" and heat to 80-85C. Slowly add to "A".
3. Mix "AC" for five minutes. Add "B" to "AC" and homogenize for fifteen minutes on medium speed while cooling.
4. At 60C transfer batch to tank equipped with a double-action side-sweep agitator. Mix and cool to room temperature.
5. Adjust solids level to 59.1+-0.1%. Solids were run in a 110C oven for one hour.

SOURCE: Inolex Chemical Co.: Formulation 383-185

Overnight Facial Moisturizer

A powerful moisturizer designed to restore the normal state of healthy skin. The high substantivity towards skin helps to provide moisture regulation.

Part A:	% by Weight
Phospholipid EFA	4.00
PEG-32	2.00
Glycerin	2.00
Water	73.00

Part B:	
Steareth-2	2.50
Cetearyl Alcohol	4.00
Cetyl Palmitate	4.00
Myristyl Myristate	4.00
Isopropyl Palmitate	3.00
Dimethicone (100cS)	1.50

Blend and heat both phases separately to 70C. Homogenize (B) into (A) with continued heat for an appropriate time while avoiding aeration. Stir-cool to 45-50C, then add fragrance, coloring, or preservative as required and fill.

Moisturizing Make-Up Foundation

An elegant product containing Phospholipid EFA which provides smooth feel and coverage while eliminating the normal drying effects of cosmetic pigments on skin.

Part A:	% By Weight
Phospholipid EFA	3.00
0.5% Kelzan AR in 0.1% NaCl	72.50
Pigment	15.00
Steareth-20	1.80
Methyl Paraben	0.25

Part B:	
Isopropyl Myristate	2.00
Hexyl Laurate	2.00
Steareth-2	2.40
Dimethylpolysiloxane (200cS)	1.00
Propyl Paraben	0.25

Combine ingredients in both phases separately and heat to 65C. Homogenize Part A thoroughly. Add (B) to (A) and continue to homogenize. Stir-cool, with minimal aeration, to 40C, then fill.

SOURCE: Mona Industries, Inc.: PHOSPHOLIPID EFA: Suggested Formulations

O/W Cleansing Milk

Ingredients:	%W/W
A Arlatone 985	4.00
Brij 721	2.00
Arlamol HD	10.00
B Atlas G-2330	3.00
*Preservative	
Water	80.90
C Perfume	0.10

*q.s. these ingredients.

Procedure:
Heat (A) and (B) to 70C separately. Slowly add (B) to (A) stirring thoroughly. Homogenize mixture for one minute. Allow to cool to 30C while stirring. Add (C) and while stirring. Package.

W/O Cleansing Milk

Ingredients:	%W/W
A Arlacel 780	6.00
Arlamol S7	6.00
Arlamol HD	10.00
B Arlas G-2330	4.00
MgSO4.H2O	0.70
*Preservative	
Water	73.15
C Perfume	0.15

*q.s. these ingredients.

Procedure:
Heat (A) and (B) to 70C separately. Slowly add (B) to (A) stirring thoroughly. Homogenize mixture for one minute. Allow to cool to 30C while stirring. Add (C) and while stirring. Package.

SOURCE: ICI Surfactants: Suggested Formulations

O/W Cleansing Milk

Ingredients:	%W/W
A Paraffin oil perliquidum	15.00
Arlamol E	3.00
B Arlatone 2121	2.00
Propylene glycol	1.25
Glycerol	1.25
*Preservative	
C Water	72.50
Carbomer 934 (3% w/w mastergel)	5.00

D *NaOH (10% w/w solution) to adjust pH to 6.5-7

E *Perfume

*q.s. these ingredients.

Procedure:
    Mix the Arlatone 2121 in the water phase at 80C under moderate
stirring until a homogenous dispersion is formed. Disperse the
hydrocolloid in the heated water phase at 75C with moderate stirring. Add the oil phase (RT) to the water phase while stirring
intensively. Homogenise the mixture between 75 and 65C. Energy
input is related to final formulation viscosity. Allow to cool
to 40C while stirring. Homogenise again if adding additional
ingredients. e.g. perfume, vitamins,......

Dry Skin Oil with Alcohol

Ingredients:	%W/W
Arlacel 186	30.00
Arlamol E	30.00
Ethanol, SDA-40, (190 Proof)	30.00
Cyclomethicone	10.00

Procedure:
    Mix all ingredients well at room temperature. If necessary
warm the Arlacel 186 to achieve clarity before mixing.
Formula CP 1049

SOURCE: ICI Surfactants: Suggested Formulations

Petroleum Oil/Petroleum Wax Free Lipstick

This lipstick is a creamy stick that doesn't use any petroleum oils or waxes in the formula nor does it use any synthetic
raw materials that has petroleum base products in it.

Oil Phase I:
1. Kester Wax 100 (Koster Keunen)	6.00
2. Candelilla (Koster Keunen)	7.00
3. Hydroxy Polyester (Koster Keunen)	3.70
4. Hexanediol Behenyl Beeswax (Koster Keunen)	4.00
5. Cocoa Butter	3.00
6. Shea Butter (Koster Keunen)	4.50
7. Squalane (Centerchem)	1.50
8. Rice Bran Oil (Koster Keunen)	4.00
9. Octyl Palmitate (Inolex)	3.00
10.Coconut Oil (Alnor)	5.50
11.Vitamin E (BASF)	0.10
12.Vitamin A Palmitate (BASF)	0.10
13.Wheat Germ Oil (Lipo)	1.50
14.Sweet Almond Oil (Lipo)	1.50
15.Isopropyl Palmitate (Unichema)	15.50
16.Oyster Nut Oil (Koster Keunen)	1.50
17.Castor Oil (Caschem)	20.00
18.Ethoxylated Carnauba (Koster Keunen)	1.30
19.Orange Wax (Koster Keunen)	5.50
20.TiO2 (WC&D)	2.50
21.Z-Cote (SunSmart)	5.00
22.Pigment	2.80
23.Beta Glucan 15% Coarse (Koster Keunen)	0.50

Procedure:
Add 20,21 and 22 together. To the mixture of pigments add
17, 18, and 19, heat till everything is dispersed. Add the rest
of the ingredients except the Beta Glucan. The Beta Glucan will
be added last to the batch, just before pouring into the molds.

Adaption of formula and its influence on the product:
With the increase or decrease of oils and/or waxes one could
create a stick that is creamier or harder.

SOURCE: Koster Keunen, Inc.: Suggested Formulations

Pressed Powder Eye Shadow
(Pearlescent Type)

Ingredients: % by Weight
(1) Bentonite (NF grade)   5.0
(2) Mearlin-AC   35.0
(3) Kaopolite TLC   29.0
(4) Zinc Stearate (fine ground)   8.0
(5) Magnesium Carbonate   1.0
(6) Acetol   3.0
(7) Polysorbate 20   9.0
(8) Deionized Water   10.0
    Preservative   q.s.

Procedure:
   Dry mix (1), (2), (3), (4), and (5). In separate tank,
thoroughly mix (6), (7), and (8). Then slowly add this mixture
to the dry mix. Screen through a No. 16 sieve and press.
   Follow recommended handling practices of the supplier of each
product used.

Face Mask

Ingredients: % by Weight
(1) Deionized Water   36.7
(2) Bentonite (NF grade)   8.0
(3) Kaolin USP   40.0
(4) Isopropyl Alcohol   10.0
(5) Lanolin (PEG 50)   0.5
(6) Glyceryl Stearate (Emerest 2000)   1.5
(7) Dimethicone (Dow Corning 200)   1.5
(8) Phenoxyethanol (Emmenessence)   1.5
(9) Methyl Paraben   0.2
(10)Propyl Paraben   0.1

Procedure:
   Add (2) to (1) and heat to 90C, add (9) and (10) until
dissolved. Add (5), (6), (7), and (8). Cool and add (4) with
good mixing until dispersed. Add (3) and continue high speed
mixing until completely uniform.
   Follow recommended handling practices of the supplier of
each product used.
   Good industrial practices should be used when handling
flammable ingredients.

SOURCE: Kaopolite, Inc.: Suggested Formulations

Pressed Powder Eyeshadow

Phase A:	%Wt.
Talc	40.40
Zinc Stearate	4.00
Carmine	2.00
Gemtone Mauve Quartz	19.00
Flamenco Super Blue	12.60
Antimicrobial	q.s.

Phase B:	
Antioxidant	q.s.
Squalane	5.00
Carnation Mineral Oil	7.00

Phase C:	
Flemenco Super Blue	10.00

Procedure:
Thoroughly blend and disperse A. Add B into a support vessel, heat and mix until uniform. Spray B into premixed A and continue blending. Pulverize and return to blender. Add C to A-B and mix until uniform.

Molded Eye Shadow

	%Wt.
Polytetrafluorethylene-treated mica	50.0
Mica	27.0
Titanium-dioxide-coated mica	10.0
Colorant	3.0
Dimethylpolysiloxane	4.0
Carnation Mineral Oil	3.0
Glyceryl trioctanoate	3.0
Fragrance	0.1
Ethanol	q.s.100.0

Cosmetic Aerosol Foam

PEG-2 stearyl ether	4.50%
PEG-10 stearyl ether	1.50
Carnation Mineral Oil	7.00
Isopropyl palmitate	7.00
Octyldodecanol	6.50
Cetyl Alcohol	0.50
Water	qs to 100.00
Propellant=N2O	

SOURCE: Witco Corp.: Petroleum Specialties Group: Suggested Formulations

Pressed Powder with 1% Eldew

Ingredient:	% by Weight
Talc Supra A	53.54
Nylon 12	10.00
Lecithin Treated Sericite	10.00
Lecithin Treated Mica	10.00
Lauroyl Lysine/Amihope LL	5.00
Zinc Stearate	3.00
Isopropyl Lanolate	2.70
Hydroxylated Lanolin	1.50
Isopropyl Isostearate	1.20
Cholesteryl/Behenyl/Octyldodecyl Lauroyl Glutamate/ Eldew CL-301	1.00
Butyl Stearate	0.60
Germall II	0.30
Methylparaben	0.20
Propylparaben	0.10
Cosmetic Brown 3277	0.34
Cosmetic Brown 1985	0.32
Cosmetic Brown 1654	0.18
D & C Red 30 Lake	0.02

Pressed Powder with 3% Eldew

Ingredient:	% by Weight
Talc Supra A	52.49
Nylon 12	9.80
Lecithin Treated Sericite	9.80
Lecithin Treated Mica	9.80
Lauroyl Lysine/Amihope LL	4.90
Zinc Stearate	2.94
Isopropyl Lanolate	2.65
Hydroxylated Lanolin	1.47
Isopropyl Isostearate	1.18
Cholesteryl/Behenyl/Octyldodecyl Lauroyl Glutamate/ Eldew CL-301	2.95
Butyl Stearate	0.59
Germall II	0.29
Methylparaben	0.20
Propylparaben	0.10
Cosmetic Brown 3277	0.33
Cosmetic Brown 1985	0.31
Cosmetic Brown 1654	0.18
D & C Red 30 Lake	0.02

SOURCE: Ajinomoto USA, Inc.: Suggested Formulations

Prolonged Skin Moisturizer

Emulsion Phase:

Spermaceti	3.00%
Beeswax	2.00
Carnation Mineral Oil	15.00
Behenyl alcohol	5.00
Preservatives	0.30
Elastin hydrolysate	2.00
1,3 butylene glycol	10.00
Carboxymethylchitin	0.50
Water	53.20

Liposome Phase:

Soy lecithin	2.00
Sphingoglycolipids	1.00
Phytosterol	0.50
Glycerine	5.00
Vitamin A	

Skin Lightening Composition

Polyoxyethylene-polyoxypropylene cetyl ether	1.00%
Silicone oil	2.00
Blandol Mineral Oil	3.00
Propylene glycol	5.00
E-Aminocaproic acid	1.00
Glycerin	2.00
Ethyl alcohol	5.00
Carbomer	0.30
Hydroxypropyl cellulose	0.10
2-Aminomethylpropanol	0.10
Ascorbic acid disulfate-sodium salt	1.00
Preservative, antioxidant, fragrance	q.s.
Water	q.s. to 100.00

Three Layer Facial Moisturizer

Sodium pyrrolidone carboxylate (50%)	10.00%
Glycerin	20.00
Polyoxyethylene glyceryl monooleate	20.00
Carnation Mineral Oil	50.00

**SOURCE: Witco Corp.: Petroleum Specialties Group: Suggested
      Formulations**

Purifying Clay Face Mask

This product removes imbedded impurities which smoothes skin, cleans pores, brings a glow and removes dulling cells for a finer texture so skin has a radiant glow.

Phase A:

Magnabrite S (Whittaker & D)	4.5%
Xanthan Gum (Kelko)	0.2%
Water (Distilled)	63.2%
Glycerin (Unichema)	4.0%
Aloe Vera Gel (Active Organics)	0.5%
Propylene Glycol (Arco)	1.5%
Tween 60 (Gallard Schlesinger)	0.6%
Methyl Paraben (Sutton)	0.5%

Phase B:

Emulsifying Wax NF (Koster Keunen)	5.0%
Cetylstearyl Alcohol (P&G)	0.5%
Kester Wax K-48 (Spermaceti, Koster Keunen)	2.0%
Propyl Paraben (Sutton)	0.5%

Phase C:

Cera Bellina (Pg-3 Beeswax: Koster Keunen)	4.5%
Green Chromium Oxide (Sun Chemical)	0.5%
Titanium Dioxide (Whittaker C&D)	2.0%

Phase D:

Titanium Dioxide (Whittaker C&D)	4.0%
Vanclay (Vanderbilt)	6.0%

Procedure:

Heat and mix till homogeneous, Phase A to 80C. Heat and mix Phase B to 80C. Heat and mix Phase C till pigments are dispersed. Mix Phase B with Phase C, add BC to A at 80C under agitation. When emulsified, sprinkle a mixed Phase D into ABC and continue mixing till thoroughly dispersed, allow to cool while mixing. Pour at 55-60C.

Adaption of formula and its influence on the product:

Changes in clay concentration will reduce or increase drying time on the skin, which also changes the respective purifying qualities toward the skin. Moisturizing agents and or other actives can be added to enhance the properties of this product.

SOURCE: Koster Keunen, Inc.: Suggested Formulations

Slenderizing Cream with Green Tea Extract

This slenderizing cream incorporates Phyt'iod and Iodobio 45.
This lipolitic product develops an elective metabolic action at
the level of adipocytes through the inhibition of the phospho-
diesterase. This inhibition blocks the degradation of the cyclic
AMP which is indispensable to the protein kinase activation.
The kinase proteins activate the lipase which trigger a degradation of the triglycerides in the adipose cell in the form of
fatty acids and glycerol and then elminated.

Ingredients:	%
A. Unitina KD-16	16.00
Eumulgin B-1	1.00
Lexol IPM	8.00
Carnation white mineral oil	6.00
Tocopherol	0.01
Trisept P	0.10
B. Deionized Water	48.49
Trisept M	0.20
C. Ivy Extract	3.00
Seaweed Extract	2.00
Organic Silicon	1.00
D. Deionized Water	6.00
Iodobio 45	2.00
Green Tea Extract	1.00
E. Deionized Water	2.00
Trisept IU	0.20
F. Phyt'iod	3.00

Procedure:
    Heat phase A and B to 75-80C. Under proper agitation, add
A to B. At 50C, add phase C. Dissolve Iodobio 45 in water.
At 40C, add phase D, E and F.

Amidroxy Sugar Cane Cream

Ingredients:	%
A. Cirami N 1	11.50
Jojoba Oil	10.00
Sunflower Oil	5.00
Trisept P	0.10
B. Deionized Water	61.94
Trisept M	0.15
C. Tensami 3/06	3.00
D. Tocopherol	0.01
E. Amidroxy Sugar Cane	8.00
F. Grapefruit Essential Oil	0.30

Procedure:
    Heat phase A and phase B to 70C. Add phase A to phase B with
proper agitation. Add phase C. At 35C-40C add phase D, phase E
and phase F. Mix to room temperature. Adjust pH to 3.50.

SOURCE: TRI-K Industries, Inc.: Code AMI Suggested Formulations

Soothing Body Gel

A clear, viscous gel with microencapsulated Vitamin E designed to provide skin conditioning and soothing properties.

Ingredient/Trade Name:	Weight%
A Mineral Oil (and) Hydrogenated Butylene/Ethylene/ Styrene Copolymer (and) Hydrogenated Ethylene/ Propylene/Styrene Copolymer//Geahlene 750	78.82
B Cyclomethicone/Dow Corning 344 Fluid	7.00
C12-15 Alkyl Benzoate/Finsolv TN	10.00
Bisabolol, synthetic	0.30
Methoxypropanediol/Cooling Agent No. 10	0.50
Fragrance	0.15
Menthol	0.03
Corn Oil (and) Retinyl Palmitate/Vitamin A Palmitate type PIMO/BH	0.10
Sunflower Oil (and) Chamomile Extract/Chamomile LS	0.50
C Macademia Nut Oil/Refined Macadamia Nut Oil	2.00
Propylparaben	0.10
D Tocopherol Acetate (and) Acacia (and) Gelatin (and) Mica (and) Titanium Dioxide (and) Iron Oxides//Microencapsulated Vitamin E HC487	0.50

Procedure:

Premix B. Add B to A while mixing at moderate speed. Heat C to 60C until clear. Add C to A. Mix until completely homogeneous. Add D (pre-washed with anhydrous ethanol) and gently mix into A until uniformly dispersed.

SOURCE: Penreco: Suggested Formulation

Anhydrous Deep Penetrating Gel

Simple but effective is a way to describe this quick penetrating product. Use of this gel will diminish the signs of aging by conditioning and treating the skin with anti-inflammatory agents. The activity comes from phytosterols and bioflavonides in orange wax. These class of chemicals are also strong anti-oxidants which will scavenge free radicals thereby preventing potential premature aging.

Silicone Oil 556	56.0%
IPP (Unichema)	5.0%
Squalane (Polyesther)	10.0%
Cera Albalate 103 (Koster Keunen)	14.0%
Apricot Oil (Arista)	2.0%
Orange Wax (Koster Keunen)	10.0%
Vit. A Palmitate (Roche)	0.5%
Vit. E (BASF)	0.5%
Candelilla Wax (Koster Keunen)	2.0%

Procedure:

Melt and mix all components to 70C, cool to 60C and pour into container.

Adaption of formula and its influemce on the product:

Triglyceride oils can only be use sparingly, since incorporation at any significant concentration will destroy the gelling properties of the Hydroxy-Hexanyl-Behenyl-Beeswaxate (Cera Albalate 103). Cera Albalate 103 is a highly effective gelling agent for very low viscosity oils, as is indicated above.

SOURCE: Koster Keunen, Inc.: Suggested Formulation

Stick Type Lip Gloss

Materials:	% by Weight
1. Ozokerite White 170	7.5
2. Candelilla Wax Light Refined	9.5
3. Carnauba Wax Yellow USP #1	2.5
4. Ceraphyl 50	3.0
5. Lanogene	5.0
6. Acetol 1706	11.5
7. Flexricin 9	10.0
8. Bentone Gel CAO	27.5
9. Lexol PG 8/10	12.8
10. Isopropyl Myristate	5.0
11. Propyl Paraben	0.1
12. Tenox 4	0.1
13. Pigment Concentrate*	5.0
14. Fragrance	0.5

*Pigment Concentrate:	
Castor Oil	75.00
D&C Red #21 Aluminum Lake	15.00
Cosmetic Brown C-33 115	10.00

Manufacturing Directions:
1. Weigh items 1 through 12 into a stainless steel jacketed kettle. Heat to 85C until melted clear.
2. Bring temperature down to 80C. Add and mix in item 13 using homomixer at medium speed. Mix for 10 minutes at 80C.
3. Continue mixing at medium speed, lower temperature to 70C, and add item 14. Remove heat and stir slowly until congealed, or pour into trays.
4. When casting, melt and pour molten mass into mold at 75C with slow agitation (avoid air entrapment).
   Note: For a shiny polished surface on the lip-gloss, the sticks should be surface flamed.

Pigment Concentrate Preparation:
1. Weigh dry powder and mix to the castor oil using a slow speed Hobart type mixture until uniform.
2. Give this concentrate two passes over a 3 roller mill at room temperature.

SOURCE: Rheox, Inc.: Formula TS-286

Thickening Mascara

This is a long wearing, non-flaking and thickening mascara formula. This formulation applies effortlessly, separates lashes well and does not smear or clump. The formulation does not flake during wear and has the added benefit of washing off easily without the need of special eye make-up removers or harsh scrubbing.

	%w/w
A. Deionized Water	56.95
Colloidal Magnesium Aluminum Silicate (Veegum HV)	1.80
Hydroxyethylcellulose (QP 15,000-H)	2.70
Xanthan Gum (Keltrol T)	4.45
Pigment (Pur Oxy Yellow BC and Pur Oxy Black BC)	8.95
Methylparaben (Lexgard M)	0.18
Propylparaben (Lexgard P)	0.09
B. PVP/VA Copolymer (Luviskol 73W)	10.00
Imidazolidinyl Urea (Germall 115)	0.20
Tetrasodium EDTA (Hamp-ene Na4)	0.08
C. Glyceryl Stearate (Lexemul 515)	2.25
Beeswax NF (S&P White NF 422P)	5.25
Glyceryl Stearate (and) PEG-100 Stearate (Lexemul 561)	3.50
Glycerin/Diethylene Glycol/Adipate Crosspolymer (Lexorez 100)	3.50
Sorbitan Stearate NF (Arlacel 60)	0.10

Procedure:
1. Combine "A" except for the pigments and heat to 80-85C. When temperature is reached, homogenize mixture slowly adding pigments.
2. Combine "C" and heat to 80-85C. Slowly add to "A".
3. Mix "AC" for five minutes. Add "B" to "AC" and homogenize for fifteen minutes on medium speed while cooling.
4. At 60C transfer batch to tank equipped with a double-action side-sweep agitator. Mix and cool to room temperature.
5. Adjust solids level to 53.2%+-0.1%. Solids were run in a 110C oven for one hour.

SOURCE: Inolex Chemical Co.: Formulation 383-198

Thigh Slimming Cream

The following formula is used for developing thigh slimming and anti-cellulite creams. The final formulation is 0.50% Natural Cocoa Extract and 0.50% Caffeine which are encapsulated in unilamellar and multilamellar liposomes. They are formulated to target subcutaneous fat tissue. A similar formula can utilize aminophyline as the active ingredient.

Ingredient:	%
A. Deionized Water	69.06
Carbopol 940	0.40
Glycerin	1.50
Kelate 220	0.10
Trisept M	0.20
Tween 80	1.00
Dow Corning 2501	1.50
Dow Corning 193	1.00
B. Lanette 16	1.25
Kukui Nut Oil	3.50
Emersol 132	2.00
Butyl Stearate	0.50
Butylated Hydroxy Toluene	0.10
Emerest 2717	1.50
Lexol IPP	1.00
Dow Corning 200 (50 cps)	1.00
C. Triethanolamine	0.80
Deionized Water	5.00
D. Tristat IU	0.25
E. Liposome of 6% Cocoa Extract & 6% Caffeine Concentrate	8.34

Procedure:
Using moderate propellor agitation, spinkle Carbopol 940 into Phase A and mix until fully hydrated. Add remaining Phase A ingredients. Heat Phase A and Phase B to 65C-70C. With propellor agitation, add Phase B to Phase A. When uniform, add Phase C. At 55C-60C, add Phase D. Add Phase E at 35C, mixing slowly until uniform.

SOURCE: TRI-K Industries, Inc.: Code Biozone

Ultra Rich Hand and Body Nourisher

Veegum Ultra thickens and stabilizes this creamy, oil-in-water emulsion prepared using a nonionic emulsifier system. It features the "dry touch" application properties typical of formulas containing Veegum. Palmitoyl hydrolyzed animal protein functions as an anti-irritant that complements the group of oil phase emollients. Veegum Ultra whitens and brightens this cosmetic formula and adjusts the pH to approximate that of normal skin.

Ingredient:	% by Wt.
A Veegum Ultra (Magnesium Aluminum Silicate)	1.50
Deionized Water	79.00
B Glycerin	3.00
Aloe Vera Gel (2)	2.00
C Palmitoyl Hydrolyzed Animal Protein (3)	1.00
Cetyl Esters (4)	1.00
Glyceryl Stearate SE (5)	2.50
Isopropyl Palmitate	5.00
Sorbitan Palmitate	2.25
Polysorbate 40	2.75
D Preservative, Fragrance	q.s.

(2) Veragel Liquid
(3) Lipacide PCO
(4) Crodamol SS
(5) Kessco GMS-24SE

Procedure:
Heat the water to 55C. Add Veegum Ultra slowly while mixing at 700 rpm with a propeller stirrer. Increase mixer to 1500-1700 rpm and mix for 30 minutes while maintaining temperature at 55C. Add B to A and mix until uniform. Heat C to 60C and add to (A and B). Mix (A,B and C) for 30 minutes. Avoid air entrapment. Slow mixer to 1000 rpm and mix while cooling to 35C. Add D and mix until uniform. Package.

Product Characteristics:
Viscosity: 1900-2500 cps
pH: 5.3+-0.2

Comments:
This prototype formula is designed to serve as a guide for the development of new products or the improvement of existing ones.

SOURCE: R.T. Vanderbilt Co., Inc.: Formulation No. 447

Vegetable Based Moisturizing Cream

Ingredients:	Percent
Part A:	
Super refined almond oil	20.0
"Crodesta" F10	3.0
"Syncrowax" HGLC	2.0
Part B:	
"Hydrosoy" 2000 SF	1.0
"Crodesta" F110	3.0
"Sorbo" 70	5.0
"Germaben" 11	1.0
"Veragel" 200	0.5
Deionized Water	64.5

Procedure:
   Combine Part A. Heat to 70C+-2C. Combine Part B. Heat to 70C+-2C. Combine Part B to Part A under agitation. Cool and fill off.

SOURCE: Dr. Madis Laboratories Inc.: Suggested Formulation

Liquifying Cream Makeup Remover

	%
Penreco Mineral Oil #9	51.08
Petrolatum Alba	32.8
Rosswax 60-0254	9.1
Ross Ceresine Wax 1160/7	7.0
Beta Carotene 30%	0.02
Fragrance	q.s.
Preservative	q.s.

Procedure:
   Melt ingredients one thru four in a steam jacketed kettle to 170F with good agitation. When fully mixed cool, add the rest of the ingredients and pack at about 130F.

SOURCE: Frank B. Ross Co., Inc.: Suggested Formulation

Water Resistant Liquid Foundation

The resulting product is a thin lotion that feels like a fine powder.

	%w/w
A. Deionized Water	69.25
Colloidal Magnesium Aluminum Silicate (Veegum)	1.00
Xanthan Gum (Keltrol T)	0.50
Methylparaben (Lexgard M)	0.20
Propylene Glycol NF	5.00
Triethanolamine NF	QS
B. Octyl Stearate (Lexol EHS)	8.50
Glyceryl Stearate (and) PEG-100 Stearate (Lexemul 561)	3.50
Glycerin/Diethylene Glycol/Adipate Crosspolymer (Lexorez 100)	2.00
Propylparaben (Lexgard P)	0.05
C. Talc (J-13-AT)	5.00
Iron Oxides (AT Series)	5.00

Procedure:
1. Combine "A" heating to 80-85C. Add all ingredients except the Veegum and xanthan gum. Dry blend the gums and sift in.
2. Combine "B" and heat to 80C.
3. Move "A" to homogenizer. Turn to a low speed.
4. Add "B" to "A", homogenize at low speed for 5 minutes. Begin cooling.
5. Add "C" to "AB",homogenize at low speed for 10 minutes or until particles are dispersed.
6. Remove from homogenizer and transfer to propeller mixer.
7. At 45-50C adjust pH to 7.5-7.7 with TEA. Adjust for water loss
Formula 383-188

Lip Gloss

The resulting product is a cake lip gloss which could be hot poured into a pan or directly into a SAN compact. The lip gloss applies easily. This product is very wear resistant and exhibits a high gloss due to Lexorez 100.

	%w/w
Castor Oil NF (Crystal O)	60.90
Beeswax NF (S&P White 422P)	10.00
Lanolin Anhydrous	1.30
Caprylic/Capric Triglyceride (Lexol GT-865)	14.80
Stearyl Stearate (Liponate SS)	4.30
Glycerin/Diethylene Glycol/Adipate Crosspolymer (Lexorez 100)	8.70

Procedure:
1. Combine ingredients, heat to 75C. Mix slowly so that air is not entrapped. 2. Pour at 50C.
Formula LP-100
SOURCE: Inolex Chemical Co.: Suggested Formulations

Section V
Creams

All Purpose Cream

		Wt%
(O)	Liquid Paraffin (#70)	10.0
	Squalane	4.0
	Nikkol WCB	5.0
	Glyceryl Monostearate (Self Emulsifying Type)	8.0
	Cetyl Alcohol	4.0
	Polyoxyethylene (20) Glyceryl Monostearate	2.7
	Sorbitan Monostearate	1.3
(W)	Ajidew N-50	2.0
	Sorbitol (70%)	2.0
	Water	61.0
	Preservative	q.s.

Procedure:
1. Heat (O) and (W) to 80C.
2. Add (W) to (O) slowly with agitation.
3. Cool to 40C with stirring.
   pH: 5.6

All Purpose Cream

		Wt%
(O)	Liquid Paraffin (#70)	21.6
	Paraffin Wax (mp 42-44C)	8.0
	Cetyl Alcohol	5.0
	Nikkol WCB	10.0
	Cetyl Alcohol	4.0
	Tocopherol Acetate	0.2
	Allantoin	0.2
	Sorbitan Monostearate	2.0
	Polyoxyethylene (20) Sorbitan Monooleate	3.0
(W)	Ajidew N-50	3.0
	Propylene Glycol	2.0
	Water	45.0
	Preservative	q.s.

Procedure:
1. Heat (O) and (W) to 80C.
2. Add (W) to (O) slowly with agitation.
3. Cool to 40C with stirring.
   pH: 5.4

SOURCE: Ajinomoto USA, Inc.: Suggested Formulations

All Purpose Cream

	Wt%
(O) Amiter LGOD	10.0
Squalane	9.0
Spermaceti	2.0
Propylene Glycol Monostearate	4.0
Polyoxyethylene (20) Glyceryl Monostearate	3.0
Glyceryl Monostearate (Self Emulsifying Type)	10.0
Sorbitan Monostearate	1.5
(W) Ajidew N-50	3.0
Sorbitol (70%)	4.0
Sodium Carbonate (10% aq soln)	0.1
Water	53.1
Preservative	0.3

Procedure:
1. Heat (O) and (W) to 80C.
2. Add (W) to (O) slowly with agitation.
3. Cool to 40C with stirring.
    pH: 5.5

Vanishing Cream

	Wt%
(O) Stearic Acid	8.0
Liquid Paraffin (#70)	3.0
Isopropyl Myristate	3.0
Hydrogenated Lanolin	2.0
Nikkol WCB	2.0
Glyceryl Monostearate (Self Emulsifying Type)	1.3
Polyoxyethylene (20) Cetyl Ether	1.7
(W) Ajidew N-50	3.0
Propylene Glycol	3.0
Glycerin	3.0
Triethanolamine	0.5
Water	69.5
Preservative	q.s.

Procedure:
1. Heat (O) and (W) to 85-86C.
2. Add (W) to (O) slowly with agitation.
3. Cool slowly to 30C with stirring.
    pH: 7.2

SOURCE: Ajinomoto USA, Inc.: Suggested Formulations

All Purpose Dry Skin Cream

Formula:	%Wt.
Phase A:	
Glucquat 100	1.0
Deionized Water	83.0
Phase B:	
Glucam P-20 Distearate	2.0
Glucate DO	0.5
Promulgen D	4.5
Acetulan	2.0
Cetal	1.0
Carnation Mineral Oil	5.0
Cetyl Palmitate	1.0
Perfume and Preservative	q.s.

Procedure:
   Dissolve Glucquat 100 into deionized water and heat to 80C
with adequate agitation. Combine B ingredients and heat to 80C
with propeller mixing. Slowly add A to B and mix until uniform.
When material begins to thicken during cooling, change to slow
sweep agitation.

Dry Skin Cream

	Wt.%
Water	61.37
Protopet 1S	20.00
Dry Flo Starch	5.00
Silicone 344	5.00
Finsolv TN	5.00
Cetyl Alcohol	1.00
Sodium Stearyl 2 Lactalate	1.00
Propylparaben	0.15
Methylparaben	0.10
Carbomer 934	0.30-0.50
Triethanolamine	QS to Neutralize
Fragrance & Color	QS

SOURCE: Witco Corp.: Petroleum Specialties Group: Suggested
         Formulations

Almond Vanishing Cream with Collagen

Phase 1:	%
Rosswax 63-0412	5.9
Rosswax 573	8.9
Amerlate P	0.7
Emerest 2314	0.7
Emerest 2316	0.7
Glyceryl Monostearate SE	0.37
Almond Oil-Lipovol ALM	1.0
Phase 2:	
Emery 916 Pure Glycerine	6.0
Water	13.46
Triethanolamine	0.9
Phase 3:	
Collasol (Collagen)	0.37
Phase 4:	
Germaben II	1.0

Procedure:
In separate steam jacketed kettles heat both Phase (1) and (2) to a temperature of 170F with agitation. When the temperature is reached, add Phase (1) to Phase (2) with continued agitation. Next add Phase (3) and then Phase (4) both with agitation, cool to 120F and package.

Dry Skin Cream

Part (A):	%
Modulan	3.7
Amerchol L-101	4.2
Isopropyl Myristate	2.7
Sodium Stearate Pure	10.0
Glyceryl Mono Stearate SE	1.8
Ross Spermaceti Wax Sub. 573	5.5
Ross Jojoba Oil	1.8
Part (B):	
Water	59.7
Emery 916 Glycerine Pure	9.2
Triethanolamine	1.4
Part (C):	
Preservative	q.s.
Part (D):	
Fragrance	q.s.

Procedure:
Melt Part (A) and Part (B) in separate vessels to 170F under agitation. When temperature is reached, mix Part (A) to Part (B) and cool. Package in containers at below 120F.

SOURCE: Framk B. Ross Co., Inc.: Suggested Formulations

Alpha Hydroxy Acid Cream

Resulting product is a light cream that could be dispensed from a flexible tube or a jar. The alpha hydroxy acid (AHA) contained in this formula is buffered lactic acid with a use level of approximately 10%.

	%w/w
A. Deionized Water	65.80
Methylparaben (Lexgard M)	0.20
2,4-Dichlorobenzyl Alcohol (Myacide SP)	0.20
Glycerin USP (99.7%)	3.00
Lactic Acid USP (88%)	0.60
B. Glyceryl Stearate (and) Stearamidoethyl Diethylamine (Lexemul AR)	2.50
Glyceryl Stearate (and) PEG-100 Stearate (Lexemul 561)	4.00
Stearamidopropyl Dimethylamine (Lexamine S-13)	1.00
Cetyl Alcohol (Adol 52)	1.00
Caprylic/Capric Triglyceride (Lexol GT-865)	6.00
Isobutyl Stearate (Lexol BS)	2.00
Propylparaben (Lexgard P)	0.10
C. Lactic Acid (88%)	5.60
Sodium Lactate (Purasal S/SP 60%)	8.00

Procedure:
1. Combine section "A" heating to 75-80C. Use propeller mixer for agitation.
2. Combine section "B" heating to 80-85C. Agitate slowly with a propeller mixer.
3. When sections "A" & "B" are homogeneous and at the designated temperatures slowly add "B" to "A". Mix with rapid agitation for 5 minutes then begin cooling.
4. Reduce mixing speed during cooling to prevent vortexing.
5. At 55-60C, add section "C" to sections "AB".
6. When the batch is homogeneous and at 35-45C adjust for water loss.
7. Mix to 30C.
8. Adjust final pH to 3.50-3.75 with Lactic Acid or Sodium Lactate, as required.

Physical Properties:
Viscosity: 10,000 cPs @ 24C (Brookfield RVT, TA @ 10 rpm). 24 hour sample.
pH: 3.5 @ 24C

SOURCE: Inolex Chemical Co.: Formulation 398-79-3

Alpha Hydroxy Acid Cream with 10% Aloe Vera Gel

Phase A:	%w/w
Deionized Water	50.45
Propylene Glycol	4.00
Xanthan Gum (Ticaxan)	0.50
Phenoxyethanol	0.30

Phase B:	
Squalane NF	4.00
PPG-12/SMDI (Polyolprepolymer-2)	3.00
Hydrogenated Phospholipid (Lecinol S-10)	1.00
Caprylic/Capric Triglyceride (Miglyol 812)	5.00
Stearic Acid	2.50
Caprylic/Capric/Stearic Triglyceride (Softisan 378)	2.00
Cyclomethicone (DC 344 Fluid)	4.00
Dimethicone (DC 200 Fluid; 50 vis.)	1.00
Cetearyl Alcohol (and) Ceteareth-20 (Cosmowax J)	2.00
Glyceryl Stearate (and) PEG-100 Stearate (Arlacel 165)	1.50
Steareth-2 (Brij 72)	0.50

Phase C:	
Lactic Acid, 88% (Biolac)	7.00

Phase D:	
Terry Aloe Vera Gel, 1X, Dec	10.00
Propylene Glycol (and) Diazolidinyl Urea (and)	
Methylparaben (and) Propylparaben (Germaben II)	1.00

1. Slowly disperse Xanthan gum to water while stirring vigorously. Continue mixing at rapid speed until completely
   dispersed and hydrated. Heat to 70C. Add Propylene Glycol
   followed by Phenoxyethanol.
2. Combine all ingredients in Phase B. While slowly mixing
   commence heating to 70-75C. Continue mixing until uniform.
3. Slowly introduce Phase B into Phase A. Continue to mix for
   5-10 minutes.
4. Cool to 40C while continuing to mix. With batch at 35C, add
   Phase C and continue to mix.
5. Combine all Phase D ingredients, stirring until dissolved.
   Add to batch while mixing until uniform. Cool to 25C.

SOURCE: Terry Laboratories, Inc.: Suggested Formulations

Aloe Vera Night Cream

Ingredients:	Percent
A. Deionized Water	49.225
Tetrasodium EDTA	0.075
Propylene Glycol	3.50
Methylparaben	0.20
B. Cetyl Alcohol (Adol 52 NF)	2.00
Cetearyl Alcohol (and) Polysorbate 60 (and)	
PEG-150 Stearate (and) Steareth-20 (Ritachol 1000)	2.00
Polysorbate-40	2.00
Sorbitan Palmitate	0.70
Mineral Oil (and) Lanolin Alcohol (Ritachol)	1.00
Mineral Oil	7.00
Aloe Vera Lipoid 1:1	3.00
Petrolatum (and) Lanolin (and) Sodium PCA (and)	
Polysorbate 85 (Ritaderm)	3.00
Dimethicone 200	1.00
BHA	0.10
Propylparaben	0.10
C. Sodium Borate	0.20
D. Aloe Veragel	20.50
Fragrance	0.15
Germall II	0.25

Procedure:
Heat Phase A and Phase B separately with agitation to 75C.
Add Phase A to Phase B and mix 30 minutes. Add Phase C and cool
with agitation until temperature reaches 50C. Add Phase D and
agitate until temperature reaches room temperature.

Night Cream

Ingredients:	Percent
A. Water	q.s.
Promulgen D	1.0
Triethanolamine 99	1.2
Methylparaben	0.35
Propylparaben	0.1
B. Stearic Acid	8.0
Aloe Veragel Lipoid	6.0
Glycol Stearate	3.5
Mineral Oil	10.0
Magnesium Stearate	0.5

Procedure:
Heat phases to 80C. Add phase B to A. Mix until 40C.; and
shut off agitation.

SOURCE: Dr. Madis Laboratories Inc.: Suggested Formulations

Anti-Cellulite Cream

Actives specifically developed to reduce the signs and quantity of cellulite are incorporated into this quick pene- trating, non-greasy cream. The use of Cera Albalate 103 allows for a stable emulsion using a high concentration of low viscosity oils. Orange Wax in this formula is also an active due to similar chemistry. The claims for Centerchem's anti- cellulite products are based on groups of chemicals which are also found in Orange Wax, for example; Steroids, Flavonones, Flavonols and Cinnamates.

Phase A:

Cera Albalate 103 (Koster Keunen)	3.0%
Glycol Stearate (Koster Keunen)	4.0%
Glycerol Monostearate (Henkel)	3.0%
Orange Wax (Koster Keunen)	4.0%
Isopropyl Palmitate (Unichema)	2.0%
Silicone Oil 556 (Dow)	2.0%
Emulsifying Wax NF (Koster Keunen)	2.0%
Ceteareth-20 (Croda)	1.5%
Eucalyptus Oil (Crompton & Knowles)	0.5%
Caprylic/Capric Triglyceride (Bernel)	2.5%

Phase B:

Water (Distilled)	50.1%
Carbopol 940 (BF Goodrich)	0.2%
Glycerin (Unichema)	3.0%
Pronalan Anticellulite (Centerchem)	10.0%
Ivy Extract (Alban Muller)	5.0%
Seaweed Extract (Centerchem)	5.0%
Triethanolamine (Dow)	0.2%
Aloe Vera Gel (Aloe Corp.)	1.0%
Germaben II (Sutton)	1.0%

Procedure:

Heat Phase A to 75 to 80C. Heat and mix Phase B to 75C. Emulsify by adding Phase A to Phase B. Add the preservative and mix while cooling. Pour into containers at 40C.

Adaption of formula and its influence on the product:

There is room for changing this product by the use of other actives which produce similar claims. The use of other plant oils will not dramatically alter the finished product. Worth mentioning is the ability of Orange Wax to mask the aroma of the eucalyptus oil producing a product of low fragrance.

SOURCE: Koster Keunen, Inc.: Suggested Formulation

Antiwrinkle Daycream

Ingredients:	% w/w
A) Cremophor A6	2,00
Cremophor A25	2,00
Cutina GMS	3,00
Cetylalcohol	3,00
Luvitol EHO	5,00
Paraffin Oil	5,00
Bisabolol	0,20
Siliconoil AK 500	0,50
Phenonip	0,30
Parsol 1789	1,00
Parsol MCX	2,00
B) Water demineralized	61,50
Euxyl K-200	0,20
Propylenglycol	4,00
C) Fitobroside	5,00
Phytaluronate	5,00
D) Fragrance/Chiara 0/238927	0,30

Procedure:
Heat the ingredients of fatty phase A) to 70C.
Heat the ingredients of water phase B) to 70C.
Under stirring add phase B) to phase A), cool to 50C,
homogenize and cool to 30C.
Then add phases C) and D) one after another and stir cold.
Application No. A 006.B/02.93

Daycream

Ingredients:	% w/w
A) Tween 60	3,00
Arlacel 60	2,00
Cetylalcohol	3,00
Stearic Acid	6,00
Isopropylmyristate	10,00
Miglycol 812	5,00
Parsol 1789	1,00
Parsol MCX	2,00
Phenonip	0,50
B) Water demineralized	56,90
Glycerin	4,00
Imidazolidinyl Urea	0,30
Sericin	3,00
Immucell	3,00
C) Fragrance/Timbuktu 0/186901	0,30

Procedure:
Heat the ingredients of fatty phase A) to 70C.
Heat the ingredients of water phase B) to 75C.
Under stirring add phase B) to phase A), cool to 50C,
homogenize and cool to 30C.
Then add phase C) and stir cold.
Application No. A 005.E/11.94
SOURCE: Pentapharm Ltd.: Suggested Formulations

Benzoyl Peroxide Cream

Materials:	% by Weight
Part A:	
1. D.I. Water	71.00
2. Glycerine	3.00
3. Bentone MA rheological additive	2.00
4. Preservative	0.25
Part B:	
1. Glyceryl Monostearate, SE	3.00
2. Stearyl Alcohol	1.00
3. Isocetyl Stearate	1.00
4. Preservative	0.10
Part C:	
1. Deionized Water	11.65
2. Benzoyl Peroxide	7.00

Procedure:
1. Mix D.I. Water, glycerine and preservative. Heat to 60C.
2. Add Bentone MA slowly and mix while heating to 80C.
3. In a separate vessel, mix phase B and heat to 80C.
4. Add Phase A to Phase B and mix while cooling to 50C.
5. Mix and homogenize Phase C separately and add to above
   mixture of Phase A and Phase B.
6. Homogenize the cream and fill the container.

SOURCE: Rheox, Inc.: Suggested Formulation TS-285

Day Cream O/W for Sensitive Skin

Component:	%
I. Emulgade PL 1618	7,5
Almond Oil	2,0
Cetiol 868	6,0
Isopropyl Palmitate	2,0
Baysilon M 350	0,5
II.Glycerin 86%	3,0
Water	79,0
Preservative	q.s.
Viscosity 23C mPas: 150000	

Preparation in the Laboratory:
1. Heat oil phase to 80C. Heat aqueous phase to 80C and add to
   oil phase while stirring. Emulsify for 5 minutes at this
   temperature.
2. The emulsion is cooled down while stirring; the stirring
   must be selected in such a way that the emulsion is kept in
   continual motion, without developing a so-called "stirring
   cone". Excessive stirring, in particular below 50C, can
   lead to the reduction of the final viscosity.
3. Heat-sensitive additives are added below 40C. Finish stirring
   at 30C.

SOURCE: Henkel KGaA: Formulation No.: 92/216/106

Cleansing Cream

This formulation is a dual purpose makeup remover and skin conditioner. Myvaplex 600P glyceryl stearate provides mild cleansing, emollience, and emulsion stability. Eastman vitamin E6-81, the acetate ester of vitamin E, offers improved heat and light stability.

Phase A:	%W/W
Myverol 18-06 hydrogenated soy glyceride	6.00
Stearic acid, USP/NF	4.00
Petrolatum, USP	10.00
Drakeol 9 mineral oil	10.00
Isopropyl myristate	10.00
SF 18 (350) silicone fluid (dimethicone)	2.00

Phase B:	
Distilled water	q.s. to 100
Propylene glycol, USP	5.50
Triethanolamine 99%, USP	0.70

Phase C:	
Eastman vitamin E 6-81 (vitamin E acetate)	0.30

Phase D:	
Fragrance	q.s.
Preservative	q.s.

Procedure:
1. Combine ingredients and heat Phase A and Phase B separately to 80C, mixing until each phase is uniform.
2. Add Phase B to Phase A with propeller mixing.
3. Continue mixing and slowly cool to 50C.
4. Add Phase C with mixing and when uniform add Phase D.
5. Continue mixing and cool to room temperature.
6. Adjust agitation speed throughout process as needed. Inversion will occur at 32C, and cream will become smooth and white.
   pH: 7.71

SOURCE: Eastman Chemical Co.: Formulation X21139-007

Cleansing Cream

Ingredients:	% by Weight
(1) Deionized Water	31.0
(2) Bentonite (NF grade)	2.0
(3) Glycerin	3.0
(4) Sorbitol (70% solution)	5.0
(5) Kaopolite 1147	20.0
(6) Beeswax	5.0
(7) Petrolatum	10.0
(8) Mineral Oil (light)	17.0
(9) Arlacel 186	3.0
(10)Tween 80	1.0
(11)Solulan 16	3.0
Preservative, Fragrance & Colorant	q.s.

Procedure:
    Add (2) to (1), mix at high speed. Reduce speed and add (3) and (4) until smooth. Continue mixing, and add (5) slowly until well dispersed. Heat to 70C. In a separate container, mix (6), (7), (8), (9), and (10) and heat to 70C until uniform. Add to water phase and cool to 30C with continued mixing.
    Follow recommended handling practices of the supplier of each product used.
    Good industrial practices should be used when handling flammable ingredients.

Mascara Cream

Ingredients:	% by Weight
(1) Deionized Water	27.8
(2) Bentonite (NF grade)	2.0
(3) Carboxymethyl Cellulose (low viscosity)	0.2
(4) Propylene Glycol	5.0
(5) Sorbitol (70% solution)	5.0
(6) Solulan 98	2.5
(7) Kaopolite TLC	4.0
(8) Iron Oxides (micronized)	3.5
(9) Deodorized Kerosene	35.0
(10)Candelilla Wax	5.0
(11)Carnauba Wax (No. 1 yellow)	7.0
(12)Arlacel 186	3.0
Preservative	q.s.

Procedure:
    Disperse (2) into (1) at high speed. Add (3) and continue mixing until well dispersed. Reduce speed and add (4), (5), (6), (7), and (8). Continue mixing, heat to 70C. In a separate container, mix (9), (10), (11) and (12) and heat to 70C until dissolved. Add to the aqueous phase while mixing. Cool to 30C.
    Follow recommended handling practices of the supplier of each product used.
    Good industrial practices should be used when handling flammable ingredients.

SOURCE: Kaopolite, Inc.: Suggested Formulations

Cold Cream

		Wt%
(O)	Liquid Paraffin (#70)	20.0
	Paraffin Wax (mp 42-44C)	9.0
	Isopropyl Palmitate	3.0
	Cetyl Alcohol	1.0
	Hydrogenated Lanolin	2.0
	Nikkol WCB	10.0
	Sorbitan Monostearate	2.7
	Polyoxyethylene (20) Sorbitan Monooleate	2.3
(W)	Ajidew N-50	3.0
	Water	47.0
	Preservative	q.s.

Procedure:
1. Heat (O) and (W) to 80-85C.
2. Add (W) to (O) slowly with stirring.
3. Cool to 40C with stirring.
     pH: 7.2

Cold Cream

		Wt%
(O)	Liquid Paraffin (#70)	30.0
	Paraffin Wax (mp 42-44C)	6.0
	Petrolatum	7.0
	Nikkol WCB	7.0
	Glyceryl Monostearate	2.5
	Polyoxyethylene (10) Cetyl Ether	2.5
	Aluminum Monostearate	2.0
(W)	Ajidew N-50	3.0
	Water	42.0
	Preservative	q.s.

Procedure:
1. Heat (O) and (W) to 80C.
2. Add (W) to (O) slowly with stirring.
3. Continue stirring until 35-40C.
     pH: 5.2

SOURCE: Ajinomoto USA, Inc.: Suggested Formulations

Cream

Ingredients:	%W/W
A Stearyl alcohol	10.00
Brij 721S	4.00
B Water, deionized	80.50
Carbopol 934	0.10
C Sodium hydroxide (10% W/W aqueous)	0.10
D Dowicil 200	0.10
E Fragrance	0.20
F Ethanol, SDA-40	5.00

Procedure:
Heat (A) to 70C and (B) to 75C. Add (B) to (A) slowly using blade type agitation. Add (C). Add (D) when the temperature drops below 50C. Add (E) below 40C. Add (F) and water to compensate for evaporation. Homogenize. Adjust pH to 5.5-6.5.

Facial Washing Cream

Ingredients:	%W/W
A Mineral oil	25.00
Cetyl alcohol	2.00
Arlacel 165	10.00
Tween 60	1.00
B Glycerin	6.00
Water	56.00
*Preservative	
C *Perfume	
*Color	

*q.s. these ingredients.

Procedure:
Heat (A) to 65C and (B) to 65C. Add (B) to (A) with continued stirring until the emulsion temperature drops to 45C. Add (C) and continue agitation until cool.
Formula SK-16A

SOURCE: ICI Surfactants: Suggested Formulations

Daily Facial Cream with UV Protection (O/W)
(Expected SPF 6)

Ingredients:	%(w/w)
A Arlacel 165	3.00
Eumulgin B 2	1.00
Lanette O	2.00
Finsolv TN	3.00
Cetiol OE	5.00
Isopropyl myristate	1.00
Abil 100	1.00
Bentone Gel MIO	3.00
Neo Heliopan, Type AV	1.00
Cutina CBS	2.00
B Demineralized water	67.25
Pemulen TR-2	0.40
Sodium hydroxide (10% aq. solution)	0.20
Glycerin 86%	3.00
Phenonip	0.50
Neo Heliopan, Type Hydro; used as a 15% solution	3.35
neutralized with sodium hydroxide	(active 0.50%)
Zinc Oxide Neutral H&R	3.00
C Perfume oil	0.30

Manufacturing Process:
Part A: Heat up to 75C with thorough agitation.
Part B:
   Disperse the Pemulen in the water during high speed agitation.
Then add the other ingredients, excluding Zinc Oxide neutral
H&R and heat up to 95C. Then add and disperse Zinc Oxide
neutral H&R. Add Part B to Part A while stirring. Continue stirr-
ing to room temperature.
Part C:
   At 30C add the perfume oil to Part A/B and homogenize the
emulsion.

   The pH-value of the finished emulsion should be approx. 7.5
and has to be checked.

SOURCE: Haarman & Reimer: Formulation K 2/1-21282/E

Day Cream, Oily

		%
A.	Imwitor 960 flakes	10.0
	Miglyol 812	6.0
	Miglyol 840	6.0
	Softisan 649	5.0
	Dynacerin 660	3.0
	Stearic acid	5.0
	Cetyl alcohol	3.0
B.	Water	to 100.0
C.	Triethanolamine	0.90
D.	Perfume	qs
	Preservative	qs

Preparation:
    A. and B. are separately warmed to 75C. C. is added to B.
B. and C. are emulsified in A. Perfume is introduced at approx.
30C.

Moisturizing Cream, Slightly Oily

		%
A.	Imwitor 940	10.0
	Miglyol 812	5.0
	Miglyol 840	3.0
	Lanette N	5.0
B.	Hygroplex HHG	5.0
	Karion F	3.0
	1,2-Propylene glycol	3.0
	Water	to 100.0
C.	Perfume	qs
	Preservative	qs

Preparation:
    The constituents of A. are mixed and heated to 80-85C.
B. is heated to the same temperature and emulsified in A.
C. is added at approx. 30C.

SOURCE: Huls America Inc.: Formulations for Cosmetics:
        Suggested Formulations

Deep Moisturizing Cream

This moisturizing cream uses Cromoist CS as a moisture absorbing, film forming protein to retain moisture and bind to the skin. The emulsion is formed and stabilized by the Polawax, Syncrowax, and Super Hartolan combination. The oils are a blend of Crodalan LA, Crodamol PMP, Super Refined Babassu Oil, and Petrolatum which provides a non greasy elegant feel and helps to soften the skin.

Ingredients: %
Part A:
Polawax (Emulsifying Wax NF)   2.00
Syncrowax AW1-C (C18-36 Acid)   5.50
Syncrowax HGL-C (C18-36 Acid Triglyceride)   1.00
Crodalan LA (Cetyl Acetate (and) Acetylated Lanolin
  Alcohol)   1.00
Super Refined Babassu Oil (Babassu Oil)   3.00
Super Hartolan (Lanolin Alcohol)   1.00
Crodamol PMP (PPG-2 Myristyl Ether Propionate)   7.00
Petrolatum   1.00
Propyl Paraben   0.20

Part B:
Water, deionized   66.00
Methyl paraben   0.30
Glycerin   10.00
Triethanolamine (99%)   1.00

Part C:
Cromoist CS (Chondroitin Sulfate (and) Hydrolyzed
  Protein)   1.00

Procedure:
   Combine the ingredients of Part A with mixing and heat to 75C. Combine the ingredients of Part B with mixing and heat to 75C. Add Part B to Part A with homogenizer mixing and cool to 65C. Continue mixing with a sweep blade mixer and cool to 40C. Add part C with mixing and cool to desired fill temperature.
   Viscosity: 4,000,000+-10% (Brookfield RVT, Spindle TE, 25C)
   N.A.T.C. Approved

SOURCE: Croda Inc.: Formula SC-207

Deep Moisturizing Creme

Part A:	% by Weight
Water	84.5
Potassium Hydroxide (45%)	0.4
Monafax 160	1.0
Phospholipid SV	2.5
Germaben II-E	0.6
Titanium Dioxide	0.5

Part B:	
Cetyl Alcohol	2.0
Myristyl Myristate	3.0
Isopropyl Palmitate	4.0
Dimethicone (100cs)	1.0
Lanolin Alcohol	0.5

Procedure:
   Heat Part A with agitatiion to 60C. Separately heat Part B to 60C. Homogenize Part B into Part A for a sufficient time to ensure good emulsification. Stir cool to 45C, add fragrance, and package.

SOURCE: Mona Industries, Inc.: Suggested Formulation

Apricot Vanishing Cream

Phase (1):	%
Rosswax 63-0412	6.64
Rosswax 573	9.2
Amerlate P	0.8
Emerest 2314	0.8
Emerest 2316	0.8
Glyceryl Monostearate SE	0.8
Apricot Kernal Oil	0.4
Lipovol P	1.3

Phase (2):	
Water	72.9
Emery 916 Pure Glycerine	6.2
Triethanolamine	0.96

Phase (3):	
Germaben II	1.0

Fragrance GK-19	q.s.

Procedure:
   In separate steam jacketed kettles, heat both Phase (1) and (2) to a temperature of 170F with agitation. When the temperature is reached, add Phase (1) to Phase (2) with continued agitation. Cool to 130F, add Phase (3) and fragrance. Continue to cool to 120F and package.
SOURCE: Frank B. Ross Co., Inc.: Suggested Formulation

Emollient Cream

Part A:
Water	69.12%
Magnabrite	0.50
Methyl paraben	0.20
Propyl paraben	0.10
BHA	0.08

Part B:
Carnation Mineral Oil	4.50
Propylene glycol	3.50
Sorbitol	4.00
Myristyl propionate	3.00
Cocoa butter	5.00
Cetyl alcohol	4.00
Stearic acid	4.00

Part C:
Octyl dimethyl PABA	1.00
Triethanolamine	1.00
Fragrance	q.s.

Procedure:
    Heat water to 80 degrees celsius. Dissolve parabens and BHA into water. Sift Magnabrite into the vortex. When dispersed add B ingredients and mix until melted and uniform. Begin cooling. At 50 degrees celsius add C ingredients. Continue mixing until cool.

Stabilized Urea Moisturizing Cream

Urea	10.00%
Stearyl alcohol	10.00
Glyceryl monostearate	8.00
Octyl dodecanol	5.00
Carnation Mineral Oil	6.00
Polyoxyethylene stearyl ether	2.50
Butylparaben	0.10
Ethylparaben	0.10
Propylene glycol	3.00
Sodium PCA	3.00
Ammonium chloride	1.00
Water	q.s. to 100.00

SOURCE: Witco Corp.: Petroleum Specialties Group: Suggested
        Formulations

Emollient Cream for Dry Flaking Skin

No.:	Phase:	Ingredient:	% by Weight
1	A	Deionized Water	74.65%
2	A	Triethanolamine 99%	1.00%
3	A	Propylene Glycol	3.00%
4	A	Xanthan Gum	0.30%
5	A	Tetrasodium EDTA	0.10%
6	B	Cetearyl Alcohol	1.00%
7	B	Laneth-5	1.50%
8	B	Glyceryl Monostearate	4.00%
9	B	Oils of Aloha Kukui Nut Oil	3.00%
10	B	Vitamin E Acetate	0.20%
11	B	Stearic Acid XXX	4.00%
12	B	Dimethicone	1.00%
13	B	Octyl Palmitate	5.00%
14	C	Germaben II	1.00%
15	D	Fragrance	0.25%

Manufacturing Procedure:
Heat Phase A to 75C. Heat Phase B to 75C.
Add Phase B to Phase A. Cool to 40C and add remaining phases.
Package.

A oil in water emulsion where the primary emulsifiers are stearic acid with triethanolamine and Laneth-5.
This is a fairly inexpensive formula that uses two emollients-Kukui and octyl palmitate.
An excellent example of the feel of Kukui.

Facial Cleansing Cream

No.:	Phase:	Ingredient:	% by Weight
1	A	Deionized Water	54.00%
2	A	Glycerine	6.00%
3	B	Mineral Oil	15.00%
4	B	Oils of Aloha Macademia Nut Oil	5.00%
5	B	Oils of Aloha Kukui Nut Oil	5.00%
6	B	Cetyl Alcohol	2.00%
7	B	C12-15 Alcohols Benzoate	2.00%
8	B	Arlacel 165	10.00%
9	B	Polysorbate-60	1.00%
10	C	Preservative	QS

Manufacturing Procedure:
Phase A: Heat water and glycerin phase to 75C.
Phase B: Heat oil phase to 75C. Add to water phase. Cool to 40C.
Phase C: Add preservative.

SOURCE: Oils of Aloha: Suggested Formulations

Emollient Cream

Ingredients:	%W/W
A Arlamol E	20.00
Arlasolve 200L	1.10
Brij 72	7.20
Stearyl alcohol	2.00
B Water, deionized	69.30
Carbopol 934	0.20
C Sodium hydroxide (10% W/W aqueous)	0.20
D *Preservative	
*Perfume	

*q.s. these ingredients.

Procedure:
Heat (A) to 72C and (B) to 75C. Add (B) to (A) slowly with good agitation and add (C) at 35C and mix thoroughly. Add (D). Replace water lost due to evaporation. Package.
Formula AE-15

Non-Greasy Emollient Cream

Ingredients:	%W/W
A Isopropyl myristate	10.00
Stearyl alcohol	4.00
Brij 721S	5.11
Brij 72	1.89
B Water, deionized	58.10
Carbopol 934	0.40
C Sodium hydroxide (10% W/W aqueous)	0.40
D *Preservative	
E Fragrance	0.10
F Ethanol, SDA-40	20.00

Procedure:
Heat (A) to 70C and (B) to 72C. Add (B) to (A) slowly with moderate agitation. Add (C). Add (D) below 50C. Add (E) and (F) at 35C. Replace water lost by evaporation and package.
Formula SK-12

SOURCE: ICI Surfactants: Suggested Formulations

Emollient Cream

Ingredients:	%W/W
A Cetyl alcohol	3.00
Stearyl alcohol	2.00
Petrolatum	5.00
Liquid lanolin fraction	2.00
Brij 72	3.00
Brij 700	3.00
B Water, deionized	82.00
*Preservative	
*q.s. these ingredients.	

Procedure:
    Heat (A) to 70C. Heat (B) to 72C. Add (B) to (A) with agitation.

Emollient Cream

Ingredients:	%W/W
A Brij 72	3.00
Brij 721	2.00
Arlamol E	9.00
Stearyl alcohol	0.70
Cetyl alcohol	0.30
Stearic acid	1.50
Silicone fluid (GE SF 96-20)	1.00
B Propylene glycol	4.00
*Dowicil 200	
Water, deionized	78.50
*q.s. these ingredients.	

Procedure:
    Heat (A) to 70C. Heat (B) to 72C. Add (B) to (A) slowly with moderate stirring. Stir to 35C and add water lost due to evaporation.
Formula PC-9028

Vanishing Cream

Ingredients:	%W/W
Arlacel 165, acid stable g.m.s.	18.00
Lanolin	2.00
Cetyl alcohol	1.00
Mineral oil	1.00
Sorbo, 70% sorbitol solution, U.S.P.	2.00
Water, deionized	76.00

Procedure:
    Heat all ingredients to 90C with moderate agitation. Cool to room temperature.

SOURCE: ICI Surfactants: Suggested Formulations

European O/W Day Cream

Ingredients:	%W/W
A Isohexadecane	10.00
Caprylic/capric triglyceride	4.00
Avocado oil	2.00
Sunflower Seed Oil	2.00
Tocopherol Acetate	2.00
Behenyl alcohol	2.00
Antioxidant	0.05
B Arlatone 2121	5.50
Glycerol	4.00
Panthenol	1.00
Allantoin	0.20
C *Preservative	
Xanthan gum	0.15
Water	67.10

*q.s. these ingredients.

Procedure:
Melt Arlatone 2121 and add to remainder of (B) and (C) mixture at 80C. Add heated (A) to the (B) and (C) mixture at 75C with vigorous agitation. Homogenize mixture at 65C.

Cream

Ingredients:	W/W
A Cetyl alcohol	6.00
Brij 721S	5.00
Silicone oil, 350 cs	0.50
B Water, deionized	66.30
C Hydrogen peroxide, 27% dilution grade	22.20

Procedure:
Heat (A) to 60C and (B) to 65C. Add (B) to (A) slowly with moderate agitation. Add (C) below 35C. Replace water lost by evaporation and adjust pH to 3.5-4.0 with dilute phosphoric acid (10% C.P.). Package in suitable container for possible evolution of oxygen.
Formula PC-8200

SOURCE: ICI Surfactants: Suggested Formulations

Exfoliating Cream

The use of natural exfoliating agent (easily dispersed) in this cream formula giving a marvelous, non-irritating, abrasive quality for the removal of spent surface cells. This will renew and revitalize worn and damaged skin, leaving a barrier which contains natural anti-oxidants inherent in the beeswax and orange wax.

Phase A:

Orange Wax (Koster Keunen)	3.5%
NF White Beeswax (Koster Keunen)	1.0%
Isostearic Acid (Unichema)	3.0%
Almond Oil (Arista)	3.5%
Light Mineral Oil (Witco)	4.0%
Glycerol Monostearate (Koster Keunen)	1.5%
Cetylstearyl Alcohol (P&G)	1.0%
Octyl Palmitate (Unichema)	3.0%

Phase B:

Water (Distilled)	64.6%
Propylene Glycol (Dow)	2.5%
Triethanolamine (Dow)	1.0%
Polysorbate 60 (Gallard & Schlesinger)	0.2%
Carbopol 940 (BF Goodrich)	0.2%
Germaben II (Sutton)	1.0%

Phase C:

Microgranulated Carnauba, 20-60 mesh (Koster Keunen)	10.0%

Procedure:
Add a mixed and uniform Phase A to a mixed and uniform Phase B at 75C under agitation. Continue mixing and cool to 50C. Add Phase C and mix until homogeneous.

Adaption of formula and its influence on the product:
Viscosity of this product can easily be altered by substituting stearic acid for isostearic acid, to accomodate your preferred container. For sensitive skin, reduce the orange wax, and replace with Kester Wax 62 and Ceresine 130/135. A low viscosity will allow this product to be packaged in a bottle fitted with a pump and a product with high viscosity allows packaging in an open jar. The formulating advantages remain unaltered by the changes in viscosity.

SOURCE: Koster Keunen, Inc.: Suggested Formulation

Facial Moisturizing Cream

Ethylhexyl p-methoxycinnamate	4.90%
Butylmethoxydibenzoylmethane	2.10
Ethylene-acrylate copolymer	0.50
Hydroxypropyl glyceryl ether	5.00
Protopet 1S Petrolatum	2.00
Dimethicone	0.40
Steareth-100	0.70
Glyceryl monostearate	0.30
Cetyl alcohol	1.20
Stearic Acid	0.52
Carbomer 934	0.10
Carbomer 941	0.10
Methylparaben	0.20
Propylparaben	0.10
Imidazolidinylurea	0.10
Tetrasodium EDTA	0.10
Tyrosine	0.10
Potassium hydroxide	0.37
Titanium dioxide	0.40
Fragrance	0.15
Water	q.s. to 100.00

Moisturizing Hand & Body Cream

Ingredients:	%Wt.
Deionized Water	82.00
Carbomer 934	0.50
Glycerine	5.00
Polyaldo 10-1-S	1.50
Petrolatum	2.00
Carnation Mineral Oil	2.00
Lonzest 143/S	2.00
Aldo HMS	1.50
Stearic Acid	2.50
Triethanolamine 99%	1.00
Glydant Plus	q.s.

SOURCE: Witco Corp.: Petroleum Specialties Group: Suggested Formulations

Facial Night Cream

Part A:
Deionized Water	53.7%
Sorbitol	1.0
Methylparaben	0.2

Part B:
C14-16 alcohols benzoate	15.0
Lanolin	1.0
Protopet 1S Petrolatum	5.0
Beeswax	8.0
Polysorbate 80	4.0
Glyceryl stearate and PEG 100 stearate	2.0
Stearic acid	3.0
Triethanolamine	0.6

Part C:
Glycosphingolipids	5.0

Part D:
Dimethicone	1.0
Diazolidinyl urea	0.3
Fragrance	0.2

Procedure:
   Begin heating water to 80 celsius, add rest of Part A. Mix well. Add Part B ingredients in order. Mix until homogeneous. Begin cooling to room temperature. Slowly add Part C; mix until smooth. Add rest of Part D ingredients. Mix well.

Age Spot Treatment Cream

Diethanolamine cetyl phosphate	2.00%
Ethylene glycol monostearate	12.00
Isopropyl Palmitate	12.00
Cetearyl octanoate	2.00
Protopet 1S Petrolatum	2.50
Lanolin alcohols	3.00
Propylparaben	0.10
Methylchloroisothiazolinone and methylisothiazolinone	0.05
Imidazolidinyl urea	0.10
Methylparaben	0.10
BHA	0.20
BHT	0.30
Cysteine	0.50
Reduced glutathione	0.50
Pyrocatechol	2.00
Ascorbyl palmitate	0.20
Acetic Acid	0.20
Octylmethoxy cinnamate	1.00
Butylmethoxydibenzoylmethane	1.00
Fragrance	0.60
Water	q.s. to 100.00

SOURCE: Witco Corp.: Petroleum Specialties Group: Suggested
        Formulations

General Purpose Cream

		Wt%
1.	A-C Copolymer 540	2.0
2.	A-C Polyethylene 617	2.0
3.	Amerchol L-101	5.0
4.	Mineral Oil	10.2
5.	Hexadecyl Stearate	10.0
6.	Emerest 2452	5.5
7.	Propyl-p-Hydroxybenzoate	0.1
8.	Methyl-p-Hydroxybenzoate	0.2
9.	Sorbitol (70%)	5.0
10.	Borax	0.3
11.	Magnesium Sulfate	0.3
12.	Water	59.7

Procedure:
   Weigh 1-7 and heat to 85C with slow agitation. The blend has
a cloud point of approximately 80C. Above the cloud point all
waxes will eventually dissolve in the blend. If a higher solvating temperature is used, solvation can be much faster. Hold
the wax blend at 85C. Heat 8-12 to 85-90C and stir gently until
all has dissolved. Hold at 85C.
   Place wax blend in mixing container, add aqueous phase to it
and shear with homomixer or colloid mill. At 78C the crude
dispersion inverts and a thick creamy emulsion forms. Continue
shearing while scraping the sides of the container and make
sure the whole content is properly sheared. Add perfume, de-
aerate and package.

General Purpose Water/Oil Cream

		Wt%
1.	A-C Polyethylene 617	2.0
2.	Beeswax	4.0
3.	Mineral Oil	15.0
4.	Arlacel 83	5.5
5.	Propyl-p-Hydroxybenzoate	0.1
6.	Borax	0.3
7.	Sorbitol	5.0
8.	Methyl-p-Hydroxybenzoate	0.2
9.	Germall 115	0.3
10.	Water	67.6

Procedure:
   Weigh 1-4 and heat to 85C with slow agitation. This wax blend
has a cloud point of approximately 80C. Above the cloud point
all waxes will eventually dissolve in blend. If a higher temperature is used, the solvation can be much faster. Hold the wax
blend at 80-85C. Heat 5-10 to 85-90C and stir gently until all
is dissolved. Hold at 85C.
   Place wax blend in mixing container, add aqueous phase to it
and shear with homomixer or colloid mill. At approximately 48C
the crude dispersion inverts. Some water may not be completely
taken up at this point but continued shearing will incorporate
it all. Add perfume, de-aerate and package.
SOURCE: Allied Signal Inc.: 5011-19-5/5011-20-1

General Purpose O/W Cream

	Wt%
1. A-C Copolymer 540	2.0
2. Mineral Oil, 70 vis.	5.0
3. Dow Fluid 556	1.0
4. Emerest 2388	10.5
5. Amerchol 400	2.0
6. Solulan 25	1.0
7. Arlacel 60	2.0
8. Sorbitol (70%)	5.0
9. Tween 60	1.0
10.Carbopol 940	0.75
11.Germall 115	0.4
12.Triethanolamine	0.75
13.Water	68.6

Procedure:
    Disperse Carbopol in water. Weigh 1-7 and heat to 80-90C with
slow agitation. Add remaining ingredients, except Triethanolamine, to the Carbopol/water dispersion and heat to 80-90C.
Add the aqueous phase to the wax phase and shear in homomixer
for five minutes. Add Triethanolamine and continue to shear
while cooling to 40-50C. Add perfume, de-aerate and package.

General Purpose o/w Cream

	Wt%
1. A-C Copolymer 540	2.0
2. Mineral oil, 70 vis.	5.0
3. Dow Fluid 556	1.0
4. Emerest 2388	10.5
5. Hydroxyol	2.0
6. Ethoxyol 24	1.0
7. Arlacel 60	2.0
8. Sorbitol (70%)	5.0
9. Tween 60	1.0
10.Carbopol 940	0.75
11.Germall 115	0.4
12.Triethanolamine	0.75
13.Water	68.6

Procedure:
    Disperse Carbopol in water. Weigh 1-7 and heat to 80-90C with
slow agitation. Add remaining ingredients, except Triethanolamine, to the Carbopol/water dispersion and heat to 80-90C.
Add the aqueous phase to the wax phase and shear in homomixer
for five minutes. Add Triethanolamine and continue to shear while
cooling to 40-50C. Add perfume, de-aearate and package.

SOURCE: Allied Signal Inc.: Ref: 5011-25-2/Ref: 5011-25-2A

General Purpose O/W Cream

		Wt%
1.	A-C Copolymer 540	2.0
2.	Mineral oil, 70 vis.	5.0
3.	Dow Fluid 556	1.0
4.	Isopropyl Myristate	10.5
5.	Amerchol 400	2.0
6.	Solulan 25	1.0
7.	Arlacel 60	2.0
8.	Sorbitol (70%)	5.0
9.	Tween 60	1.0
10.	Carbopol 940	0.75
11.	Germall 115	0.4
12.	Triethanolamine	0.75
13.	Water	68.6

Procedure:
    Disperse Carbopol in water. Weigh 1-7 and heat to 80-90C with
slow agitation. Add remaining ingredients, except Triethanolamine, to the Carbopol/water dispersion and heat to 80-90C.
Add the aqueous phase to the wax phase and shear in homomixer
for five minutes. Add Triethanolamine and continue to shear
while cooling to 40-50C. Add perfume, de-aerate and package.

General Purpose O/W Cream

		Wt%
1.	A-C Polyethylene 617	1.0
2.	A-C Copolymer 540	1.0
3.	Mineral oil, 70 vis.	5.0
4.	Dow Fluid 556	1.0
5.	Emerest 2388	10.5
6.	Amerchol 400	2.0
7.	Solulan 25	1.0
8.	Arlacel 60	1.3
9.	Sorbitol (70%)	5.0
10.	Tween 60	1.8
11.	Carbopol 940	0.75
12.	Germall 115	0.4
13.	Triethanolamine	0.75
14.	Water	68.5

Procedure:
    Disperse Carbopol in water. Weigh 1-8 and heat to 80-90C with
slow agitation. Add remaining ingredients, except Triethanolamine, to the Carbopol/water dispersion and heat to 80-90C.
Add the aqueous phase to the wax phase and shear in homomixer
for five minutes. Add Triethanolamine and continue to shear
while cooling to 40-50C. Add perfume, de-aerate and package.

SOURCE: Allied Signal Inc.: Ref: 5011-25-2B/Ref: 5011-25-3

General Purpose O/W Cream

	Wt%
1. A-C Polyethylene 617	1.0
2. A-C Copolymer 540	1.0
3. Mineral oil, 70 vis.	5.0
4. Dow Fluid 556	1.0
5. Emerest 2388	10.5
6. Hydroxyol	2.0
7. Ethoxyol 24	1.0
8. Arlacel 60	1.3
9. Sorbitol (70%)	5.0
10.Tween 60	1.8
11.Carbopol 940	0.75
12.Germall 115	0.4
13.Triethanolamine	0.75
14.Water	68.5

Procedure:
    Disperse Carbopol in water. Weigh 1-8 and heat to 80-90C
with slow agitation. Add remaining ingredients, except Triethanolamine, to the Carbopol/water dispersion and heat to 80-90C.
Add the aqueous phase to the wax phase and shear in homomixer
for five minutes. Add Triethanolamine and continue to shear
while cooling to 40-50C. Add perfume, de-aerate and package.

General Purpose O/W Cream

	Wt%
1. A-C Polyethylene 617	1.0
2. A-C Copolymer 540	1.0
3. Mineral oil, 70 vis.	5.0
4. Dow Fluid 556	1.0
5. Isopropyl Myristate	10.5
6. Amerchol 400	2.0
7. Solulan 25	1.0
8. Arlacel 60	1.3
9. Sorbitol (70%)	5.0
10.Tween 60	1.8
11.Carbopol	0.75
12.Germall 115	0.4
13.Triethanolamine	0.75
14.Water	68.5

Procedure:
    Disperse Carbopol in water. Weigh 1-8 and heat to 80-90C
with slow agitation. Add remaining ingredients, except Triethanolamine, to the Carbopol/water dispersion and heat to 80-90C.
Add the aqueous phase to the wax phase and shear in homomixer
for five minutes. Add Triethanolamine and continue to shear
while cooling to 40-50C. Add perfume, de-aerate and package.
SOURCE: Allied Signal Inc.: Ref. 5011-25-3A/5011-25-3B

General Purpose O/W Cream

	Wt%
1. A-C Polyethylene 617	1.0
2. A-C Copolymer 540	1.0
3. Mineral Oil (70 ss)	5.0
4. Dow 556 Fluid	1.0
5. Propylene Glycol Dipelargonate	10.5
6. Hydroxyol	2.0
7. Ethoxyol 24	1.0
8. Arlacel 60	1.3
9. Tween 60	1.8
10. Propyl-P-Hydroxybenzoate	0.1
11. Sorbitol (70%)	5.0
12. Carbopol 940	0.75
13. Germall 115	0.4
14. Methyl-P-Hydroxybenzoate	0.2
15. Triethanolamine	0.75
16. Water	68.3

Procedure:
   Disperse Carbopol in water. Weigh 1-10 and heat to 80-90C
with slow agitation. Add remaining ingredients to Carbopol/water
dispersion, except triethanolamine, and heat to 80-90C. Add the
wax phase to the aqueous phase and shear in homomixer. Continue
shearing while cooling to 40C, then add triethanolamine, mixing
well. Cool to 30C, add perfume, de-aerate and package.
Ref: 5011-25-4

Acne Scrub Cream
For acne scrub cream, add 10% A-C Polyethylene 9A to this
formulation, after cooling.

Emollient Cream, Water/Oil

1. A-C Polyethylene 617A	1%
2. A-C Copolymer 540	4
3. Lanolin Wax	10
4. Mineral Oil	30
5. 2 Ethyl Hexyl Stearate	4.5
6. Arlacel 20	4.6
7. Tween 20	1.5
8. P Paraben	0.1
9. M Paraben	0.15
10. Water	43.85
11. Borax	0.15
12. Carbopol 941	0.15

Procedure:
1. Heat 1-5 to 90C until all has dissolved.
2. Add 6, 7 and 8 to 1-5. Hold at 85C.
3. Heat 10 and 11 to 90-95C until all has dissolved; add 12
   gradually with shear using the Gifford-Wood Homomixer or
   equivalent.
4. Add 1-8 to mixer and shear for 10 minutes at 85-90C. Shear
   to 65C, add perfume and package.
SOURCE: Allied Signal Inc.: Suggested Formulations

General Purpose Cream

	Wt%
1. A-C Polyethylene 617A	3.0
2. Beeswax	2.0
3. Lanolin Alcohol	1.0
4. Isopropyl Palmitate	8.2
5. 2-ethyl Hexyl Palmitate	10.0
6. Dow 200 Fluid, 350 cs	1.0
7. Triglyceral Diisostearate	5.5
8. Propyl-P-Hydroxybenzoate	0.1
9. Sorbitol (70%)	5.0
10.Aloe Extract	5.0
11.Sodium Borate, Anhydrous	0.3
12.Methyl-P-Hydroxybenzoate	0.2
13.Germall 115	0.3
14.Butyl-P-Hydroxybenzoate	0.1
15.Water	58.3

Procedure:

Weigh 1-8 and and heat to 85C with slow agitation. The blend has a cloud point of approximately 80C. Above the cloud point all waxes will eventually dissolve in the blend. If a higher solvating temperature is used, solvation can be much faster. Hold the wax blend at 85C. Heat 9-15 to 85-90C and stir gently until all has dissolved. Hold at 85C.

Place wax blend in mixing container, add aqueous phase to it and shear with homomixer or colloid mill. At 67C the crude dispersion inverts and a thick creamy emulsion forms. Continue shearing while scraping the sides of the container to make sure the whole content is properly sheared. Add perfume, de-aerate and package.

Ref: 5189-4-10A

Water-In-Oil Cream

	%
1. A-C 617	3.0
2. Mineral Oil (70-75 SS)	15.0
3. Isopropyl Palmitate	6.0
4. Escalol 507	4.0
5. Lanogene	5.0
6. Sorbitan Sesquioleate (Arlacel 83)	5.0
7. Sorbo	5.0
8. Borax	0.3
9. Germaben IIE	1.0
10.Water	Balance
11.Perfume	Q.S.

Procedure:

Heat 1-6 with mild agitation until all the A-C 617 has dissolved. Maintain it at 100-110C. Heat 7-10 with vigorous agitation with a colloid mill or homofixer until the solution is at 85C. Continue vigorous agitation and pour in 1-6. A crude oil-in-water emulsion is formed. Cool with cold water bath to about 55-60C when it inverts into a smooth cream. At 50-55C add perfume and package.

SOURCE: Allied Signal Inc.: Suggested Formulations

General Purpose Water/Oil Cream

	Wt.%
1. A-C Polyethylene 617	2.0
2. A-C Copolymer 540	2.0
3. Mineral Oil (70 ss)	10.2
4. 2 Ethyl Hexyl Stearate	10.0
5. Lanolin Alcohol (L 101)	5.0
6. Triglycerol Di-isostearate	5.5
7. Propyl-P-Hydroxybenzoate	0.1
8. Methyl-P-Hydroxybenzoate	0.2
9. Sorbitol (70%)	5.0
10.Borax	0.3
11.Water	59.7

Procedure:
   Weigh 1-5 and heat to 85C with slow agitation. This wax blend has a cloud point of 79C. Above the cloud point all waxes will eventually dissolve in blend. If a higher solvating temperature is used, solvation can be much faster. Hold the wax blend at 70-80C and add 6 and 7 to it with stirring until all has dissolved. Heat 8-11 to 85-90C and stir gently until all has dissolved. Hold at 70-80C.
   Place wax blend in mixing container, add aqueous phase to it and shear with homomixer or colloid mill. At 53C the crude dispersion inverts and a thick creamy emulsion forms. Continue shearing while scraping the sides of the container to make sure the whole content is properly sheared. Add perfume, de-aerate and package.
Ref: 5011-18-6

General Purpose Water/Oil Cream

	Wt.%
1. A-C Polyethylene 617	2.0
2. Beeswax	4.0
3. Mineral Oil	15.0
4. Arlacel 83	5.5
5. Propyl-P-Hydroxybenzoate	0.1
6. Methyl-P-Hydroxybenzoate	0.2
7. Sorbitol (70%)	5.0
8. Borax	0.3
9. Water	67.9

Procedure:
   Weigh 1-4 and heat to 85C with slow agitation. The blend has a cloud point of approximately 80C. Above the cloud point all waxes will eventually dissolve in the blend. If a higher solvating temperature is used, solvation can be much faster. Hold the wax blend at 85C. Heat 5-9 to 85-90C and stir gently until all has dissolved. Hold at 85C.
   Place wax blend in mixing container, add aqueous phase to it and shear with homomixer or colloid mill. At 50C the crude dispersion inverts and a thick creamy emulsion forms. Continue shearing while scraping the sides of the container to make sure the whole content is properly sheared. Add perfume, de-aerate and package.
Ref: 5011-18-5
SOURCE: Allied Signal Inc.: **Suggested Formulations**

General Purpose W/O Cream

	Wt%
1. A-C Polyethylene 617	2.0
2. Beeswax	3.0
3. Rewomid S-280	1.0
4. Amerchol CAB	5.0
5. Mineral Oil	11.0
6. Isopropyl Isostearate	8.8
7. Triglycerol Diisostearate	3.0
8. Sorbitol (70%)	5.0
9. Borax	0.3
10.Germall 115	0.2
11.Magnesium Sulfate	0.3
12.Water	60.0

Procedure:
Weigh 1-7 and heat to 85C with slow agitation. The blend has a cloud point of approximately 80C. Above the cloud point all waxes will eventually dissolve in the blend. If a higher solvating temperature is used, solvation can be much faster. Hold the wax blend at 85C. Heat 8-12 to 85-90C and stir gently until all has dissolved. Hold at 85C.

Place wax blend in mixing container, add aqueous phase to it and shear with homomixer or colloid mill. At 76C the crude dispersion inverts and a thick creamy emulsion forms. Continue shearing while scraping the sides of the container to make sure the whole content is properly sheared. Add perfume, de-aerate and package.

General Purpose W/O Cream

	Wt%
1. A-C Polyethylene 617	2.0
2. A-C Copolymer 540	3.0
3. Rewomid S-280	1.0
4. Amerchol Cab.	5.0
5. Mineral Oil	11.0
6. Isopropyl Isostearate	8.8
7. Triglycerol Diisostearate	3.0
8. Sorbitol (70%)	5.0
9. Borax	0.3
10.Germall 115	0.2
11.Magnesium Sulfate	0.3
12.Water	60.0

Procedure:
Weigh 1-7 and heat to 85C with slow agitation. The blend has a cloud point of approximately 80C. Above the cloud point all waxes will eventually dissolve in the blend. If a higher solvating temperature is used, solvation can be much faster. Hold the wax blend at 85C. Heat 8-12 to 85-90C and stir gently until all has dissolved. Hold at 85C.

Place wax blend in mixing container, add aqueous phase to it and shear with homomixer or colloid mill. At 76C the crude dispersion inverts and a thick creamy emulsion forms. Continue shearing while scraping the sides of the container to make sure the whole content is properly sheared. Add perfume, de-aerate and package.

SOURCE: Allied Signal Inc.: Formula 5011-19-3/5011-19-4

General Purpose W/O Cream

	Wt%
1. A-C Polyethylene 617	2.0
2. A-C Copolymer 540	2.0
3. Amerchol L-101	5.0
4. Mineral Oil	10.0
5. Dow Fluid 556	3.0
6. 2-Ethyl Hexyl Stearate	5.0
7. Arlacel 83	5.5
8. Sorbitol (70%)	5.0
9. Borax	0.3
10.Germall 115	0.5
11.Water	61.7

Procedure:
Weigh 1-7 and heat to 85C with slow agitation. The blend has a cloud point of approximately 80C. Above the cloud point all waxes will eventually dissolve in the blend. If a higher solvating temperature is used, solvation can be much faster. Hold the wax blend at 85C. Heat 8-11 to 85-90C and stir gently until all has dissolved. Hold at 85C.

Place wax blend in mixing container, add aqueous phase to it and shear with homomixer or colloid mill. At 79C the crude dispersion inverts and a thick creamy emulsion forms. Continue shearing while scraping the sides of the container to make sure the whole content is properly sheared. Add perfume, de-aerate and package.

General Purpose W/O Cream

	Wt%
1. A-C Polyethylene 617	3.0
2. Beeswax	2.0
3. Amerchol L-101	5.0
4. Mineral Oil	8.2
5. Dow Fluid 200-350 cps	1.0
6. 2-Ethyl Hexyl Stearate	10.0
7. Triglycerol Diisostearate	5.5
8. Propyl-P-Hydroxybenzoate	0.1
9. Sorbitol (70%)	5.0
10.Borax	0.3
11.Methyl-P-Hydroxybenzoate	0.2
12.Water	59.7

Procedure:
Weigh 1-8 and heat to 85C with slow agitation. The blend has a cloud point of approximately 80C. Above the cloud point all waxes will eventually dissolve in the blend. If a higher solvating temperature is used, solvation can be much faster. Hold the wax blend at 85C. Heat 9-12 to 85-90C and stir gently until all has dissolved. Hold at 85C.

Place wax blend in mixing container, add aqueous phase to it and shear with homomixer or colloid mill. At 68C the crude dispersion inverts and a thick creamy emulsion forms. Continue shearing while scraping the sides of the container to make sure the whole content is properly sheared. Add perfume, de-aerate and package.
SOURCE: Allied Signal Inc.: Formulas Ref. 5011-20-4/5011-22-1

General Purpose W/O Cream

	Wt%
1. A-C Polyethylene 617	3.0
2. Beeswax	2.0
3. Amerchol L-101	5.0
4. Mineral Oil, 70 vis.	8.2
5. Dow Fluid 200, 350 cs.	1.0
6. 2-Ethyl Hexyl Stearate	10.0
7. Triglycerol Diisostearate	5.5
8. Propyl-P-Hydroxybenzoate	0.1
9. Sorbitol (70%)	5.0
10.Sodium Borate, Anhydrous	0.3
11.Methyl-P-Hydroxybenzoate	0.2
12.Germall 115	0.3
13.Water	59.4

Weigh 1-8 and heat to 85C with slow agitation. The blend has a cloud point of approximately 80C. Above the cloud point all waxes will eventually dissolve in the blend. If a higher solvating temperature is used, solvation can be much faster. Hold the wax blend at 85C. Heat 9-13 to 85-90C and stir gently until all has dissolved. Hold at 85C.

Place wax blend in mixing container, add aqueous phase to it and shear with homomixer or colloid mill. At 67C the crude dispersion inverts and a thick creamy emulsion forms. Continue shearing while scraping the sides of the container to make sure the whole content is properly sheared. Add perfume, de-aerate and package.

General Purpose W/O Cream

	Wt%
1. A-C Polyethylene 617	3.5
2. A-C Copolymer 540	1.5
3. Beeswax	2.0
4. Lantrol	9.5
5. Mineral Oil, 70 vis.	15.0
6. Isopropyl Stearate	10.0
7. 2-Ethyl Hexyl Stearate	10.0
8. Arlacel 83	5.5
9. Propyl-P-Hydroxybenzoate	0.1
10.Sorbitol (70%)	7.0
11.Borax	0.3
12.Magnesium Sulfate	0.3
13.Methyl-P-Hydroxybenzoate	0.2
14.Germall 115	0.2
15.Water	34.9

Weigh 1-9 and heat to 85C with slow agitation. This wax blend has a cloud point of approximately 80C. Above the cloud point all waxes will eventually dissolve in blend. If a higher temperature is used, the solvation can be much faster. Hold the wax blend at 80-85C. Heat 10-15 to 85-90C and stir gently until all is dissolved. Hold at 85C.

Place wax blend in mixing container, add aqueous phase to it and shear with homomixer or colloid mill. At approximately 80C the crude dispersion inverts. Some water may not be completely taken up at this point but continued shearing will incorporate it
SOURCE: Allied Signal Inc.: Formula Ref:5011-24-2/Ref: 5011-24-4

General Purpose W/O Cream

1. A-C Polyethylene 617	3.0%
2. A-C Copolymer 540	2.0
3. Beeswax	2.0
4. Lantrol	9.5
5. Mineral Oil, 70 vis.	15.0
6. Isopropyl Stearate	10.0
7. 2-Ethyl Hexyl Stearate	10.0
8. Arlacel 83	5.5
9. Propyl-P-Hydroxybenzoate	0.1
10.Sorbitol (70%)	7.0
11.Borax	0.3
12.Magnesium Sulfate	0.3
13.Methyl-P-Hydroxybenzoate	0.2
14.Germall 115	0.2
15.Water	34.9

Weigh 1-9 and heat to 85C with slow agitation. The blend has a
cloud point of approximately 80C. Above the cloud point all
waxes will eventually dissolve in the blend. If a higher solvating temperature is used, solvation can be much faster. Hold
the wax blend at 85C. Heat 10-15 to 85-90C and stir gently until
all has dissolved. Hold at 85C.
Place wax blend in mixing container, add aqueous phase to it and
shear with homomixer or colloid mill. At 80C the crude dispersion
inverts and a thick creamy emulsion forms. Continue shearing
while scraping the sides of the container to make sure the whole
content is properly sheared. Add perfume, de-aerate and package.

General Purpose W/O Cream

1. A-C Polyethylene 617	2.0%
2. A-C Copolymer 540	3.0
3. Beeswax	2.0
4. Lantrol	9.5
5. Mineral Oil, 70 vis.	15.0
6. Isopropyl Stearate	10.0
7. 2-Ethyl Hexyl Stearate	10.0
8. Arlacel 83	5.5
9. Propyl-P-Hydroxybenzoate	0.1
10.Sorbitol (70%)	7.0
11.Borax	0.3
12.Magnesium Sulfate	0.3
13.Methyl-P-Hydroxybenzoate	0.2
14.Germall 115	0.2
15.Water	34.9

Weigh 1-9 and heat to 85C with slow agitation. This wax blend has
a cloud point of approximately 80C. Above the cloud point all
waxes will eventually dissolve in blend. If a higher temperature
is used, the solvation can be much faster. Hold the wax blend at
80-85C. Heat 10-15 to 85-90C and stir gently until all is dissolved. Hold at 85C.
Place wax blend in mixing container, add aqueous phase to it and
shear with homomixer or colloid mill. At approximately 80C the
crude dispersion inverts. Some water may not be completely taken
up at this point but continued shearing will incorporate it all.
Add perfume, de-aerate and package.
SOURCE: Allied Signal Inc.: Ref: 5011-24-5/Ref: 5011-24-6

General Purpose W/O Cream

	Wt%
1. A-C Polyethylene 617A	3.0
2. Beeswax	2.0
3. Nimlesterol D	5.0
4. Mineral Oil, 70 vis.	8.2
5. Dow Fluid 200-350 cps	1.0
6. 2-Ethyl Hexyl Stearate	10.0
7. Emerest 2452	5.5
8. Propyl-P-Hydroxybenzoate	0.1
9. Sorbitol (70%)	5.0
10. Borax	0.3
11. Methyl-P-Hydroxybenzoate	0.2
12. Water	59.7

Weigh 1-8 and heat to 85C with slow agitation. The blend has a cloud point of approximately 80C. Above the cloud point all waxes will eventually dissolve in the blend. If a higher solvating temperature is used, solvation can be much faster. Hold the wax blend at 85C. Heat 9-12 to 85-90C and stir gently until all has dissolved. Hold at 85C.

Place wax blend in mixing container, add aqueous phase to it and shear with homomixer or colloid mill. At 68C the crude dispersion inverts and a thick creamy emulsion forms. Continue shearing while scraping the sides of the container to make sure the whole content is properly sheared. Add perfume, de-aerate and package.

Mineral Oil Cream - W/O

	Wt%
1. A-C Polyethylene 617	3.0
2. A-C Copolymer 540	1.0
3. Beeswax	2.0
4. Mineral Oil, 70 vis.	25.0
5. Isopropyl Stearate	3.0
6. 2-Ethyl Hexyl Stearate	9.4
7. Triglycerol Diisostearate	3.5
8. Propyl-P-Hydroxybenzoate	0.2
9. Sorbitol (70%)	7.0
10. Sodium Borate, Anhydrous	0.3
11. Methyl-P-Hydroxybenzoate	0.3
12. Magnesium Sulfate	0.3
13. Water	45.0

Weigh 1-3 and heat to 85C with slow agitation. The blend has a cloud point of approximately 80C. Above the cloud point all waxes will eventually dissolve in the blend. If a higher solvating temperature is used, solvation can be much faster. Hold the wax blend at 85C. Heat 9-13 to 85-90C and stir gently until all has dissolved. Hold at 85C.

Place wax blend in mixing container, add aqueous phase to it and shear with homomixer or colloid mill. At 77C the crude dispersion inverts and a thick creamy emulsion forms. Continue shearing while scraping the sides of the container to make sure the whole content is properly sheared. Add perfume, de-aereate and package.
SOURCE: Allied Signal Inc.: Formula Ref 5011-22-1A/5011-22-4

General Purpose W/O Cream

		Wt%
1.	A-C Polyethylene 617A	2.0
2.	A-C Copolymer 540	3.0
3.	Beeswax	2.0
4.	Mineral Oil, 70 vis.	15.0
5.	Dow Fluid 334	9.5
6.	Isopropyl Stearate	10.0
7.	2-Ethyl Hexyl Stearate	10.0
8.	Arlacel 83	5.5
9.	Propyl-P-Hydroxybenzoate	0.1
10.	Sorbitol (70%)	7.0
11.	Borax	0.3
12.	Magnesium Sulfate	0.3
13.	Methyl-P-Hydroxybenzoate	0.2
14.	Germall 115	0.2
15.	Water	34.9

Procedure:

Weigh 1-9 and heat to 85C with slow agitation. The blend has a cloud point of approximately 80C. Above the cloud point all waxes will eventually dissolve in the blend. If a higher solvating temperature is used, solvation can be much faster. Hold the wax blend at 85C. Heat 10-15 to 85-90C and stir gently until all has dissolved. Hold at 85C.

Place wax blend in mixing container, add aqueous phase to it and shear with homomixer or colloid mill. At 80C the crude dispersion inverts and a thick creamy emulsion forms. Continue shearing while scraping the sides of the container to make sure the whole content is properly sheared. Add perfume, de-aerate and package.

Ref: 5011-31-2

Oil-In-Water Cream and Its Scrub Derivative

1.	A-C 617 Polyethylene	2.0%
2.	Stearic Acid	0.5
3.	Lanogene (lanolin alcohol & mineral oil)	6.0
4.	Isopropyl palmitate	12.5
5.	Sorbitan Monostearate	1.3
6.	Polyoxyethylene 20 Sorbitan Monostearate	1.8
7.	Sorbo (sorbitol 70%, water 30%)	5.0
8.	Carbopol 940	1.0
9.	Germaben II	0.8
10.	Water	68.35
11.	Triethanolamine	0.75
12.	Perfume	Q.S.

Procedure:

Weigh 1-6, then weigh 7-10. Heat 7-10 with agitation using a homomixer or colloid mill. When the aqueous solution reaches 85C, heat oil phase to 90-95C with slow agitation until all the wax has dissolved. Combine 1-10 and shear until mixture is homogeneous. Add 11 and shear well. Cool to 55C and add 12.

SOURCE: Allied Signal Inc.: Suggested Formulations

General Purpose W/O Hand Cream

	Wt%
1. A-C Polyethylene 617A	3.0
2. Beeswax	2.0
3. Amerchol L-101	5.0
4. Dow 200 Fluid, 350 cs	1.0
5. 2-Ethyl Hexyl Stearate	10.0
6. Triglycerol Diisostearate	5.5
7. Isopropyl Palmitate	8.2
8. Propyl-P-Hydroxybenzoate	0.1
9. Sorbitol	5.0
10.Sodium Borate, Anhydrous	0.3
11.Methyl-P-Hydroxybenzoate	0.2
12.Germall 115	0.3
13.Water	59.4

Procedure:
    Weigh 1-8 and heat to 85-90C with slow agitation. The blend
has a cloud point of approximately 80C. Above the cloud point,
all waxes will eventually dissolve in the blend. If a higher
solvating temperature is used, solvation can be much faster.
Hold the wax blend at 85-90C. Heat 9-13 to 85-90C and stir
gently until all dissolved. Hold at 85-90C.
    Place wax blend in mixing container, add aqueous phase to it
and shear with homomixer or colloid mill. At 75C, the crude
dispersion inverts and a thick creamy emulsion forms. Continue
shearing while scraping the sides of the container to make sure
the whole content is properly sheared. Add perfume, de-aerate
and package.
Ref: 5189-4-10

Oil-In-Water Cream

	%
1. A-C 540	2.0
2. Dow Corning Silicone 344	3.0
3. Lanogene	5.0
4. Sorbitan Monostearate (Arlacel 60)	1.3
5. Polyoxyethylene 20 Sorbitan Monostearate (Tween 60)	1.8
6. Isopropyl Palmitate	7.0
7. Escalol 507	4.0
8. Sorbo (Sorbitol 70%)	5.0
9. Carbopol 941	1.0
10.Germaben II	0.8
11.Water	Balance
12.Triethanolamine (TEA)	0.75
13.Perfume	Q.S.

Procedure:
    Weigh 1-7, then weigh out 8-11. Heat 7-10 with agitation
using a colloid mill or homomixer. When the aqueous solution
reaches 85C, heat 1-6 to 95C. Combine 1-10 until well mixed,
then add TEA and shear until all appears to be homogeneous.
Cool to 55C and add perfume and package.
SOURCE: Allied Signal Inc.: Suggested Formulations

Gentle Night Repair Cream

This product encapsulates an active ingredient which can decrease the oxidation of a precious compound thereby increasing shelf life and reducing the chance of rancidity. The emulsion is very stable, quickly penetrates the skin and leaves a silky feel.

Oil Phase:

Cera Bellina (Pg-3 Beeswax, Koster Keunen)	7.10%
Ozokerite Wax 160/164 (Koster Keunen)	2.20%
Light Mineral Oil (Witco)	7.60%
Isopropyl Palmitate (Unichema)	3.00%
Glycerine Monostearate (Henkel)	1.20%
PEG-100 Stearate (Lipo)	1.00%
Emulsifying Wax NF (Koster Keunen)	1.00%
Incropol SC-20 (Croda)	4.00%
Isostearic Acid (Unichema)	4.00%
Vitamin E (Roche)	0.10%
Vitamin A Palmitate (BASF)	0.10%
Safflower Oil (Lipo)	2.00%
Liquapar (Sutton)	0.50%
Wheat Germ Oil (Lipo)	3.00%
Squalane (Centerchem)	2.00%
Propylene Glycol Dioctanate (Inolex)	4.00%
Silicone 200/100 (Dow)	2.00%

Water Phase:

Water (Distilled)	35.8%
Methyl Paraben (Sutton)	0.30%
Sodium Borate (Borax)	0.60%
Carbopol 940 2%	10.00%

Active Phase:

Phenonip (Nipa)	0.50%
Hyaluronic Acid (Active Organics)	1.00%
Firming Liposome (Centerchem)	4.00%
APT (Centerchem)	3.00%

Procedure:

Heat the components of the oil phase, mix and maintain at a temperature of 82C. In a separate vessel dissolve the components of the water phase while mixing and increasing the temperature to 80C. Add the oil phase to the water phase under moderate agitation, and cool slowly while mixing. When a temperature of 40C is mixed, add phenonip. Cool to 35C and add the three actives. Cool to room temperature.

Adaption of formula and its influence on the product:

You can substitute your preferred secondary emulsifier and some oils with only slight changes to viscosity, gloss, etc. The concentration of Cera Bellina (CTFA=Pg-3 Beeswax) can only be slightly altered since it helps control consistency and the micelle size and shape. The water soluble biologically active ingredients are well protected in this type of emulsion allowing you to incorporate a wide range of active compounds. A formulation of this type is not affected by ionic strength.

SOURCE: Koster Keunen, Inc.: Suggested Formulation

Glossy Cream

Ingredients:	%W/W
A Stearyl alcohol	10.00
Brij 721S	4.00
B Water, deionized	85.50
Carbopol 934	0.10
C Sodium hydroxide (10% W/W aqueous)	0.10
D Dowicil 200	0.10
E Fragrance	0.20

Procedure:
Heat (A) to 70C and (B) to 75C. Add (B) to (A) slowly using blade type agitation. Add (C). Add (D) when the temperature drops below 50C. Add (E) below 40C. Add water to compensate for evaporation. Homogenize. Adjust pH to 5.5-6.5.

Glossy Cream

Ingredients:	%W/W
A Isopropyl myristate	12.00
Stearyl alcohol	4.00
Brij 721S	2.80
Brij 72	2.20
B Water, deionized	78.80
C Dowicil 200	0.10
D Fragrance	0.10

Procedure:
Heat (A) to 70C and (B) to 75C with good propeller agitation. Add (C) at 50C and mix thoroughly. Add (D) at 35C. Add water to compensate for loss due to evaporation.

SOURCE: ICI Surfactants: Suggested Formulations

Glycolic Acid Cream

A rich cream with the benefits of an alpha-hydroxy acid.
Incorporation of Geahlene 750 helps moisturize the skin and
gives it a rich, silky after feel.

Ingredient/Trade Name:	Weight%
A Deionized Water	61.00
Magnesium Aluminum Silicate/Veegum Ultra	1.00
Xanthan Gum/Keltrol	0.30
Methylparaben	0.20
Butylene Glycol	5.00
Tetrasodium EDTA/Hamp-Ene 220	0.10
B Mineral Oil (and) Hydrogenated Butylene/Ethylene/ Styrene Copolymer (and) Hydrogenated Ethylene/ Propylene/Styrene Copolymer//Geahlene 750	9.00
DEA-Cetyl Phosphate/Amphisol	2.00
C12-15 Alkyl Benzoate/Finsolv TN	5.00
Propylparaben	0.10
Cetyl Alcohol/Lanette 16	2.00
Stearic Acid/Emersol 132	4.00
C Diazolidinyl Urea/Germall II	0.30
Glycolic Acid/Glypure 70%	8.00
D Deionized Water	1.00
Sodium Hydroxide	1.00

Procedure:
Disperse Veegum in rapidly agitated deionized water (part A).
Add Keltrol. Heat to 80C. Mix until uniform. In a separate container, heat part B to 80C and mix until all the solids are dissolved. Add part B to part A. Mix for 30 minutes until completely
homogeneous. Cool to 50C. Add part C and mix well. Add premixed
part D. Continue mixing and cooling to 30C. Add fragrance at 30C
if desired.

SOURCE: Penreco: Suggested Formulation

Hand and Body Cream

A rich, protective hand and body cream designed to relieve dryness and leave the skin soft and silky smooth. Geahlene 750 provides long-lasting moisturization, a perceptible conditioning effect, and an elegant after feel.

Ingredient/Trade Name:	Weight%
A Deionized Water	70.60
Carbomer/Carbopol 940	0.20
Propylene Glycol	7.00
Methylparaben	0.20
Panthenol/DL-Panthenol	0.10
B Mineral Oil (and) Hydrogenated Butylene/Ethylene/ Styrene Copolymer (and) Hydrogenated Ethylene/ Propylene/Styrene Copolymer//Geahlene 750	7.00
Propylene Glycol Dicaprrylate/Dicaprate//Myritol PC	5.00
Isostearyl Alcohol/Prisorine 3515	2.00
Propylparaben	0.10
Cetyl Alcohol/Lanette 16	2.00
Glyceryl Stearate (and) PEG-100 Stearate/Arlacel 165	2.50
Potassium Cetyl Phosphate/Amphisol K	1.75
Tocopheryl Acetate/Vitamin E Acetate	0.10
C Triethanolamine, 99%	0.15
D Diazolidinyl Urea/Germall II	0.20
E Soy Lecithin/Sedermasome	1.00
Fragrance	0.10

Procedure:
   Disperse Carbomer into rapidly agitated deionized water. Add remaining part A ingredients. Heat to 75-80C and mix until uniform and lump-free. Combine part B. Heat to 80C and mix until all the solids are dissolved. Add part B to part A. Mix for 30 minutes with good agitation. Add part C. Mix until completely smooth and homogeneous. Cool to 50C. Add part D. Cool to 40C. Add part E. Continue mixing and cooling to 30C.

SOURCE: Penreco: Suggested Formulation

High Content Mineral Oil Cream

Ingredients:	%W/W
A Mineral oil, (Carnation brand)	70.00
Brij 721S	3.70
Brij 72	3.30
B Water, deionized	22.60
C Sodium hydroxide (10% aqueous)	0.10
D Dowicil 200	0.10
E Herbal fragrance (SL 79-1224, PFW)	0.20

Procedure:
   Heat (A) to 65C and (B) to 60C. Add (A) to (B) slowly with
moderate anchor type agitation and add (C). Add (D) at about
50C. Add (E) at 35C and add water to compensate for loss due
to evaporation.
Formula PC-9026

O/W Petrolatum Cream

Ingredients:	%W/W
A Petrolatum, white	35.00
Brij 721, POE 21 stearyl ether	1.00
Brij 72, POE 2 stearyl ether	4.00
Dimethicone (350 cs.)	3.00
B Water, deionized	56.70
Carbomer 934	0.10
C NaOH, 10% aqueous solution	0.10
D Germall II	0.10

Procedure:
   Heat (A) to 70C and (B) to 72C. Add (B) to (A) with moderate
anchor stirring. Add (C). Add (D) below 50C. Stir to 35C and
add any water lost due to evaporation.
Formula CP 1090

SOURCE: ICI Surfactants: Suggested Formulations

Hydrocortisone Cream

Ingredients:	%w/w
Phase A:	
Propylene Glycol	5.0
Glycerin	2.5
Water	10.0
Hydrocortisone USP	(0.5/1.0)
Phase B:	
Water	Q.S.
Preservatives	Q.S.
Phase C:	
Polyglyceryl-3 Methylglucose Distearate (Tego Care 450)	3.0
Octyl Stearate (Tegosoft OS)	6.0
Octyl Palmitate (Tegosoft OP)	5.0
Mineral Oil	5.5
Glyceryl Stearate (Tegin M)	1.8
Stearyl Alcohol	0.8
Cetyl Dimethicone (Abil Wax 9801)	1.0
Phase D:	
Fragrance	Q.S.

Procedure:
1. Mix Phase A to dissolve the active.
2. Add the balance of the water of Phase B plus the preservatives. (Allow for water loss during manufacturing). Heat A/B to 70C.
3. Add the ingredients of Phase C. Mix. Maintain temperature of 70C.
4. Begin cooling when fully dispersed. Begin homogenization at 60C.
5. Homogenize while cooling to 35C.
6. Add fragrance with sweep mixer.

All Natural Cream
(W/O type)

Ingredients:	%w/w
Phase A:	
Polyglyceryl-3 Oleate (Isolan GO 33)	4.0
Hydrogenated Castor Oil	1.5
Beeswax	1.5
Octyl Stearate (Tegosoft OS)	11.5
Cetearyl Octanoate (Tegosoft Liquid)	11.5
Phase B:	
Glycerin	2.5
D-panthenol	0.5
Magnesium Sulfate	0.5
Water	66.5
Phase C:	
Fragrance	Q.S.

Procedure:
1. Add the ingredients of Phase A to a mix tank. Heat to 80-85C and disperse the waxes. Cool to 65-70C.
2. Mix the ingredients of Phase B. Heat to 65-70C.
3. Add Phase B to Phase A. Homogenize and cool to 40C with mixing
4. Add the fragrance. Cool and fill.
SOURCE: Goldschmidt Chemical Corp.: Suggested Formulations

Hydrogen Peroxide Cream

Ingredients:	%W/W
A Stearic acid, triple pressed	6.00
Arlacel 60	2.00
Tween 60	3.00
B Water, deionized	71.72
Phenacetin	0.04
Disodium ethylenediaminetetra-acetate dihydrate	0.10
C Hydrogen peroxide, 35% dilution grade	17.14
D Phosphoric acid, 10% C.P., as required	

Procedure:
Heat (A) to 70C. Heat (B) to 72C. Add (B) to (A) slowly with thorough agitation. Cool to 45C. with agitation. Compensate for lost water due to evaporation. Add (C). Adjust pH to 3.5-4.0 by adding (D).

Hydrogen Peroxide Cream

Ingredients:	%W/W
A Cetyl alcohol	10.00
Arlacel 40	2.00
Tween 40	3.00
B Water, deionized	67.86
C Hydrogen peroxide (10% dilution grade)	17.14
D Phosphoric acid (10% C.P.) as required	

Procedure:
Heat (A) to 70C. Heat (B) to 72C. Add (B) to (A) slowly with thorough agitation. Cool to 45C. with agitation. Compensate for lost water due to evaporation. Add (C). Adjust pH to 3.5-4.0 by adding (D).

Hydrogen Peroxide Cream

Ingredients:	%W/W
A Cetyl alcohol	10.00
Arlacel 165	2.50
B Water, deionized	70.36
C Hydrogen peroxide (10% dilution grade)	17.14
D Phosphoric acid (10% C.P.) as required	

Procedure:
Heat (A) to 70C. Heat (B) to 72C. Add (B) to (A) slowly with thorough agitation. Cool to 45C. with agitation. Compensate for lost water due to evaporation. Add (C). Add (C). Adjust pH to 3.5-4.0 by adding (D).

SOURCE: ICI Surfactants: Suggested Formulations

Hydroxy Acid Cream
(W/O Emulsion)

Ingredients:	%w/w
Phase A:	
Polyglyceryl-4 Isostearate (and) Cetyl Dimethicone Copolyol	
(and) Hexyl Laurate (Abil WE-09)	5.5
Cetyl Dimethicone (Abil Wax 9801)	1.0
Cyclomethicone	5.0
Isohexadecane	4.0
Octyl Stearate (Tegosoft OS)	5.0
Stearyl Heptanoate (Tegosoft SH)	2.5
Beeswax	0.6
Hydrogenated Castor Oil	0.6
Phase B:	
Water	67.7
Sodium Chloride	0.6
Lactic Acid (44% Solution)	5.0
Sodium Salicylate	1.5
Preservatives	Q.S.
Sodium Lactate (and) Sodium PCA (and) Glycine (and)	
Fructose (and) Inositol (and) Sodium Benzoate (and)	
Lactic Acid (Lactil)	1.0
Phase C:	
Fragrance	Q.S.

Procedure:
1. Heat Phase A in a closed kettle to 85C, mixing until all components are fully dispersed or solubilized. When uniform, cool to 45-50C.
2. Mix the ingredients of Phase B together. Heat to 40-45C.
3. Add B to A slowly with soft propeller agitation (150-250 RPM's). Mix until uniform.
4. Homogenize at 35-40C. 5. Add fragrance.

Moisturizing Cream
Cold Process
W/O

Ingredients:	%w/w
Phase A:	
Polyglyceryl-4 Isostearate (and) Cetyl Dimethicone	
Copolyol (and) Hexyl Laurate (Abil WE-09)	5.0
Mineral Oil	5.0
Caprylic/Capric Triglycerides (Tegosoft CT)	5.0
Isopropyl Myristate (Tegosoft M)	5.0
Phase B:	
Water	79.2
Sodium Chloride	0.8
Preservatives	Q.S.
Perfume	Q.S.
Color	Q.S.

Procedure:
1. Mix the oils of Phase A together. Mix well. 2. Dissolve the sodium chloride in the water. Mix until uniform.
3. Add Phase B slowly into Phase A with agitation. 4. Homogenize
5. Preservatives, perfume and color can be added at any time.
SOURCE: Goldschmidt Chemical Corp.; Suggested Formulations

Light Texture Hand Creme

This high humectant creme provides a non-greasy, long lasting soothing feel.

Part A:	% By Weight
Phospholipid SV	3.00
Steareth-20	0.45
Glycerin	5.00
Methyl Paraben	0.25
Water	77.75

Part B:	
Steareth-2	0.80
Cetearyl Alcohol	3.50
Myristyl Myristate	3.50
Finsolv TN	1.50
Isopropyl Palmitate	3.00
Dimethicone (100 cS)	1.00
Propyl Paraben	0.25

Procedure:
Heat both phases to 65C, and homogenize the oil phase into the water phase. Stir-cool to 40C and add fragrance, coloring or preservative as required.

Facial Moisture Creme

This elegant formulation provides high moisturization, excellent rub off resistance, and is ideally suited for overnight skin care.

Part A:	% By Weight
Phospholipid SV	3.00
Steareth-20	0.20
Methyl Paraben	0.25
Water	81.50

Part B:	
Steareth-2	1.30
Cetearyl Alcohol	4.00
Myristyl Myristate	4.00
Isopropyl Myristate	4.00
Dimethicone (100cS)	1.00
Lanolin Alcohol	0.50
Propyl Paraben	0.25

Procedure:
Heat both phases to 65C, and homogenize the oil phase into the water phase. Stir-cool to 40C and add fragrance, coloring or preservative as required.

SOURCE: Mona Industries, Inc.: Phospholipid SV: Suggested Formulations

Mixed Alpha Hydroxy Acid Cream

Resulting product is a viscous firm cream, most appropriately dispensed from a jar. The alpha hydroxy acids (AHA's) contained in this formula consists of Malic, Citric and Lactic Acids. The total use level of AHA's contained in the formula is approximately 8.5%. This formulation utilizes a combination anionic (Lexemul AS) and nonionic (Lexemul 561) emulsion system.

	%w/w
A. Deionized Water	64.70
Methylparaben (Lexgard M)	0.20
2,4-Dichlorobenzyl Alcohol (Myacide SP)	0.20
Propylene Glycol USP	1.00
B. Glyceryl Stearate (and) Sodium Lauryl Sulfate (Lexemul AS)	4.00
Glyceryl Stearate (and) PEG-100 Stearate (Lexemul 561)	2.10
Glyceryl Stearate (Lexemul 515)	3.00
Cetyl Alcohol NF (Adol 52)	1.00
Caprylic/Capric Triglyceride (Lexol GT-865)	3.00
Octyl Stearate (Lexol EHS)	2.00
Avocado Oil	2.00
Tocopherol (Copherol 1300)	0.10
Butylparaben (Lexgard B)	0.10
C. Deionized Water	4.50
Malic Acid (Granular)	0.60
Sodium Citrate USP-FCC	3.00
Lactic Acid USP (88%)	0.60
Sodium Lactate (Purasal S/SP 60%)	7.90

Procedure:
1. Combine section "A" heating to 75-80C. Use a propeller mixer for agitation.
2. Combine section "B" heating to 80-85C. Agitate slowly with a propeller mixer.
3. When sections "A" & "B" are homogeneous and at the designated temperatures slowly add "B' to "A" then begin cooling.
4. Reduce mixing speed during cooling to prevent vortexing.
5. At 55-60C, add section "C" to sections "AB".
6. When the batch is homogeneous and at 35-45C adjust for water loss.
7. Mix to 30C.
8. Adjust final pH to 5.00-5.50 with Lactic Acid or Sodium Lactate.

Procedure:
   Viscosity: 10,000 cPs @ 24C (Brookfield RVT TC @ 10 rpm).
              24 hour sample.
   pH: 4.8 @ 24C.

SOURCE: Inolex Chemical Co.: Formulation 398-104-5

Multi-functional Day Cream

This Bee's Milk formulation creates a barrier that effectively replenishes moisture, softens and imparts radiance. This will combat dry skin caused by variations in humidity and will help to minimize the signs of aging.

Water Phase I:

Bermocol E 481 (Whittaker)	0.4%
Glycerine (Unichema)	2.7%
Water (Distilled)	52.1%
Triethanolamine	0.2%

Oil Phase:

Nikkol Lecinol S-10-M (Barnet)	2.1%
Squalane (Polyesther)	7.9%
Macademia Nut Oil (Tri-K)	4.3%
Borage Oil (Tri-K)	2.9%
Vitamin E (BASF)	0.3%
Vitamin A Palmitate (BASF)	0.3%
Kester Wax-62 (Koster Keunen)	1.0%
Glycerol Monostearate (Koster Keunen)	0.2%
Ozokerite 158/160 (Koster Keunen)	0.3%
Cera Albalate 103 (Koster Keunen)	0.3%
Phytoglycolipid (Barnet)	4.0%
Liquipar (Sutton)	1.0%

Water Phase II:

Bee's Milk (Koster Keunen)	20.0%

Procedure:
Mix and heat water phase I components to 75C. Add all the oil phase components, heat till 75C and mix. Add slowly the water phase I to the oil phase under agitation (approx. 700 rpm) maintaining mixing a temperature of 75C for 5 minutes (make sure that mixing does not exceed 800 rpm's, as a phase inversion will occur). Allow to cool to 50-55C, add water phase II at room temperature, under moderate agitation (approx. 200 rpm's), continue mixing for 5 minutes and pour into container.

Adaption of formula and its influence on the product:
Sunscreens are easily incorporated to give this product an SPF of 6-8 by using Escalol 507 (Van Dyk) at approximately 5% and reducing the concentration of water and or oils.

SOURCE: Koster Keunen, Inc.: Suggested Formulation

Multi-Purpose Cream

Ingredients:	%W/W
A Mineral oil, Carnation	15.00
Stearyl alcohol	5.00
Brij 721S	2.50
Brij 72	2.50
B Water, deionized	74.70
C Dowicil 200	0.10
D Fragrance	0.20

Procedure:
Heat (A) to 70C and (B) to 75C with good propeller agitation. Add (C) at 50C and mix thoroughly. Add (D) at 35C. Add water to compensate for loss due to evaporation.

Versatile Cream

Ingredients:	%W/W
A Cetyl alcohol	10.00
Brij 721S	4.00
B Water, deionized	85.50
Carbopol 934	0.10
C Sodium hydroxide (10% W/W aqueous)	0.10
D Dowicil 200	0.10
E Fragrance	0.20

Procedure:
Heat (A) to 70C and (B) to 75C. Add (B) to (A) slowly using blade type agitation. Add (C). Add (D) when the temperature drops below 50C. Add (E) below 40C. Add water to compensate for evaporation. Homogenize. Adjust pH to 5.5-6.5.

SOURCE: ICI Surfactants: Suggested Formulations

Night Cream (15% Aloe)

Ingredients:	Percent
A. Water	67.625
Methylparaben	0.2
Promulgen-D	2.0
Triethanolamine	0.75
B. Ceraphyl-368	10.0
Kessco-653	3.0
Emerson-1323	6.0
Light Mineral Oil	6.0
Glyceryl monostearate	0.05
Propylparaben	2.0
Vybar-5013	2.0
C. Aloe Veragel Liquid Concentrate 1:40	0.375
D. Fragrance	q.s.

Aloe Bath Soap

Ingredients:	Percent
Sodium Laureth Sulfate (60%)	12.0
Disodium Laureth Sulfosuccinate	7.0
Cocamidopropyl Betaine	6.0
Disodium Oleamido MEA Sulfosuccinate	6.0
Aloe Veragel Liquid 1:1	5.0
Citric Acid	q.s. to pH 6.0
Sodium Chloride	q.s. to 1M cps
Water, Dye, Preservative, Fragrance	q.s. to 100.0

Procedure:
    Add components to water and heat to 40C. Adjust viscosity with sodium chloride and pH with citric acid. Add dye, preservative and fragrance and cool to room temperature.

SOURCE: Dr. Madis Laboratories Inc.: Suggested Formulations

Oil/Water Cold Cream

Ingredients:	%W/W
A Stearic acid, triple pressed	10.00
Mineral oil	6.00
Petrolatum	4.00
Cetyl alcohol	1.00
Arlacel 60	3.00
Tween 60	1.50
B Glycerin	1.00
Triethanolamine	0.60
Water, deionized	72.90
*Preservative	

*q.s. these ingredients

Procedure:
Heat (A) to 70C and (B) to 72C. Add (B) to (A) with constant agitation. Pour at slightly more than room temperature.

O/W Cold Cream (Soap Free)

Ingredients:	%W/W
A Mineral oil	50.00
Beeswax	7.00
Tween 40	2.00
Atlas G-1726, beeswax derivative	8.00
B Water, deionized	33.00
*Preservative	
C *Perfume	

*q.s. these ingredients.

Procedure:
Heat (A) to 75C. Heat (B) to 77C. Add (B) to (A) slowly with moderate but thorough agitation. Add (C) at 45C. Stir until room temperature and package.

SOURCE: ICI Surfactants: Suggested Formulations

Oil/Water Moisturizing Cream

Ingredients:	%W/W
A Octyl dimethyl PABA	5.00
Mineral oil	5.00
Stearyl alcohol	0.50
Brij 721S	2.00
Brij 72	2.00
Dimethicone	0.50
B Water, deionized	84.60
Carbomer 940	0.20
C Sodium hydroxide (10% W/W aqueous)	0.20
D *Preservative	
*Fragrance	

*q.s. these ingredients.

Procedure:
Disperse Carbomer in water and heat (B) to 60C. Heat (A) to 65C and add (B) to (A) with propeller agitation. Slowly add (C) and stir until uniform. Cool to 50C. Add (D) and any water lost due to evaporation.

Oil/Water Moisturizing Cream

Ingredients:	%W/W
A Octyl dimethyl PABA	7.00
Benzophenone-3	3.00
Mineral oil	5.00
Stearyl alcohol	0.50
Brij 721S	2.00
Brij 72	2.00
Dimethicone	0.50
B Water, deionized	79.60
Carbomer 940	0.20
C Sodium hydroxide (10% W/W aqueous)	0.20
D *Preservative	
*Fragrance	

*q.s. these ingredients.

Procedure:
Disperse Carbomer in water and heat (B) to 60C. Heat (A) to 65C and add (B) to (A) with propeller agitation. Slowly add (C) and stir until uniform. Cool to 50C. Add (D) and any water lost due to evaporation.

SOURCE: ICI Surfactants: Suggested Formulations

Oil/Water Night Cream

Ingredients:	%W/W
A Stearic acid, triple pressed	3.00
Mineral oil	3.50
Cetyl alcohol	3.00
Beeswax	2.50
Amerchol H-9	4.00
Isopropyl lanolate	1.50
Arlacel 165	6.00
B Triethanolamine	1.00
Propylene glycol	2.50
Water	73.00
C *Fragrance	
*Preservative	

*q.s. these ingredients.

Procedure:
    Heat (A) to 80C. Heat (B) to 85C. Add (B) to (A) slowly with moderate but thorough agitation. Add fragrance at 50C and continue to stir until room temperature.

W/O Night Cream

Ingredients:	%W/W
A Arlacel 1689	3.50
Arlamol HD	6.00
Arlamol M812	2.00
Arlamol DOA	2.00
Paraffin oil	8.00
Aerosil R972	0.50
B Glycerol	4.00
MgSO4.7H2O	0.50
*Preservative	
Water	73.35
C Perfume Rocelia 74475	0.15

*q.s. these ingredients.

Procedure:
    Heat phase (A) (without Aerosil R972) to 75C. Slowly add Aerosil R972 while stirring. Heat phase (B) to 75C. Slowly add phase (B) to phase (A) while stirring thoroughly. Homogenise the mixture intensively. Allow to cool down while stirring and add phase (C) and allow to cool to 35C while stirring.

SOURCE: ICI Surfactants: Suggested Formulations

Oil/Water Pigmented Cream

Ingredients:	%W/W
A Stearic acid, triple pressed	12.00
Isopropyl myristate	1.00
Arlacel 60	2.00
Tween 60	1.00
B Sorbo	3.00
Propylene glycol	12.00
Titanium dioxide	2.00
Talc	8.00
C Iron oxide	1.00
Water	58.00
D *Preservative	

*q.s. these ingredients.

Procedure:
   Heat (A) to 90C and (B) to 95C. Add (B) and (C) to (A) slowly with constant agitation. Add (D). Homogenize if necessary.

O/W All-Purpose Cream

Ingredients:	%W/W
A Stearic acid	15.00
Lanolin	4.00
Beeswax	2.00
Mineral oil	23.00
Tween 85	1.00
Arlacel 85	1.00
B Sorbo 70% sorbitol solution	12.20
Water	41.80
*Preservative	
C *Perfume	

*q.s. these ingredients.

Procedure:
   Heat (A) to 75C. Heat (B) to 77C. Add (B) to (A) slowly with moderate but thorough agitation. Add (C) at 45C. Stir until room temperature and package.

SOURCE: ICI Surfactants: Suggested Formulations

Oil-in-Water-in-Oil Cream

Oil-in-Water Phase:
Oil Phase:

Carnation Mineral Oil	15.00%
Paraffin	2.00
Cetyl alcohol	3.00
Polyoxyethylene sorbitan tristearate	1.00
Diglycerin monooleate	2.00

Water Phase:

Bentonite	0.50
Dextrin Palmitate	1.00
Glycerin	5.00
Methylparaben	0.10
Water	29.80

Add oil-in-water emulsion to the following oil phase:

Paraffin	5.00
Cetyl Alcohol	3.00
Beeswax	2.00
Sorbitan sesquioleate	3.00
Sorbitan monostearate	1.00
Butylparaben	0.10
Bentonite	0.50
Dextrin palmitate	1.00
Carnation Mineral Oil	25.00

Transparent Cleansing Cream

Di(polyoxyethylene lauryl ether) phosphate sodium salt	5.00%
Glyceryl tri-2-ethylhexanoate	40.00
Carnation Mineral Oil	20.00
Glycerin	30.00
Water	5.00

SOURCE: Witco Corp.: Petroleum Specialties Group: Suggested Formulations

O/W Protective Hand Cream

Ingredients:	%W/W
A Arlacel 165	6.00
Stearic acid	2.00
Petrolatum	7.00
Arlamol HD	10.00
Arlamol E	3.00
B Atlas G-2330	4.00
*Preservative	
Water	66.35
C Fomblin HC/R	1.50
D *Perfume, Bouquet Eau de Mer PC 916.315	0.15

*q.s. these ingredients.

Procedure:
   Heat phases (A) and (B) separately to 75C. Add phase (A) slowly to (B) while stirring intensively. Add phase (C) and homogenise for 1 minute at 75C. Allow to cool while stirring. Homogenise again at 40C after the addition of phase (D). Allow to cool down to 30C while stirring and package.

Hand Cream

Ingredients:	%W/W
A Arlamol E	8.00
Brij 721S	2.40
Brij 72	2.60
Stearyl alcohol	4.00
B Water	82.80
C *Preservative	
D Fragrance	0.20

*q.s. these ingredients.

Procedure:
   Heat (A) to 70C and (B) to 75C. Add (B) to (A) slowly using blade type agitation. Add (C). Add (D) when the temperature drops below 50C. Add (E) below 40C. Add water to compensate for evaporation. Homogenize. Adjust pH to 5.5-6.5.
Formula SK-5

SOURCE: ICI Surfactants: Suggested Formulations

Modified Oil-in-Water Stearic Acid Cream

Ingredients:	%W/W
A Stearic acid, triple pressed	8.00
Arlamol E	2.00
Arlacel 165	5.00
B Sorbo, sorbitol solution USP	10.00
Water	75.00
*Preservative	

*q.s. these ingredients.

Procedure:
Heat (A) to 70C. Heat (B) to 72C. Add (B) to (A) with rapid agitation. Cool to room temperature with stirring.
Formula SK-2B

Skin Cream

Ingredients:	%W/W
A Arlacel 165, acid stable g.m.s.	12.00
Lanolin	1.00
Cetyl alcohol	3.00
Mineral oil	4.00
B Propylene glycol	1.00
Water, deionized	79.00

Procedure:
Heat (A) to 70C. Heat (B) to 72C. Add (B) to (A) slowly with moderate stirring. Stir to 35C and add water lost due to evaporation.

Hand Cream

Ingredients:	%W/W
A Stearic acid, triple pressed	20.00
Isopropyl myristate or mineral oil	2.00
Arlacel 165	5.00
B Sorbo	20.00
Water, deionized	53.00
*Preservative	

*q.s. these ingredients.

Procedure:
Heat (A) to 70C. Heat (B) to 72C. Add (B) to (A) rapid agitation. Stir until set.

SOURCE: ICI Surfactants: Suggested Formulations

Moisturizer O/W Cream

This product utilizes the minor components of natural products to deliver mild anti-microbial activity, which allows the formulator to reduce the concentration of synthetic preservatives. The light yellow coloured cream quickly penetrates the skin, leaving a non-greasy feel and diminishes the signs of aging by laying down a flexible barrier.

Oil Phase:

Orange Wax (Koster Keunen)	4.5%
NF Yellow Beeswax (Koster Keunen)	2.5%
Isostearic Acid (Unichema)	1.0%
Almond Oil (Arista)	3.0%
Mineral Oil (Witco)	3.5%
Glycerol Monostearate (Henkel)	6.0%
Cetylstearyl Alcohol (P&G)	6.0%
Octyl Palmitate (Unichema)	2.5%

Water Phase:

Water (Distilled)	67.8%
Propylene Glycol (Dow)	1.5%
Triethanolamine (Dow)	1.0%
Polysorbate 60 (Gallard & Schlesinger)	0.2%
Germaben II (Sutton)	0.5%

Procedure:
Add a mixed and uniform water phase to a mixed and uniform oil phase at 75C under agitation. Continue mixing till cool.

Adaptation of formula and its influence on the product:
Actives are easily incorporated into a product such as this without altering the aesthetics. Changing the types of oils should only alter the viscosity, if maintaining similar concentrations of oils. Viscosity changes are achieved by reducing the concentration of cetylstearyl alcohol and glycerol monostearate, by approximately 30-40%, however this may cause instability.

SOURCE: Koster Keunen, Inc.: Suggested Formulations

Moisturizing Cream

This formulation demonstrates a polymeric approach to occlusive moisturization eliminating undesirable feel properties of petrolatum or other traditional occlusive ingredients.

Epolene N-34 wax is a non-emulsifiable polyethylene which provides a film that is resistant to wash-off. It is incorporated into this emulsion with Eastman AQ 55 water-dispersible polyester which also contributes a protective film. The wash-off resistance is due to the inability of Epolene N-34 to be emulsified by soap or surfactants. The dual film-forming action of this polymer combination has been shown to reduce moisture loss. In vitro occlusivity data is available.

Phase A:	% W/W
Distilled water	q.s. to 100
Propylene glycol, USP	4.00

Phase B:	
Polawax emulsifying wax, NF	3.00
Arlacel 165 glyceryl stearate and PEG-100 stearate	3.00
Myverol 18-06 distilled monoglyceride	3.00
Isopropyl myristate	5.00
Robane squalane, NF	5.00
Epolene N-34 polyethylene wax	1.00

Phase C:	
Distilled water	10.00
Propylene glycol	3.00
Eastman AQ 55S polymer	1.00

Phase D:	
Eastman vitamin E TPGS (20%)	1.00

Phase E:	
Fragrance	q.s.
Preservative	q.s.

Procedure:
1. Prepare Phase C by adding propylene glycol to water and heating to 95C and then adding Eastman AQ 55S with mixing. Allow to cool to 70C.
2. Heat Phase A with mixing to 95C.
3. Heat Phase B with mixing to 105-110C.
4. Cool Phase B to 95C and add to Phase A with propeller mixing.
5. Continue mixing and cool to 70C, avoiding air entrapment.
6. At 70C, add Phase C.
7. Continue mixing and cool to 50C; then add Phase D and Phase E.
8. With mixing, force cool to room temperature. Product will thicken and form a cream at 44C.
   pH: 5.5

SOURCE: Eastman Chemical Co.: Formulation X20491-069

Neck Firming Cream

The neck is the most neglected part of the body and shows the first visible signs of aging. The product delivers actives to that area in a cost effective product which has emollient and moisturizing properties that firms tiny lines. Continued use helps promote the look of firm, resilient, soft, youthful-looking skin.

Phase A:

Orange Wax (Koster Keunen)	3.0%
Emulsifying Wax NF (Koster Keunen)	5.0%
Shea Butter (Koster Keunen)	3.0%
Stearic Acid (Unichema)	3.0%
Mineral Oil (Witco)	5.0%
Isopropyl Palmitate (Unichema)	5.0%
Glycerol Monostearate (Koster Keunen)	1.5%
Squalane (Barnet)	3.0%
Phytoglycolipid (Barnet)	2.0%
Cetyl Stearyl Alcohol (P&G)	0.5%

Phase B:

Water (Distilled)	57.9%
Carbopol 940 (BF Goodrich)	0.2%
Triethanolamine (Dow)	0.4%
Glycerin (Unichema)	2.0%
Aloe Vera Gel (Active Organics)	0.5%
Propylene Glycol (Dow)	2.0%
Germaben II (Sutton)	1.0%

Phase C:

Elastosol Animal Collagen & Elastin (Croda)	5.0%

Procedure:
Weigh out materials for Phase A, heat to 80C and mix till homogeneous. Heat Phase B to 75C. while mixing. Add Phase A to Phase B under rapid agitation. Cool to 45C and add Phase C. Continue mixing till 35 to 40C.

Adaption of formula and its influence on the product:
Sunscreens are easily incorporated into a formula of this type. By reducing the concentration of water and or the oils, Escalol 507 (Van Dyk) can be substituted at approximately 5% to produce a SPF of 6-8.

SOURCE: Koster Keunen, Inc.: Suggested Formulation

Night Cream, Non-Greasy

	%
A. Miglyol 812	18.0
Imwitor 900	6.0
Miglyol 840	5.0
Imwitor 370	6.0
Imwitor 375	1.0
Cetyl alcohol	1.0
Fluilan	1.0
Emulan ODE 50	1.5
B. Glycerol	6.0
Water	to 100.0
C. Perfume	qs
Preservative	qs

Preparation:
The constituents of A. are mixed and heated to 80-85C. B. is
heated to the same temperature and emulsified in A. C. is
incorporated at approx. 30C.

O/W Cream, with Bacteriostat

	%
A. Softisan 601	20.0
Imwitor 960 flakes	8.0
Miglyol 829	5.0
Softigen 701	5.0
Imwitor 312	5.0
Softisan 649	3.0
Silicon oil 344 fluid	1.0
B. Water	to 100.0
C. Perfume	qs
Preservative	qs

Preparation:
The constituents of A. are heated to 75-80C. Those of B.
are heated to the same temperature and emulsified in A. This
is followed by stirring until cold and incorporating the perfume.

SOURCE: Huls America Inc.: Formulations for Cosmetics:
Suggested Formulations

Protective Face and Body Cream

Ingredients:	Percent
A. Cocoa butter	5.00
Petrolatum and lanolin alcohol (Amerchol CAB)	3.00
Stearic acid XXX	5.00
Glycol stearate	8.50
Ethyl dihydroxy propyl PABA (Amerscreen P)	5.00
Methyl gluceth-20 sesquistearate (Glucamate SS 20)	2.00
Propyl paraben	0.10
B. Aloe Veragel (Veragel Liquid 1:1)	50.00
Water	17.50
Methyl gluceth-10 (Glucose E-10)	1.00
Quaternium-6 (Merquat 100)	1.50
Triethanolamine	1.00
C. Methyl paraben	0.10
Diazolidinyl ura (Germall II)	0.30
Fragrance	q.s.
Color (Optional)	q.s.

Product Characteristics:
Non-tacky cream with good rub-in quality and nice sun
protection.

Procedure:
Heat phase (A) and (B) separately to 75C. Add (B) to (A) with
mixing. Let cool to approximately 45-50C. Add preservative system
and fragrance. SPF is approximately 8-10.

Eye Cream

Ingredients:	Percent
A. Water	q.s.
Carrageenan Gum	0.1
Propylene Glycol	5.0
Sorbic Acid	0.05
Glycamate SSE-20	1.5
B. Petrolatum	15.5
Glyceryl Mono-Stearate	4.0
Soybean Oil	2.0
Steareth-20	0.5
Cetyl Alcohol	0.5
Stearate-2	0.2
Glucate-SS	2.0
C. Aloe Veragel Liquid 1:1	5.0

Procedure:
Heat Phases to 80C. Add Phase B to A at 80C. Let mix until
batch is at 55C. Add aloe gel to batch slowly.. Mix and cool to
below 40C.

SOURCE: Dr. Madis Laboratories Inc.: Suggested Formulations

Protective Face and Body Cream

Ingredients:	Percent
A. Oil phase	13.50
Mineral oil	3.00
Sweet Almond oil	13.50
White Petrolatum	5.50
Glyceryl stearate	3.50
White beeswax	3.50
Tocopherol (Vitamin E)	2.00
B. Aloe veragel (Veragel Liquid)	45.00
Magnesium aluminum silicate (Veegum F)	0.10
Deionized water	10.00
Octyl dimethyl PABA (Escalol 507)	3.00
Diazolidinyl urea (Germall II)	0.35
Methyl Paraben	0.15
Fragrance, color and preservative	q.s.

Product Characteristics:
This is a good rub-in cream with a SPF 6 (approximately), allowing tanning with some protection.

Procedure:
Decrease 0.1 (%w/w) Veegum in 10 (%w/w) water at boiling temperature with mixing, then let cool to approximately 70C. and add rest of B (except for fragrance) and maintain temperature. Heat (A) to 70-75C., then add (B) to (A) with mixing. Let cool to 45C. Then add fragrance, let cool further and package.

Moisturizing Cream

Ingredients:	Percent
A. Mineral oil and lanolin oil (Amerchol L-101)	1.00
Laneth-5	1.00
Isopropyl lanolate	1.00
Stearic acid XXX	3.00
Glyceryl monostearate	2.00
Mineral oil	8.00
B. Aloe veragel (Veragel Liquid 1:1)	78.00
Propylene glycol	4.50
Triethanolamine	0.50
Propylene glycol (and) diazolidinyl urea (and)	
methyl paraben (and) propyl paraben (Germaben II)	1.00

Procedure:
Heat (A) and (B) separately to 75 degrees C. Mix (A) until uniform. Add (A) to (B) with constant stirring and cool to room temperature.

SOURCE: Dr. Madis Laboratories Inc.: Suggested Formulations

Rejuvenating Cream

Ingredients:	% w/w
A) Cremophor A-6	2,50
Cremophor A-25	2,50
Cutina GMS	4,00
Lanette-O	3,00
Stearic Acid	1,00
Paraffin oil	10,00
Cetiol SN	5,00
Vaseline white	3,00
Abil-350	0,40
B) Water demineralized	55,60
Imidazolidinyl urea	0,20
Phenonip	0,50
Glycerin	3,00
Glycolic acid	2,00
Malic acid	1,00
Pentavitin	3,00
Immucell	3,00
C) Fragrance/Chiara 0/238927	0,30

Procedure:
    Heat the ingredients of fatty phase A) to 70C.
    Incorporate item's 11-15 in water (10), adjust the pH to 4.5.
Then incorporate items 16 + 17.
    Heat phase B) to 75C.
    Under stirring add phase B) to phase A), cool to 50C, homogen-
ize and cool to 30C.
    Then add phase C) and stir cold.
Application No. A 032.0/11.94

W/O Night Repair Cream
"cold procedure"

Ingredients:	% w/w
1 A) Pionier KWH-soft	30,00
2 B) Water demineralized	53,70
3 Glycerin	5,00
4 Magnesiumsulfat-7H2O	0,50
5 Revitalin	5,00
6 Hyasol	5,00
7 Phenonip	0,30
8 Euxyl K-200	0,20
9 C) Fragrance: 0/232511 Black Dragon II	0,30

Procedure:
    Dissolve items 3-8 in water (2).
    Under very good stirring add phase B) slowly to phase A).
    Finally incorporate phase C).
Application No. B 006.0/04.93

SOURCE: Pentapharm Ltd.: Suggested Applications

Silk Protein Skin Cream

	Weight,%
1. Mineral Oil	10.0
2. Coco Butter	2.0
3. Cetearyl Alcohol & Ceteareth 20	4.0
4. Emulsifying wax N.F.	6.0
5. Stearic Acid	1.0
6. Glyceryl Monostearate	2.8
7. Glycerin	2.0
8. Propylene Glycol	2.0
9. Acetamide MEA 100%	0.5
10.Triethanolamine	0.2
11.Mackpro NSP (Oleyl/Palmityl/Palmitoleamidopropyl/Silkhydroxy-propyl Dimonium Chloride)	
12.Mackstat DM (DMDM Hydantoin)	qs
13.Fragrance	qs
14.Deionized Water	qs

Procedure:
1. Melt 1,2,3,4,5,6,7,8,9, in a separate container to 75 degrees C.
2. In the mixing tank heat the water to 78 degrees C. add 10,11.
3. Start mixing and add hot mixture of 1 thru 9 slowly with good agitation, mix for 20 minutes then start cooling.
4. While mixing add at 50 degrees C. items 12 then 13 and mix until everything is homogeneous.
5. Check pH and adjust if needed with triethanolamine or acid solution to 5.4-6.5.

SOURCE: McIntyre Group Ltd.: Personal Care Formulary

Moisturizing Cream w/Tritosol

A blend of Polawax and Incroquat Behenyl TMS provide this cream with excellent stability and mildness. The incorporation of Tritisol allows the skin to retain moisture, and acts to condition the skin.

Ingredients:	%
Part A:	
Polawax (Emulsifying Wax NF)	10.00
Incroquat Behenyl TMS (Behentrimonium Methosulfate (and) Cetearyl Alcohol)	3.00
Mineral Oil (70ssu)	5.00
Part B:	
Deionized Water	80.00
Part C:	
Tritisol (Soluble Wheat Protein)	1.00
Propylene Glycol (and) Diazolidinyl Urea (and) Methyl Paraben (and) Propyl Paraben*	1.00

Procedure:
    Combine ingredients of Part A with mixing and heat to 70C. Heat Part B to 70C. Add Part B to Part A with good mixing and cool to 45C. Add Part C with mixing and cool to desired fill temperature.

pH: 5.0+-0.5	Viscosity: 11,000+-10%
*Germaben II	N.A.T.C. Approved

SOURCE: Croda, Inc.: Formula SC-229

Silky-Smooth Lotion/Cream-A

Materials:	%W/W
Part A:	
Deionized Water	79.66
Rheolate 5000	0.3
Propylene Glycol	0.5
Part B:	
Panalene	8.0
Silicone 7207	1.0
Promulgen D	1.0
Ceraphyl 494	2.0
Part C:	
AMP	0.24
Part D:	
Euxyl K-400	0.3
Finsolv TN	2.0
Escalol 507	5.0
Approximate Viscosity, cps: 15,000	

Silky-Smooth Lotion/Cream-B

Materials:	%W/W
Part A:	
Deionized Water	83.55
Rheolate 5000	0.3
Propylene Glycol	2.5
Part B:	
Silicone 7207	2.0
Ceraphyl 494	1.0
Stearic Acid	1.75
Cetyl Alcohol	1.0
Dow 200 Fluid	5.0
Part C:	
Triethanolamine	0.6
Part D:	
Euxyl K-400	0.3
Finsolv TN	2.0
Approximate, Viscosity, cps: 45,000	

Manufacturing Directions:
1. Combine the ingredients in Part A by slowly sifting in the polymer to the water, mixing for 20 minutes, then adding the propylene glycol. Heat to 80C.
2. Combine all ingredients in Part B, then heat to 78C.
3. Add Part B to Part A while stirring. Mix for 10 minutes, then add Part C.
4. Cool to 40C before adding Part D. Package at Room Temperature.

SOURCE: Rheox, Inc.: Formulations TS-312

Silky-Smooth Lotion/Cream-C

Materials:	%W/W
Part A:	
Deionized Water	78.16
Rheolate 5000	0.3
Propylene Glycol	2.5
Part B:	
Panalene	8.0
Silicone 7207	1.0
Promulgen D	0.5
Ceraphyl 494	2.0
Part C:	
AMP	0.24
Part D:	
Euxyl K-400	0.3
Finsolv TN	2.0
Escalol 507	5.0

   Approximate Viscosity, cps: 24,500

Manufacturing Directions:
1. Combine the ingredients in Part A by slowly sifting in the polymer to the water, mixing for 20 minutes, then adding the propylene glycol. Heat to 80C.
2. Combine all ingredients in Part B, then heat to 78C.
3. Add Part B to Part A while stirring. Mix for 10 minutes, then add Part C.
4. Cool to 40C before adding Part D. Package At Room Temperature.

Formula TS-312

Silky Smooth Cream Base

Ingredient:	%W/W
Part A:	
Deionized Water	83.55
Rheolate 5000	0.3
Propylene Glycol	2.5
Part B:	
Silicone 7207	2.0
Ceraphyl 494	1.0
Stearic Acid	1.75
Cetyl Alcohol	1.0
Dow 200 Fluid	5.0
Part C:	
Triethanolamine	0.6
Part D:	
Euxyl K-400	0.3
Finsolv TN	2.0

Manufacturing Directions:
1. Combine the ingredients in Part A by slowly sifting in the polymer to the water, mixing for 20 minutes, then adding the propylene glycol. Heat to 80C.
2. Combine all ingredients in Part B, then heat to 78C.
3. Add Part B to Part A while stirring. Mix for 10 minutes, then add Part C.
4. Cool to 40C before adding Part D. Package at Room Temperature.

Formula TS-331

SOURCE: Rheox, Inc.: Suggested Formulations

Skin Cream

	Wt%
A. Squalane	4.0
Beeswax	1.0
Amiter LG-OD	2.0
Cetyl Octanoate (Emalex CC-168)*	3.0
Hydrogenated Oil (Emalex S.T.G-R)*	4.0
Behenyl Alcohol	1.5
Propyleneglycol Monostearate	1.0
Glyceryl Monostearate, Self Emulsifying	
(Emalex GMS-7CAE)*	5.0
Dimethylpolysiloxane (300 c.s.)	0.4
Butylparaben	0.1
Amihope LL	1.0
B. CAE	0.5
Glycerin	5.0
Hydroxyethylcellulose (1% aq. soln.)	10.0
Methylparaben	0.2
Water	58.3
*Nihon Emulsion Co.	

Procedure:
1. Mix (A) at 80C.
2. Mix (B) at 80C.
3. Add (A) to (B).
4. Mix them with a homomixer, and then cool slowly to 30C.

Note: This skin cream spreads well.

Skin Cream

	Wt%
A. Liquid Petrolatum	17.0
Cetanol	3.0
Propylene Glycol Monostearate	1.0
Glyceryl Monostearate, Self Emulsifying (HLB 5)	
(Emalex GMS-45RT; Nihon Emulsion Co.)	3.0
POE (10) Monostearate	2.0
POE (30) Monostearate	1.0
Butyl Paraben	0.1
Amihope LL	5.0
B. 1,3-Butylene Glycol	5.0
Acylglutamate HS-11	0.3
Methylparaben	0.2
Water	62.4

Procedure:
1. Mix (A) at 80C.   2. Mix (B) at 80C.
3. Add (B) to (A)   4. Mix them with a homomixer.
5. Cool them slowly to 30C.
Note: This skin cream has low friction touch after use.

SOURCE: Ajinomoto USA, Inc.: **Suggested Formulations**

Skin Cream with CAE and Amihope

Ingredients:	Wt%
Phase A:	
Squalane	4.0
Beeswax	1.0
Amiter LG-OD	2.0
Cetyl Octanoate* (Emalex CC-168)	3.0
Hydrogenated Oil* (Emalex S.T.G.-R)	4.0
Behenyl Alcohol	1.5
Stearic Acid	3.0
Propylene Glycol Monostearate	1.0
Glyceryl Monostearate Self Emulsifying (Emalex GMS-7 CAE)*	5.0
Methylpolysiloxane (300 c.s.)	0.4
Butylparaben	0.1
Amihope LL	1.0
Phase B:	
CAE	0.5
Glycerin	5.0
Hydroxyethylcellulose (1% Aq. Soln.)	10.0
Methylparaben	0.2
Water	58.3

Manufacturing Procedure:

Mix Phase A and heat to 80C. Mix Phase B and heat to 80C. Add Phase A to Phase B and homogenize. Cool slowly to 30C.

*Nihon Emulsion Co., Ltd., Japan

Formula No. LC-7

Cleansing Cream

	Wt%
(O) Liquid Paraffin (#70)	21.0
Petrolatum	5.0
Beeswax	10.0
Lanolin Alcohol	1.7
Polyoxyethylene (5) Glyceryl Isostearate	4.0
Polyoxyethylene (5) Stearyl Ether	2.5
Polyethylene Glycol (500) Distearate	1.6
Sorbitan Monolaurate	0.2
Sorbitan Monooleate	0.4
Aluminum Monostearate	2.0
(W) Ajidew N-50	3.0
Propylene Glycol	3.0
Water	45.6
Preservative	q.s.

Procedure:
1. Suspend aluminum monostearate to liquid paraffin.
2. Add other (O) ingredients to the suspension.
3. Heat (O) to 95C to dissolve.
4. Heat (W) to 85C.
5. Add (W) to (O) slowly with stirring.
6. Cool to 40C with stirring.
    pH: 5.1

SOURCE: Ajinomoto USA, Inc.: **Suggested Formulations**

Skin Care Cream

This product imparts good skin feel and has a high quality appearance, though the production is cost effective. There are also barrier and moisturizing properties which are highly effective hydration system for all skin types. The formula also will help reduce transepidermal water loss.

Oil Phase:

Cetylstearyl Alcohol 1618 (P&G)	4.50%
NF White Beeswax (Koster Keunen)	2.50%
Isopropyl Palmitate (Unichema)	3.00%
Light Mineral Oil (Witco)	5.00%
Propylene Glycol Dioctanate (Inolex)	1.50%
Stearic Acid (Unichema)	0.50%
Coconut Oil (CocoChem)	1.25%
Propyl Paraben (Sutton)	0.20%

Water Phase:

Water (Distilled)	74.25%
Glycerine (UniChema)	5.50%
Carboxymethyl Cellulose (CMC, Hercules)	0.30%
Carbopol 940 (BF Goodrich)	0.60%
Sodium Borate (Borax)	0.40%
Triethanolamine (Dow)	0.30%
Methyl Paraben (Sutton)	0.20%

Procedure:
Add to the water phase under agitation, in order; CMC until everything is dissolved then methyl paraben while mixing. Then add carbopol, mix till homogeneous making sure there are no agglomerations. Add the remainder of the water phase components, mix and heat to 75C. Add all the oil phase components, heat till 75C and mix. Add slowly the oil phase to the water phase under agitation maintaining a temperature of 75C. When the oil phase is added, cool and pour into container.

Adaption of formula and its influence on the product:
By reducing the concentrations of mineral oil by 2.0%, propylene glycol dioctanate by 0.5% and the addition of 2.5% Escalol 507 (Van Dyk) the cream will take on an SPF of 6-8. The product has the same appearance, skin feel and stability. Fragrances can also be added without affecting its texture.

SOURCE: Koster Keunen, Inc.: Suggested Formulation

Soft Cream

Ingredients:	%W/W
A Arlamol ISML	4.00
Stearyl alcohol	1.00
Silicone oil (350 cs)	0.50
Arlamol E	1.00
Brij 700	2.25
Brij 72	2.25
B Water, deionized	88.10
Carbopol 934	0.40
C Sodium hydroxide (10% aqueous)	0.40
D Preservative	0.10
E *Perfume	

*q.s. these ingredients.

Procedure:
Heat (A) to 70C and (B) to 72C. Add (B) to (A) with moderate agitation. Add (C). Add (D) below 50C. Add (E) at 35C and add water to compensate for loss due to evaporation.

Jojoba Oil Cream

Ingredients:	%W/W
A Jojoba Oil	10.00
Brij 72	2.70
Brij 700	1.30
Stearyl alcohol	4.00
B Water, deionized	81.90
C Dowicil 200	0.10
D *Perfume	

*q.s. these ingredients.

Procedure:
Heat (A) to 60C and (B) to 65C. Add (B) to (A) using prop-eller type agitation. Add (C) at about 50C and (D) at about 45C. Add water at 35C to compensate for loss due to evapora-tion. Package.
Formula PC-7158

SOURCE: ICI Surfactants: Suggested Formulations

Vanishing Cream

Ingredients:	%W/W
A Arlamol ISML	10.00
Stearyl alcohol	4.00
Silicone oil (350 cs)	0.50
Arlamol E	3.00
Brij 700	2.00
Brij 72	3.00
B Water, deionized	72.90
Sorbo	4.00
Carbopol 934	0.20
C Sodium hydroxide (10% aqueous)	0.20
D Germall II	0.10
E Herbal Fragrance SL 79-1224, PFW	0.10

Procedure:
   Heat (A) to 70C and (B) to 72C. Add (B) to (A) with moderate agitation. Add (C). Add (D) below 50C. Add (E) at 35C and add water to compensate for loss due to evaporation.

Vanishing Cream

Ingredients:	%W/W
A Arlamol ISML	10.00
Stearyl alcohol	4.00
Silicone oil, 350 cs.	0.50
Arlamol E	3.00
Brij 721	2.65
Brij 72	2.35
B Water	73.00
Sorbo, 70% Sorbitol solution	4.00
Carbopol 934	0.20
C NaOH (10% aqueous)	0.20
D Dowicil 200 preservative	0.10
*Perfume	

*q.s. these ingredients.

Procedure:
   Disperse Carbomer in water and heat (B) to 70C. Heat (A) to 72C and add (B) to (A) with propeller agitation. Slowly add (C) and increase speed of the agitation as needed. Add (D) and replace water lost by evaporation.

SOURCE: ICI Surfactants: Suggested Formulations

Vanishing Cream

Phase 1:	Parts by Weight
Rosswax 63-0412	8.0
Rosswax 573	12.0
Amerlate P	1.0
Emerest 2314	1.0
Emerest 2316	1.0
Glyceryl Monostearate SE	0.5

Phase 2:	
Water	99.0
Emery 916 Pure Glycerine	8.0
Triethanolamine	1.2
Fragrance	q.s.
Preservative	q.s.

Procedure:
   In separate steam jacketed kettles heat both phase 1 and 2 to temperature of 170F with agitation. When the temperature is reached add phase 1 to 2 with continued agitation cooling to 120F to package. Fragrance may be added to the product as it is cooling.

Ross Cold Cream Formulation with Jojoba Oil

Part A:	%
Ross Beeswax Substitute 628/5	11.0
Ross Fully Refined Paraffin Wax 150/160	2.0
Mineral Oil 80/90	45.5
Glycerol Monostearate S.E.	0.3
Ross Jojoba Oil	2.0

Part B:	
Borax	0.8
Water	38.4
Fragrance	q.s.
Preservative	q.s.

Procedure:
   Heat Part A to 170F and agitate. Heat Part B to 170F and agitate. Cool to 160F and add Part A to Part B at 160F with good agitation. Cool slowly with agitation and pour at 110F.

SOURCE: Frank B. Ross Co., Inc.: Suggested Formulations

W/O Cream with Eldew

Ingredients/Trade Name:	% by weight
Part A:	
Di-(Cholesteryl, behenyl, octyldodecyl)	
N-Lauroyl-L-glutamic acid ester/Eldew CL-301	2.0
Cetearyl Octanoate	8.0
C12-15 Alkyl Benzoate	5.0
Phenoxyethanol	0.60
Tocopheryl Acetate	0.05
Part B:	
Polyglyceryl-4 Isostearate (and) Cetyl Dimethicone	
Copolyol (and) Hexyl Laurate	5.00
Cetyl Dimethicone	2.00
Part C:	
Deionized Water	68.55
Sodium Chloride	0.80
Glycerin (99.5%)	5.00
Partially Deacetylated Chitin (1.0%)/Marine Dew PC-100	2.00
Part D:	
Methylparaben	0.20
Butylene Glycol	0.80

Procedure:
    Pre-melt part A at 50 degrees Centigrade. Add part B to part
A. Pre-melt part D by heating to 50 degrees C. Add to part C.
Slowly add part C and D mixture to parts A and B with high shear
mixing.
    Appearance: White, smooth, shiny lotion   pH: 6.0-6.5
    Viscosity: 20,000-20,000 (RVT #6 @ 10rpm @ 25 degrees C)

Cleansing Cream

		Wt%
(O)	Liquid Paraffin (#70)	35.0
	Paraffin Wax (mp 42-44C)	10.0
	Squalane	2.0
	Isopropyl Palmitate	3.0
	Cetyl Alcohol	1.0
	Nikkol WCB	10.0
	Sorbitan Monostearate	2.4
	Polyoxyethylene (15) Cetyl Ether	2.6
(W)	Ajidew T-50	3.0
	Water	31.0
	Preservative	q.s.

Procedure:
1. Heat (O) and (W) to 80C.
2. Add (W) to (O) slowly with stirring.
3. Cool to 40C with stirring.
     pH: 5.7

SOURCE: Ajinomoto USA, Inc.: **Suggested Formulations**

90% Water Cream

	Parts by Weight
Water	450.0Gr.
Carbomer 934	2.0Gr.
Protox T-25	1.0Gr.
Rosswax 63-0412	4.0Gr.
Rosswax 1824	16.0Gr.
GMS SE	4.0Gr.
Coconut Oil #76	16.0Gr.
Jojoba Oil	4.0Gr.
Triethanolamine	4.0Gr.
Germaben IIE	6.0Gr.
Fragrance GK-21	q.s.

Procedure:
  Disperse the Carbomer 934 in the water, on a stainless steel
vessel. In a separate vessel melt the Oil Phase. When the Oil
Phase is melted add it to the Water Phase with agitation. Next
add the fragrance, the preservative and last add the Triethanolamine with increased agitation.

Soft & Silky Vanishing Cream

	Parts by Weight
Part (A):	
Rosswax 63-0412	8.0
Rosswax 573	10.0
Ross Lotion Oil 2745	8.0
GMS-SE	0.5
Part (B):	
Water	97.0
Propylene Glycol	8.0
Triethanolamine	2.0
Germaben II	1.2
Part (C):	
Fragrance	q.s.

Procedure:
  Heat Part (A) and Part (B) to 170F in separate steam jacketed
kettles under agitation. When fully heated add Part (A) to Part
(B) under agitation. Cool to 130F, Fragrance and package.

SOURCE: Frank B. Ross Co., Inc.: Suggested Formulations

Section VI
Hair Care Products

"Arctic Mist" Spray Gel

This water white, clear, sprayable gel uses Diaformer Z-301 to provide excellent hold, clear films, and no flaking. It exhibits good viscosity characteristics. Sandoxylate SX-424 is used as a fragrance solubilizer.

Ingredients:	%W/W
Deionized Water	91.80
Carbopol 980	0.30
NaOH 10%	0.40
Diaformer Z-301	6.95
Glycerin	0.10
Disodium EDTA	0.05
Sandoxylate SX-424	0.25
Fragrance	0.05
Preservative	0.10

Procedure:
Add Carbopol 980 to water with rapid agitation and mix until homogeneous. Add NaOH 10% to neutralize the Carbopol 980. Mix well. Add Diaformer Z-301 and Glycerin one at a time with mixing. Add Disodium EDTA. Presolubilize fragrance in Sandoxylate SX-424 and add to batch. Mix well. Adjust pH as needed with NaOH 10%.

Properties:
pH: 6.5
Viscosity: 12,000-14,000 cps.
Appearance: Water white, clear gel
Formulation CHF-20

Super Hard Hold Hair Spray

Diahold A-503 is used to create this super hard hold hair spray that meets the 80% VOC requirements. Its ability to form hard, crystal clear films result in a hair spray that has good gloss and excellent hold on the hair. Diahold A-503 is also easily removed from the hair by shampooing.

Ingredients:	%W/W
Diahold A-503	17.50
Deionized Water	12.70
Dow Corning 190	0.10
SD 40 Alcohol	69.50
Lauramide DEA	0.10
Fragrance	QS

Procedure:
Add Diahold A-503 to alcohol with mixing. Add water and mix well. Add remaining ingredients in order with mixing.

Properties:
Appearance: Pale Yellow Liquid
pH: 8.2-8.6
Formulation CHF-18

SOURCE: Sandoz Chemicals Corp.: Suggested Formulations

Balsam Conditioner

	Weight,%
Mackine 301 (Stearamidopropyl Dimethylamine)	1.6
Mackol 16 (Cetyl Alcohol)	1.8
Phosphoric Acid (85%)	0.9
Sodium Chloride	0.3
Mackstat DM (DMDM Hydantoin)	qs
Balsam of Peru	qs
Water, Dye qs to	100.0

Procedure:
1. Add the first four components to water and heat to 70 degrees C.
2. Blend until homogeneous.
3. Cool to 45 degrees C. and add Mackstat DM and Balsam of Peru.
4. Cool to room temperature and fill.

Clear Conditioner with Wheat Germ Cationic

	Weight,%
Mackalene 716 (Wheat Germamidopropyl Dimethylamine Lactate)	1.0
Natrosol 250 HHR	1.0
Mackstat DM (DMDM Hydantoin)	qs
Water, Fragrance, Dye, qs to	100.0

Procedure:
1. Completely disperse Natrosol in water.
2. Heat to 45 degrees C. and add Mackalene 716.
3. Adjust pH to 5.0 with lactic acid.
4. When product is clear, add remaining components.
5. Cool and fill.

Clear Leave-On Conditioner

	Weight,%
Mackalene 426 (Isostearamidopropyl Morpholine Lactate)	6.0
Natrosol 250 HHR	1.0
Mackstat DM (DMDM Hydantoin)	qs
Deionized Water, Dye, Fragrance qs to	100.0

Procedure:
1. Completely dispense Natrosol in water.
2. Add Mackalene 426 and blend until clear.
3. Heat to 40 degrees C. and add remaining components.

SOURCE: McIntyre Group Ltd.: **Personal Care Formulary**

Birch Hair Lotion

	Wt%
	Wt%
Ethanol 96%	50.0
Birch (water) Special	5.0
Softigen 767	5.0
Marlazin KC 30/50	0.5
Allantoin	0.2
Vitamin F	0.2
Panthenol	0.2
Water	to 100.0

Preparation:
  The components are mixed together in sequence and stirred until homogeneous. The pH is adjusted to 5.5-6.5.

Invigorating Hair Lotion

	Wt%
Isopropanol	45.0
Softigen 767	3.0
Marlowet R 40/K	1.0
Menthol	0.2
Camphor	0.05
Stinging Nettle Special	1.0
Perfume	qs
Water	to 100.0

Preparation:
  The components are mixed together in sequence and stirred until homogeneous. The pH is adjusted to 5.5-6.5.

Hair Conditioner

	Wt%
A. Marlazin KC 30/50	6.0
Cetyl alcohol	3.0
Marlamid M 1218	1.5
Cellosize QP 100 MH	qs
Water	to 100.0
B. Perfume	qs
Colour	qs
Preservative	qs

Preparation:
  The constituents of A. are added together in sequence and stirred while warm until homogeneous. The ingredients of B. are added to A. at approx. 30C. The pH is adjusted to 5.5.

SOURCE: Huls America Inc.: Formulations for Cosmetics: Formulas

Brushing Gel

	Wt%
A. Amihope LL	0.5
POE (20) Sorbitan Monolaurate	0.5
POE (20) Sorbitan Monostearate	0.5
POE (25) Glyceryl Monopyroglutamate Monoisostearate	1.0
Propylene Glycol	2.0
Grape Seed Oil	0.5
B. Carboxyvinyl Polymer (Carbopol 940) (1.0wt% solution)	50.0
Deionized Water	balance
Preservatives	0.2
C. 10% wt. NaOH Solution	2.0

Procedure:
1. Weigh each ingredient (A) in glass vessel and mix.
2. Add (B) to the former mixture and heat to 70-80C with stirring.
3. After dissolution, cool down to room temperature. Then add (C) to the mixed solution and it turns to gel state.

Note:
   This brushing gel reduces an electostatic charge produced by combing and leads to smooth combing.

Hair Brushing Lotion

	Wt%
Amihope LL	1.0
POE (20) Sorbitan Monolaurate (Polysorbate 20)	1.0
Water	50.0
Ethanol	48.0

Procedure:
   Mix all components at room temperature.

Note:
   This hair brushing lotion has good antistatic effect and smoothness for the hair.
   Amihope LL acts as hair conditioning agent instead of the cationic surfactant.

Usage:
   Spray the hair before brushing.

SOURCE: Ajinomoto USA, Inc.: Suggested Formulations

Brushing Lotion

		Wt%
A.	Amihope LL	0.5
	POE (20) Sorbitan Monolaurate	1.5
	POE (25) Glyceryl Monopyroglutamate Monoisostearate	1.0
	Propylene Glycol	2.0
B.	Stearyl Alcohol	0.5
	Carboxyvinyl Polymer (Carbopol 941)	15.0
	(0.5wt% solution; neutralized by NaOH)	
	Deionized Water	balance
	Preservatives	0.2
C.	Prodew 100	0.2
	Ethanol	5.0

Procedure:
1. Weigh each ingredient (A) in glass vessel and mix.
2. Add (B) to the former mixture and heat to 70-80C with stirring.
3. After dissolution, cool down to 50C. Then add (C) to the mixed solution.
4. With stirring, cool down to room temperature.

Note:
   This brushing lotion reduces an electrostatic charge produced by combing and leads to a smooth combing.

Hair Rinse

		Wt%
A.	Cetanol	3.0
	Amiter LGOD-2	5.0
	Glyceryl Monostearate, Self Emulsifying (HLB 11)	2.0
	Amihope LL	3.0
B.	1,3-Butyleneglycol	5.0
	Stearyltrimethylammonium Choride	3.0
	Methylparaben	0.2
	Water	the rest

Procedure:
1. Dissolve (A) at 80C.
2. Dissolve (B) at 80C.
3. Add (B) to (A).
4. Mix them with a homomixer, and cool to 30C.

Note:
   This creamy hair rinse has light finishing touch.

SOURCE: Ajinomoto USA, Inc.: **Suggested Formulations**

Clear Conditioner

Conditioning plus film clarity maximizes the natural beauty of the hair.

Ingredients:	%W/W	Function
1. Hydroxyethyl Cellulose (Natrosol 250 HHR)	0.80	Viscosity
2. Distilled/Deionized Water	84.32	---------
3. Propylene Glycol	5.00	Humectant
4. EDTA	0.10	Clarity
5. Ammonyx KP (Olealkonium Chloride)	6.00	Conditioning
6. Polyquta 400 (Polyquaternium-10)	1.00	Combing
7. Laneto 50 (PEG-75 Lanolin)	0.75	Conditioning
8. Ritapan DL (dl-Panthenol)	1.00	Body
9. Ritabate 20 (Polysorbate 20)	0.80	Clarity
10. Fragrance #189-724	0.20	Odor
11. Kathon CG	0.03	Preservative

Compounding Procedure:
In 60% of the water dissolve item 1. Add item 3, mix until clear. Add item 4 and mix. In a separate beaker disperse item 6 in remaining water. Mix until clear. Add items 5,7 and 8 in order. Combine and add item 9. Add perfume and preservative.

Ref. No. 118-125

Light Hair Conditioner

A conditioner designed for normal/oily hair to maximize combability and manageability. Helps keep hair cleaner longer.

Ingredients:	%W/W	Function
1. Rita CA (Cetyl Alcohol)	3.50	Emulsifier,Thickener
2. Rita-CTAC (Cetrimonium Chloride)	2.00	Conditioner
3. Dow Corning 344 Fluid	2.00	Silky Feel, Lubrication
4. Ritapro 200 (R.I.T.A. Blend)	2.00	Emulsifier
5. Citric Acid @ 100%	0.05	pH Adjuster
6. Glydant	0.20	Preservative
7. Distilled/Deionized Water	90.25	------------

Compounding Procedure:
Heat water and Citric Acid to 65C. Pre-mix Rita CA, Rita-CTAC and Ritapro 200 and heat to 70C. Slowly add pre-mix to water phase and agitate. While mixing cool to 45-50C. Add Glydant and Dow Corning.

Ref. No. 116-166

SOURCE: R.I.T.A. Corp.: Suggested Formulations

Clear Gel Hairdressing

Ingredients:	%W/W
A Mineral oil, Naphthenic, Drakol 10B	13.70
Brij 97	15.50
Arlatone G	15.50
Propylene glycol	8.60
Sorbo	6.90
B Water, deionized	39.80
C *Perfume #44575, Fritzsche Brothers	

*q.s. these ingredients.

Procedure:
Heat (A) and (B) to 90C. Add (B) to (A) with gentle stirring. Cool to 70C and add (C). Stir until uniform and pour while still fluid.

Water/Oil Hair Dressing

Ingredients:	%W/W
A Petrolatum	7.50
Mineral oil	37.50
Lanolin	3.00
Arlacel 83	3.00
Beeswax	2.00
Zinc stearate	1.00
B Borax	0.50
Water, deionized	45.50
C *Fragrance and Preservative	

*q.s. these ingredients.

Procedure:
Heat (A) to 75C. Heat (B) to 77C. Add (B) to (A) slowly with moderate but thorough agitation. Add (C) at 45C. Stir until room temperature and package.

SOURCE: ICI Surfactants: Suggested Formulations

Clear 'N' Natural Conditioner

This clear conditioner acquires its viscosity from the
superior thickening properties of Crothix*.

Ingredients:	%
Part A:	
Deionized water	87.5
Crovol PK-70 (PEG-45 Palm Kernel Glycerides)	3.0
Part B:	
Procetyl AWS (PPG-5 Ceteth 20)	3.0
Crothix* (PEG-150 Pentaerythrityl Tetrastearate)	2.0
Part C:	
Incroquat O-50 (Olealkonium Chloride)	2.0
Incroquat BA-85 (Babassuamidopropalkonium Chloride)	0.5
Part D:	
Triethanolamine 99%	qs to pH 6
Part E:	
Propylene Glycol (and) Diazolidinyl Urea (and) Methyl	
Paraben (and) Propyl Paraben**	1.0
Hydrotriticum WAA (Wheat Amino Acids)	1.0

Procedure:
    Combine ingredients of Part A and heat to 80-85C. When Part
A is homogeneous, add Part B with mixing until uniform. Add
Part C with mixing and cool batch to 30-35C. Adjust pH to
6.0+-0.5 using Part D. Add Part E with mixing.
    **Germaben II   N.A.T.C. Approved
    *The use of Crothix in cosmetic and other formulations is
covered under U.S. Patent #5,192,462.
    Formula HP-165

Cream of Wheat Conditioner

This conditioner gives excellent conditioning/moisturizing
characteristics.

Ingredients:	%
Part A:	
Polawax (Emulsifying Wax NF)	5.00
Incroquat Behenyl TMS (Behentrimonium Methosulfate	
(and) Cetearyl Alcohol)	3.00
Super Refined Wheat Germ Oil (Wheat Germ Oil)	2.50
Part B:	
Deionized Water	85.50
Part C:	
Cropeptide W (Hydrolyzed Wheat Protein (and) Wheat	
Oligosaccharides)	3.00
Germaben II	1.00

Procedure:
    Combine ingredients of Part A with mixing and heat to 75C.
Heat Part B to 75C. Add Part B to Part A with mixing and cool
to 45C. Add ingredients from Part C sequentially with mixing
and cool to desired fill temperature.
    pH: 4.5+-0.5   Viscosity: 50,000cps+-10%
    N.A.T.C. Approved   Formula HP-164

SOURCE: Croda Inc.: Suggested Formulations

Conditioner and Setting Lotion

	Weight,%
Mackalene 316 (Stearamidopropyl Dimethylamine Lactate)	4.0
Gafquat 755	8.0
Mackol 16 (Cetyl Alcohol)	0.5
Mackstat DM (DMDM Hydantoin)	qs
Water, Dye, Fragrance qs to	100.0

Procedure:
1. Completely disperse Gafquat 755 in water.
2. Add Mackalene 316 and Mackol 16 and heat to 70 degrees C.
3. Blend until completely homogeneous.
4. Cool to 45 degrees C. and add remaining components.
5. Cool and fill.

Curl Conditioner and Oil Sheen

	Weight,%
Glycerine	47.0
Propylene Glycol	3.0
Mackpro NLP (Quaternium-79 Hydrolyzed Animal Protein) (Natural Lipid Protein)	4.0
Mackanate DC-30 (Disodium Dimethicone Copolyol Sulfosuccinate)	3.0
Mackstat DM (DMDM Hydantoin)	qs
Deionized Water qs to	100.0

Procedure:
   Add components in order and blend until clear.

Foaming Conditioner

	Weight,%
Mackam 35 (Cocamidopropyl Betaine (Via Glyceride)	10.0
Mackalene 116 (Cocamidopropyl Dimethylamine Lactate)	15.0
Mackpro NLP (Quaternium-79 Hydrolyzed Animal Protein) (Natural Lipid Protein)	4.0
Natrosol 250 HHR	0.7
Mackstat DM (DMDM Hydantoin)	qs
Water, Dye, Fragrance qs to	100.0

Procedure:
1. Thoroughly disperse the Natrosol in water and heat to 45 degrees C.
2. Add Mackam 35, Mackalene 116 and Mackpro NLP.
3. Blend until clear.
4. Add Mackstat DM, fragrance and dye.
5. Cool and fill.

SOURCE: McIntyre Group Ltd.: Personal Care Formulary

Conditioner with Croquat HH

This creamy white conditioner offers the substantivity benefits of quaternizing with the permanant conditioning potential of a cysteine-containing protein.

Ingredients:	%
Part A:	
Incroquat S-85 (Stearalkonium Chloride)	1.50
Polawax (Emulsifying Wax NF)	4.50
Crodacol CS-50 (Cetearyl Alcohol)	5.60
Crodamol W (Stearyl Heptanoate)	6.50
Incrocas 40 (PEG-40 Castor Oil)	2.75
Part B:	
Deionized Water	75.15
Part C:	
Croquat HH (Cocodimonum Hydroxypropyl Hydrolyzed Hair Keratin)	3.00
Part D:	
Propylene Glycol (and) Diazolidinyl Urea (and) Methyl Paraben (and) Propyl Paraben*	1.00

Combine ingredients of Part A with mixing and heat to 75-80C. Heat Part B to 75-80C. Add Part B to Part A with mixing, while avoiding aeration, and cool to 55C. Add Part C with mixing and cool to 40C. Add Part D with mixing and cool to desired fill temperature.

pH: 4.0+-0.5
Viscosity: 35,000+-10% (@ 25C)
* Germaben II
N.A.T.C. Approved
Formula HP-172

High Performance Creme Rinse Conditioner

This formulation gives excellent conditioning to hair.

Ingredients:	%
Part A:	
Crodafos CES (Cetearyl alcohol (and) Cetearyl phosphate)	6.00
Incroquat Behenyl TMS (Behentrimonium Methosulfate (and) Cetearyl Alcohol)	1.00
Propyl paraben	0.10
Volpo S-2 (Steareth-2)	0.50
Crodacol C-70 (Cetyl Alcohol)	2.00
Crodamol PTIS (Pentaerythrityl Tetraisostearate)	1.00
Part B:	
Deionized Water	87.98
Incromectant LAMEA (Acetamide MEA (and) Lactamide MEA)	1.00
Methyl paraben	0.10
Part C:	
TEA 99%	0.32

Combine ingredients of Part A with mixing and heat to 65-70. Combine ingredients of Part B with mixing and heat to 65-70C. Add Part B to Part A with mixing and cool to 40C. Continue mixing and add Part C. Cool to desired fill temperature.

pH: 4.50+-0.5   Viscosity: 56,000+-10%
N.A.T.C. Approved   Formula HP-178-1
SOURCE: Croda Inc.: Suggested Formulations

Cream Curl Activator

Formula:	%Wt.
Phase A:	
Deionized Water	80.43
Hydroxypropyl Methylcellulose	0.20
Triethanolamine	0.02
Panthenol	1.00
Hydrolyzed Silk Protein (Ikeda)	1.00
Quaternium-15	0.30
Phase B:	
Macol CPS	6.00
Solulan 16	1.00
Carnation Mineral Oil	2.00
Masil 656 Fluid	3.00
Masil SF-V Fluid	3.00
Masil 280	2.00
Phase C:	
Citric Acid 50%	0.05
Fragrance	q.s.

Procedure:
    Disperse Hydroxypropyl Methylcellulose in the water; add TEA to initiate hydration. After 20 mins. add remaining A ingredients, heat to 55C and stir. Separately blend B and heat to 55C. Add B to A, maintain agitation while cooling to 40C. Adjust pH and add fragrance.

Hair Relaxer Emulsion

	Percent
Stearyl alcohol	14.00
PEG-75 lanolin	4.00
White Protopet 1S Petrolatum	18.00
Emulsifying wax	5.50
Propylene glycol	5.00
Hydrolyzed animal protein	5.00
Stearic acid	1.50
Sodium hydroxide, 50%	6.00
Fragrance	0.30
Water	q.s.100.00

SOURCE: Witco Corp.: Petroleum Specialties Group: Suggested
        Formulations

Cream Hair Conditioner

	Weight,%
A. Oleyl Alcohol	10.0
Mackol 16 (Cetyl Alcohol)	2.5
Mackester SP (Glycol Stearate Modified)	3.0
BHA	0.1
Propyl Paraben	0.1
B. Mackalene 316 (Stearamidopropyl Dimethylamine Lactate)	25.0
Mackstat DM (DMDM Hydantoin)	qs
Water, Dye, Fragrance qs to	100.0

Procedure:
1. Heat part A to 70 degrees C.
2. Add Mackalene 316 to water and heat to 70 degrees C.
3. Add A to B and with continuous blending cool to 45 degrees C.
4. Add remaining components and cool.

Hair Conditioner

	Weight,%
Mackadet CBC (Conditioner concentrate for viscous cream consistency)	5.0
Mackstat DM (DMDM Hydantoin)	qs
Water, Fragrance, Dye qs to	100.0

Procedure:
1. Add Mackadet CBC to water and heat to 70 degrees C.
2. With continuous mixing cool to 50 degrees C.
3. Add remaining components and cool.

Pearl Conditioner

	Weight,%
Macadet LCB (Liquid Conditioner Concentrate that can be cold blended)	10.0
Triethanol Amine	1.0
Sodium Chloride	0.5
Mackstat DM (DMDM Hydantoin)	qs
Water, Dye, Fragrance q.s. to	100.0

Procedure:
1. Warm water to 40 degrees C.
2. Add sodium chloride and TEA.
3. Add Mackadet LCB and blend slowly.
4. When completely dispersed add dye, preservative and fragrance.
5. Cool and fill.

SOURCE: McIntyre Group Ltd.: **Personal Care Formulary**

Cream Hair Rinse

Ingredients:	%W/W
Cetyl alcohol	1.50
Brij 721S	1.00
Forestall	1.40
Water, deionized	96.10

Procedure:
   Heat to 70 deg. C with stirring until uniform. Cool with stirring and add makeup water.
Formula PC-8214

Hydrogen Peroxide Emulsion

Ingredients:		%W/W
A	Cetyl alcohol	6.00
	Brij 721, Steareth-21	5.00
	Silicone oil, 350 cs.	0.50
B	Water, deionized	66.30
C	Hydrogen peroxide, 27% dilution grade	22.20

Procedure:
   Heat (A) to 60C and (B) to 65C. Add (B) to (A) slowly with moderate agitation. Add (C) below 35C. Replace water lost by evaporation and adjust pH to 3.5-4.0 with dilute phosphoric acid (10% C.P.). Package in suitable container for possible evolution of oxygen.
Formula HC-13

Permanent Wave Lotion

Ingredients:		%W/W
A	Forestall	1.40
	Brij 35 SP	2.00
	Water, deionized	80.10
B	Ethanolamine	9.50
C	Thioglycolic acid	7.00

Procedure:
   Mix (A) with gentle heat if necessary until uniform. Add (B). Add (C). Adjust pH to 9.0-9.5 with additional ethanolamine or thioglycolic acid.
Formula HC-15

SOURCE: ICI Surfactants: Suggested Formulations

Deep Conditioning Treatment

Deep conditioning formula based on a controlled deposition of Cholesterol.

Ingredients	%W/W	Function
1. Distilled/Deionized Water	75.17	---------
2. Propylene Glycol	3.00	Moisture
3. Ritachol (R.I.T.A. Blend)	2.00	Control
4. Rita-STAC (Steartrimonium Chloride)	1.50	Conditioner
5. Cholesterol NF	1.00	Repair
6. Supersat AWS-4 (PEG-20 Hydrogenated Lanolin)	2.00	Repair
7. Ritachol 1000 (R.I.T.A. Blend)	10.00	Emulsifier
8. Petrolatum	5.00	Repair
9. Fragrance	0.30	Odor
10.Kathon CG	0.03	Preservative

Compounding Procedure:
   Combine items 1 and 2 and heat to 75-80C. In a separate beaker combine items 3-8 and heat to 70C. Mix and add to water phase with good mixing. Cool to 50C and add perfume and preservative.
   Ref. No. 118-116

Intensive Conditioner with Body

A detangling conditioner with great wet/dry combing. Hair has increased body/volume.

Ingredients:	%W/W	Function
1. Rita-BTAC (Behentrimonium Chloride)	4.00	Conditioner, Anti-Static
2. Rita CA (Cetyl Alcohol)	3.00	Emulsifier, Thickener
3. Rita SA (Stearyl Alcohol)	1.00	Emulsifier, Thickener
4. Distilled/Deionized Water	80.80	-------------
5. Volatile Silicone Fluid 344 or 345	5.00	Lubricant, Emollient
6. Silicone Fluid 200 (100 cSt)	5.00	Lubricant, Emollient
7. Wheat Germ Oil	1.00	Nourishment
8. Glydant	0.20	Preservative

Compounding Procedure:
   Heat items 1-3 to 70C. Heat water to 80-85C. Add items 1-3 to water and mix until it reaches 50C. Then add items 5-7 and mix. At 45C add items 8. Mix until 35-40C.
   Ref. No. 116-170

SOURCE: R.I.T.A. Corp.: Suggested Formulations

Ethnic Hair Care
Clear Hair Rinse

Ingredients:	%w/w
Water	91.30
Hydroxypropyl Methylcellulose	1.20
Preservative	Q.S.
Propylene Glycol	3.00
Glycerin	2.00
Quaternium-80 (Abil Quat 3272)	0.50
Dimethicone Copolyol (Abil B 88183)	1.00
PEG-30 Glyceryl Laurate (Tagat L)	1.00
Fragrance	Q.S.

Procedure:
1. Add the water and preservative. Heat to 50C. Mix. Disperse the HMC. Mix until clear.
2. Add the remaining ingredients. Cool to 35C.
3. Pre blend the Tagat L and fragrance. Add to batch. Mix until clear using slow speed.

Ethnic Hair Care
Hair Dressings

Ingredients:	%w/w
Water	91.30
Hydroxypropyl Methylcellulose	1.20
Preservative	Q.S.
Propylene Glycol	3.00
Glycerin	2.00
Quaternium-80 (Abil Quat 3272)	0.50
Dimethicone Copolyol (Abil B 88183)	1.00
PEG-30 Glyceryl Laurate (Tagat L)	1.00
Fragrance	Q.S.

Procedure:
1. Add the water and preservative. Heat to 50C. Mix. Disperse the HMC. Mix until clear.
2. Add the remaining ingredients. Cool to 35C.
3. Pre blend the Tagat L and fragrance. Add to batch. Mix until clear using slow speed.

SOURCE: Goldschmidt Chemical Corp.: Suggested Formulations

Ethnic Hair Care
Curl Activation Gels
Hair Repair

Ingredients:	%W/W
Tetrasodium EDTA	0.10
Water	75.80
Glycerin	5.00
Propylene Glycol	15.00
Carbomer 940	0.75
Dimethicone Copolyol (Abil B 88183)	1.00
Dimethicone Propyl PG Betaine (Abil B 9950)	0.50
Dimethicone/Sodium PG Propyl Dimethicone Thiosulfate Copolymer (Abil S 201)	1.00
PEG-30 Glyceryl Laurate (Tagat L)	0.50
Fragrance	Q.S.
Preservatives	Q.S.
Triethanolamine (99%)	0.35

Procedure:
1. Mix the water, Glycerin, Propylene Glycol, and Tetrasodium EDTA together. Disperse the Carbomer 940.
2. Add the rest of the ingredients in order, mixing well between additions. Note: Preblend the Tagat L and fragrance.
3. Neutralize the Carbomer with Sodium Hydroxide solution or the Triethanolamine.

Ethnic Hair Care
Pump Spray-on Moisturizing/Oil-Sheens
Conditioning Sheen

Ingredients:	%W/W
Water	72.00
Tetrasodium EDTA	0.10
Propylene Glycol	10.00
Glycerin	15.00
Quaternium-80 (Abil Quat 3272)	0.40
Dimethicone Propyl PG Betaine (Abil B 9950)	0.25
Dimethicone Copolyol (Abil B 88183)	0.75
PEG-75 Lanolin	0.75
PEG-30 Glyceryl Laurate (Tagat L)	0.50
PEG-18 Glyceryl Cocoate/Oleate (Antil 171)	0.25
Fragrance	Q.S.
Preservatives	Q.S.
Citric Acid (25% solution)	To pH 5-6

Procedure:
1. Heat water to 50C. Dissolve Tetrasodium EDTA.
2. Add the Tagat L, Antil 171, and PEG-75 Lanolin. Mix until clear.
3. Add the remaining ingredients in order, mixing between additions.
4. Adjust pH.

SOURCE: Goldschmidt Chemical Corp.: Suggested Formulations

Ethnic Hair Care
Curl Activation Gel
For Shine

Ingredients:	%W/W
Tetrasodium EDTA	0.10
Water	76.65
Glycerin	15.00
Propylene Glycol	5.00
Carbomer 940	0.75
Dimethicone Copolyol (Abil B 8851)	1.00
Dimethicone Copolyol (Abil B 88183)	1.00
PEG-30 Glyceryl Laurate (Tagat L)	0.50
Fragrance	Q.S.
Preservatives	Q.S.
Sodium Hydroxide (25% Solution)	to pH 6

Procedure:
1. Mix the water, Glycerin, Propylene Glycol, and Tetrasodium EDTA together. Disperse the Carbomer 940.
2. Add the rest of the ingredients in order, mixing well between additions. Note: Preblend the Tagat L and fragrance.
3. Neutralize the Carbomer with Sodium Hydroxide solution or the Triethanolamine.

Ethnic Hair Care
Curl Activation Gel
Moisturizing

Ingredients:	%W/W
Tetrasodium EDTA	0.10
Water	76.30
Glycerin	10.00
Propylene Glycol	10.00
Carbomer 940	0.75
Dimethicone Copolyol (Abil B 88183)	1.00
Dimethicone Propyl PG Betaine (Abil B 9950)	1.00
PEG-30 Glyceryl Laurate (Tagat L)	0.50
Fragrance	Q.S.
Preservatives	Q.S.
Triethanolamine (99%)	0.35

Procedure:
1. Mix the water, Glycerin, Propylene Glycol, and Tetrasodium EDTA together. Disperse the Carbomer 940.
2. Add the rest of the ingredients in order, mixing well between additions. Note: Preblend the Tagat L and fragrance.
3. Neutralize the Carbomer with Sodium Hydroxide solution or the Triethanolamine.

SOURCE: Goldschmidt Chemical Corp.: Suggested Formulations

Ethnic Hair Care
Leave-In Moisturizing/Sheen

Ingredients:	%w/w
Phase A:	
Tetrasodium EDTA	0.1
Water	77.9
Oleth-20	2.0
Dimethicone Copolyol (Abil B 88183)	3.0
Quaternium-80 (Abil Quat 3272)	0.5
Glycerin	15.0
Phase B:	
PEG-30 Glyceryl Laurate (Tagat L)	1.5
Fragrance	Q.S.
Preservative	Q.S.

Procedure:
1. Add the water, Oleth-20, heat to 40C. Mix until clear.
2. Add the remaining ingredients of Phase A mixing each until clear.
3. Add the fragrance to the Tagat L. Mix well - add to Phase A.
4. Cool with mixing - add preservatives.

Ethnic Hair Care
Hair Cuticle Coat Conditioner

Ingredients:	%w/w
Cyclomethicone (and) Dimethiconol (and) Dimethicone	
(Abil OSW 12)	55.0
Cyclomethicone	25.0
Isohexadecane or Mineral Oil	10.0
Isopropyl Myristate (Tegosoft M)	3.0
Octyl Stearate (Tegosoft OS)	2.0
Phenyl Trimethicone (Abil AV 20)	5.0
Fragrance	Q.S.

Procedure:
    Combine all ingredients in order with mixing.

SOURCE: Goldschmidt Chemical Corp.: Suggested Formulations

Ethnic Hair Care
Pump Spray-on Moisturizing/Oil-Sheens
Moisturizing Sheen

Ingredients:	%W/W
Water	72.40
Tetrasodium EDTA	0.10
Propylene Glycol	10.00
Glycerin	15.00
Dimethicone Propyl PG Betaine (ABI1 B 9950)	0.50
Dimethicone Copolyol (Abil B 88183)	0.50
PEG-75 Lanolin	0.75
PEG-30 Glyceryl Laurate (Tagat L)	0.50
PEG-18 Glyceryl Cocoate/Oleate (Antil 171)	0.25
Fragrance	Q.S.
Preservatives	Q.S.
Citric Acid (25% solution)	To pH 5-6

Procedure:
1. Heat water to 50C. Dissolve Tetrasodium EDTA.
2. Add the Tagat L, Antil 171, and PEG-75 Lanolin. Mix until clear.
3. Add the remaining ingredients in order, mixing between additions.
4. Adjust pH.

Ethnic Hair Care
Pump Spray-on Moisturizing/Oil-Sheens
Extra Sheen

Ingredients:	%W/W
Water	66.15
Tetrasodium EDTA	0.10
Propylene Glycol	10.00
Glycerin	20.00
Dimethicone Propyl PG Betaine (Abil B 9950)	0.75
Dimethicone Copolyol (Abil B 88183)	1.50
PEG-75 Lanolin	0.75
PEG-30 Glyceryl Cocoate/Oleate (Antil 171)	0.25
Fragrance	Q.S.
Preservatives	Q.S.
Citric Acid (25% solution)	To pH 5-6

Procedure:
1. Heat water to 50C. Dissolve Tetrasodium EDTA.
2. Add the Tagat L, Antil 171, and PEG-75 Lanolin. Mix until clear.
3. Add the remaining ingredients in order, mixing between additions.
4. Adjust pH.

SOURCE: Goldschmidt Chemical Corp.: Suggested Formulations

Ethnic Hair Glosser/Conditioner

This water-in-oil formula based on a silicone polymeric emulsifier is designed for use as a leave on conditioner to give gloss and body especially for hair which has been chemically treated.

Ingredients:	%w/w
Phase A: Oil Phase:	
Cetyl Dimethicone Copolyol (Abil EM-90)	2.0
Petrolatum	6.0
Mineral Oil	10.0
Cetyl Dimethicone Copolyol (Abil Wax 9801)	2.0
Octyl Palmitate (Tegosoft OP)	3.5
Isopropyl Palmitate (Tegosoft P)	3.5
Lanolin Oil	3.0
Phenyl Trimethicone (Abil AV 20)	2.0
Phase B:	
Fragrance	Q.S.
Phase C: Water Phase:	
Water	64.3
Sodium Chloride	0.7
Glycerin	3.0
Preservatives	Q.S.

Procedure:
1. Blend the components of Phase A together, heating to 50C. Mix until fully dispersed.
2. Cool to 40-45C with agitation. Add fragrance.
3. In a separate vessel, mix the components to Phase C together.
4. Add Phase C to Phase A/B slowly with slow lightning mix. Mix until all water is incorporated into the oil phase.
5. Homogenize.

Ethnic Pump Spray Conditioner

Ingredients:	%w/w
Water	82.2
Propylene Glycol	7.5
Glycerin	7.5
Dimethicone Copolyol (Abil B 88183)	2.0
Quaternium-80 (Abil Quat 3272)	0.5
Panthenol	0.2
Tocopherol Acetate (Vitamin E)	0.1
Preservatives	Q.S.

Procedure:
Combine all ingredients in order with mixing.

SOURCE: Goldschmidt Chemical Corp.: Suggested Formulations

Ethnic Hair Glosser/Extra Conditioning

This water-in-oil formula based on a silicone polymeric emulsifier is designed for use as a leave on conditioner to give gloss and body especially for hair which has been chemically treated.

Ingredients:	%w/w
Phase A: Oil Phase:	
Cetyl Dimethicone Copolyol (Abil EM-90)	2.0
Petrolatum	6.0
Mineral Oil	10.0
Cetyl Dimethicone (Abil Wax 9801)	2.0
Octyl Palmitate (Tegosoft OP)	3.5
Isopropyl Palmitate (Tegosoft P)	3.0
Lanolin Oil	3.0
Phenyl Trimethicone (Abil AV 20)	2.0
Quaternium-80 (Abil Quat 3474)	0.5
Phase B:	
Fragrance	Q.S.
Phase C: Water Phase:	
Water	64.3
Sodium Chloride	0.7
Glycerin	3.0
Preservatives	Q.S.

Procedure:
1. Blend the components of Phase A together - heating to 50C. Mix until fully dispersed.
2. Cool to 40-45C with agitation - add fragrance.
3. In a separate vessel, mix the components to Phase C together.
4. Add Phase C to Phase A/B slowly with slow lightning mix. Mix until all water is incorporated into the oil phase.
5. Homogenize.

Ethnic Hair Care
Extra Conditioning Coat Conditioner

Ingredients:	%w/w
Cyclomethicone (and) Dimethiconol (and) Dimethicone	
(Abil OSW 12)	50.0
Cyclomethicone	30.5
Isohexadecane or Mineral Oil	8.0
Octyl Palmitate (Tegosoft OP)	3.0
Octyl Stearate (Tegosoft OS)	3.0
Phenyl Trimethicone (Abil AV 20)	5.0
Quaternium-80 (Abil Quat 3474)	0.5

Procedure:
   Combine all ingredients in order with mixing.

SOURCE: Goldschmidt Chemical Corp.: Suggested Formulations

Extra Hold Conditioning Mousse

	% by Weight
I. Water	37.35
Amphomer	3.75
Aminomethyl Propanol	0.60
Dow Corning 929 Emulsion	0.40
Phospholipid EFA	0.60
SD3A Alcohol	10.00
II. Hydroxyethyl Cellulose	0.30
Water	37.00
III.Propellant	10.00

Procedure:
  Prepare Part I and II separately. To prepare Part II,
carefully sprinkle hydroxyethyl cellulose into water with good
agitation. Heat may be applied to help solubilization. Blend
Part II to I and then aerosolize.

  Formula F-554

Leave-On Hair Conditioner

Ingredients:	% by Weight
Water	90.2
Hydroxyethyl Cellulose	0.5
Glycol Distearate	1.5
Cetearyl Alcohol	1.8
Monaquat TG	5.5
Phospholipid EFA	0.3
Methyl Paraben	0.1
Propyl Paraben	0.1

Procedure:
  Charge water, slowly add hydroxyethyl cellulose with good
agitation. Add remaining ingredients and heat to 65-70C.
Stir cool to blend 40C and add fragrance, color, etc. Package.

  White Pearled Lotion
  Viscosity: 2500 cp
  pH: 5.6

SOURCE: Mona Industries, Inc.: Suggested Formulations

Firm Hold Hairspray

Ingredients:	Parts by Weight
Resyn 28-2913	2.50
AMP	0.24
Crotein AD Anhydrous	0.20
DC-190	0.20
Armeed DM18D	0.15
Fragrance	Q.S.
Anhydrous Ethanol, SDA-40	71.71
Propellant A-46	25.00

    Valve: Seaquist NS-41   Stem: 2 x 0.020" (Acetal)
    Stem Gasket: Buna N 0.042"   Spring: 0.023" 302 SS
    Body: 0.062" (Nylon) 0.020" VT
    Mounting Cup: Regular, Epoxy Top/Bottom, Dimpled
    Actuator: 0.020" Misty, Acetal (black)

Preparation:
    Dissolve AMP in anhydrous ethanol, SDA-40. Slowly add
Resyn 28-2913 to the solution while maintaining good agitation.
Add remaining ingredients and mix until homogeneous. Filter and
fill concentrate. Charge propellant.
    Formula 6472:94

High Performance Styling Spray

Ingredients:	Parts by Weight
Resyn 28-2930	6.75
AMP	0.63
Dow Corning 556 fluid	0.15
Crotein AD Anh.	0.20
Citroflex 2	0.15
Fragrance	0.10
190 Proof Ethanol, SDA-40	92.02

Preparation:
    Dissolve AMP in the 190 proof SDA-40. While maintaining good
agitation, slowly add Resyn 28-2930 to the vortex. Continue
mixing until solution is complete. Add balance of ingredients.
When homogeneous, filter and fill.
    Formula 6472:95

SOURCE: National Starch and Chemical Corp.: Suggested Formulas

Gel Curl Activator

	%
Water	57.7
Acrylate/Steareth-20/Methacrylate Copolymer	
(Acrysol ICS-1)	2.0
Hystar CG	10.3
Glycerine 99%	23.9
Propylene Glycol	2.0
Dimethicone Copolyol 193	2.3
Germaben II	1.0
Triethanolamine	0.7
Fragrance	q.s.
Jojoba Oil	0.1

Procedure:
   Add ingredients in descending order in a stainless steel
tank, with slow agitation and mix til clear. Pack in a plastic
tube or a plastic bottle.

Alcohol Free Styling Gel
pH: 7.2

	%
Water	83.2
Propylene Glycol	12.0
Acrylate/Steareth-20/Methacrylate Copolymer	
(Acrysol ICS-1)	2.0
Germaben II	1.0
Dimethicone Copolyol 193	0.5
Fragrance	0.5
Triethanolamine	0.7
Jojoba Oil	0.1

Procedure:
   To the water add ingredients 2 thru 5 plus 7 with vary slow
agitation. Next add item 6, agitate til clear and package.

SOURCE: Frank B. Ross Co., Inc.: Suggested Formulations

Guanidine No Base Relaxer

A cream/activator system for a no base relaxer.

Ingredients:	%W/W	Function
I. Activator Solution:		
1. R.I.T.A. GC (Guanidine Carbonate)	30.00	Activator
2. Distilled/Deionized Water	70.00	---------
3. Color, Preservative	q.s.	Preservative

Compounding Procedure:
Combine ingredients at 20C with adequate mixing.

II. No Base Cream:		
1. Ritachol 5000 (R.I.T.A. Blend)	12.00	Emulsifier
2. Ritaderm (R.I.T.A. Blend)	4.00	Emollient
3. Petrolatum	10.00	Texture
4. Mineral Oil 70 wt.	16.00	Spread
5. Supersat AWS 4 (PEG-20 Hydrogenated Lanolin)	2.00	Moisture
6. Calcium Hydroxide, Dry	7.00	Relaxer
7. Propylene Glycol	5.00	Moisture
8. Distilled/Deionized Water	44.00	--------

Compounding Procedure:
Combine ingredients 1-5 and heat to 70C. Combine ingredients 6-8 and heat to 70C. Combine both phases, mix well and cool with mixing to 45C. Add remaining ingredients and cool with mixing to 40C. Package I at 1.75 fl.oz. Package II at 7.50 fl. oz.

Directions:
Mix I in II, stir well with wooden stick until color is completely dispersed. Use as per normal relaxer instructions.
Ref. No. 117-82

Potassium Relaxer

Potassium Hydroxide relaxer based on the emulsification power of Ritachol 5000.

Ingredients:	%W/W	Function
1. Ritachol 5000 (R.I.T.A. Blend)	15.00	Emulsifier
2. Mineral Oil 70 wt.	17.00	Emollient
3. Ritachol (Mineral Oil and Lanolin Alcohol)	2.00	Spread
4. Petrolatum	17.50	Spread
5. Ritahydrox (Hydroxylated Lanolin)	0.50	Stability
6. Propylparaben	0.13	Preservative
7. Rita SA (Stearyl Alcohol)	2.00	Emulsifier
8. Glycerine @ 99%	6.00	Humectant
9. Distilled/Deionized Water	29.67	---------
10. Methylparaben	0.20	Preservative
11. Potassium Hydroxide @ 25%	10.00	Relaxer

Compounding Procedure:
Combine items 1-7 in separate container and heat to 80C. Combine items 8-10 and heat to 80C and add to items 1-7. Mix in variable speed mixer for 1 minute. Add water to cooling pan and mix until 65C. Slowly add KOH. Add ice to cooling pan and cool to 25C.
Ref. No. 118-176

SOURCE: R.I.T.A. Corp.: Suggested Formulations

Hair Brushing Lotion

	Wt%
A. Amihope LL	0.5
TL-10 (PEG-20 Sorbitan Monolaurate)	0.5
1,3-Butanediol (Butylene Glycol)	0.5
Stearyl Alcohol	0.3
Carbopol 941 (Carbomer 941)*	10.0
B. Water	68.2
C. Ethanol	20.0

*0.5% wt Carbomer 941 aqueous solution (neutralized by NaOH)

Procedure:
1. Each chemical cited in (A) is weighed in a glass vessel.
   Heat it to 60-70C and stir until the solution is homogeneous.
2. Add a previously warmed water (B) [60-70C] to the former
   prepared solution and cool it to room temperature with
   stirring.
3. Add a 20.0g ethanol (C) to the cooled solution with stirring.

Note:
This hair brushing lotion has a good antistatic effect and
gives good combability. Amihope LL decreased the friction between
hair and comb or brush remarkably.

Usage: Spray the hair before brushing.

Hair Brushing Lotion

	Wt%
Amihope LL	0.5
POE (20) Sorbitan Monolaurate (Polysorbate 20)	0.5
Ethanol	99.0

Procedure:
Mix all components at room temperature.

Note:
This hair brushing lotion alcoholic type has good antistatic
effect and smoothness for the hair.
Amihope LL acts as hair conditioning agent instead of the
cationic surfactant.

Usage: Spray the hair before brushing.

SOURCE: Ajinomoto USA, Inc.: Suggested Formulations

Hair Conditioner

This formula incorporates a mixture of protein and extracts the have a good affinity for hair, especially when bleached and damaged, leaving your hair smooth, silky and more manageable.

Water Phase I:

Water (Distilled)	61.20%
Butylene Glycol (Hoechst)	3.00%
Glycerine (Unichema)	2.00%
Cellosize (Hercules)	0.30%
Methylparaben (Sutton)	0.30%
Sorbitol 70%	2.00%
Carbopol 940 2%	6.00%
Tris(hydroxymethyl)aminomethane	0.60%

Oil Phase:

Cera Bellina (Pg-3 Beeswax, Koster Keunen)	5.00%
Liquapar (Sutton)	0.40%
Glycerol Monostearate (Henkel)	2.00%
Deo. Orange Wax (Koster Keunen)	1.00%
Vitamin E (BASF)	0.10%
Vitamin A Palmitate (BASF)	0.10%
Propylparaben (Sutton)	0.20%
Stearic Acid (Unichem)	2.00%
Wheat Germ Oil (Lipo)	1.20%
Sweet Almond Oil (Lipo)	1.20%
Propylene Glycol Dioctanate (Inolex)	2.50%

Water Phase II:

Water (Distilled)	5.00%
Triethanolamine (Dow)	1.00%

Active Phase:

Ginseng Extract (Active Organics)	1.00%
Arnica Extract (Active Organics)	1.00%
Wheat Protein (Vege-Tech)	0.70%
Fragrance	0.20%

Procedure:

Disperse the cellosize in the water and when dispersed and added the rest of phase. Heat the water phase I to 80C and the oil phase to 82C. Slowly add the oil phase to water phase I under moderate agitation. Cool to 70C and add water phase II. Continue mixing and added the active phase at 35C. Cool to room temperature.

Adaptation of formula and its influence on the product:

Other active compounds and conditioning agents are easily introduced into this type of formula.

SOURCE: Koster Keunen Inc.: Suggested Formulation

Hair Conditioner with Moisturizers & Quaternium-79 Hydrolyzed Animal Protein

	Weight,%
Mackol 1618 (Cetearyl Alcohol)	3.0
Mackernium SDC-85 (Stearalkonium Chloride)	3.0
Propylene Glycol	1.0
Glycerin	1.0
Mackamide AME-100 (Acetamide MEA)	1.0
Mineral Oil	1.0
Mackpro NLP (Quaternium-79 Hydrolyzed Animal Protein)	
(Natural Lipid Protein)	2.0
Mackstat DM (DMDM Hydantoin)	qs
Deionized Water, Fragrance, Dye qs to	100.0

pH: 3.5-4.5
Viscosity (cps 25 degrees C): 1500-3000

Procedure:
1. Melt waxes and oils to 70 degrees C.
2. Separately heat water plus Mackpro NLP to 70 degrees C. and add hot water solution to hot oils and waxes.
3. Start stirring vigorously for 10 minutes and then start slow cooling while mixing and at 40 degrees C. add Mackstat DM then fragrance and dye and slow mixing down close to room temperature.
4. Stop mixing at 30 degrees C.
5. Adjust pH with citric acid.

High Quality Conditioner

	Weight,%
Mackernium SDC-25 (Stearalkonium Chloride)	10.0
Mackol 1618 (Cetearyl Alcohol)	2.0
Brij 72	2.0
Mackstat DM (DMDM Hydantoin)	qs
Water, Dye, Fragrance qs to	100.0

Procedure:
1. Add components to water and heat to 70 degrees C.
2. With mild agitation blend until homogeneous.
3. Cool to 50 degrees C. and add dye and fragrance.
4. Cool and fill.

SOURCE: McIntyre Group Ltd.: Personal Care Formulary

Hair Conditioning Gel

Ingredients/Trade Name:	% by Weight
Part A:	
Deionized Water	89.94
Hydroxyethylcellulose	0.80
Hydroxypropyl Guar	
Hydroxypropyltrimonium Chloride	0.40
Citric Acid	0.06
Methylparaben	0.02
Part B:	
Panthenol	0.30
Sodium PCA/Ajidew N-50	3.00
Partially Deacetylated Chitin (1% Solution)/Marine-Dew	3.00
Polyquaternium-11	2.00
Diazolidinyl Urea	0.30

Procedure:
    Add part A ingredients in order. Heat to 75 degrees Cent-
igrade. Mix until clear and uniform. Cool to 50 degrees Centigrade. Add part B ingredients in the given order, mixing
well after each addition. Continue mixing and cooling to
35 degrees Centigrade.
    Appearance: Clear gel
    pH: 4.5-5.5
    Viscosity: 7,000-10,000 (RVT #5 @ 10rpm @ 25 degrees C)

Hair Conditioner with CAE

Ingredients:	Wt%
Phase A:	
Cetanol	3.0
Amiter LGS-2	2.0
Nikkol MYS-55	1.0
Phase B:	
CAE	1.0
A-SM	3.0
Prodew 100	2.0
Glycerin	10.0
Methylparaben	0.2
Water	77.8

Specifications: pH: 5.0   Vis: 2500 cps
Procedure:
    Mix ingredients Phase A and heat to 75-80 degrees C. Mix
ingredients Phase B and heat to 75-80 degrees C. Add Phase A
to Phase B and mix with homomixer (1500 to 2500 RPM). Cool to
35 degrees C.
    This hair conditioner has good antistatic and conditioning
effects.
    Formula No. 1 CR-56-11
SOURCE: Ajinomoto USA, Inc.: Suggested Formulations

Hair Conditioning Lotion

Ingredients: %W/W
A Brij 721S   1.30
   Cetyl alcohol   0.90
   Stearyl alcohol   0.60
   Stearalkonium   0.50

B Water, deionized   96.70

Procedure:
Heat (A) to 60C and (B) to 62C. Add (B) to (A) with moderate agitation. Cool while agitating to 40C. Add water to compensate for loss due to evaporation.

Clear Hair Conditioner Lotion

Ingredients: %W/W
A Hydroxyethyl cellulose, Natrosol 250HR   40.00

B Forestall   1.40
   Water, deionized   58.60

Procedure:
Mix (A) in advance by dispersing hydroxyethyl cellulose in water to yield a 3% solution. Prepare solution (B). Add (B) to (A) with stirring until homogeneous.

Hair Conditioning Gel

Ingredients: %W/W
A Mineral oil   11.00
   Arlasolve 200   20.00
   Brij 93   6.00

B Water, deionized   49.60
   Propylene glycol   5.00
   Sorbo   7.00
   Forestall   1.40

Procedure:
Heat (A) and (B) to 90C. Add (B) to (A) with gentle stirring. Cool to 60C and add make-up water. Stir until uniform and pour while still fluid.

SOURCE: ICI Surfactants: Suggested Formulations

Hair Cream

		Wt%
(O)	Liquid Paraffin (#70)	38.0
	Isopropyl Myristate	3.0
	Squalane	0.5
	Nikkol WCB	0.5
	Polyoxyethylene (6300) Monostearate	0.1
	Glyceryl Monostearate (Self Emulsifying Type)	4.1
	Polyoxyethylene (20) Cetyl Ether	2.9
(W)	Ajidew N-50	3.0
	Water	47.9
	Preservative	q.s.

Procedure:
1. Heat (O) and (W) to 80C.
2. Add (W) to (O) slowly with agitation.
3. Finish stirring at 40C.

pH: 7.3
Viscosity: 900 cps

Hair Tonic

		Wt%
(O)	Pyroter GPI-25	2.7
	Pyroter CPI-40	0.3
	Ethyl Alcohol (95%)	50.0
	L-Menthol	0.4
	Camphor	0.05
	Methylparaben	0.05
	Perfume	q.s.
(W)	Ajidew T-50	4.0
	Water	42.5

Procedure:
1. Dissolve (O) and (W) separately at room temperature to be clear solutions.
2. Add (W) to (O) with stirring.

pH: 7.5
Viscosity: 5 cps

SOURCE: Ajinomoto USA, Inc.: Suggested Formulations

Hair Dressing
W/O Cold Process

Ingredients:	%w/w
Phase A:	
Cetyl Dimethicone Copolyol (Abil EM-90)	2.0
Mineral Oil	7.3
Caprylic/Capric Triglycerides (Tegosoft CT)	3.0
Cetyl Dimethicone (Abil Wax 9801)	1.0
Cyclomethicone (Abil B 8839)	5.0
Polyglyceryl-4 Isostearate (Isolan GI 34)	1.0
Cetearyl Isononoanoate (Tegosoft CI)	1.5
Cetyl Octanoate (Tegosoft CO)	0.8
Fragrance	Q.S.
Phase B:	
Water	72.9
Propylene Glycol	3.0
Quaternium-80 (Abil Quat 3270)	0.5
Glycerin	1.2
Sodium Chloride	0.8
Preservatives	Q.S.

Procedure:
1. Mix the ingredients of Phase A together.
2. Dissolve the Sodium Chloride into the water. Add the Glycerin and Propylene Glycol. Mix until clear.
3. Add the preservatives and Quaternium-80 to the water phase. Mix until fully dispersed.
4. Add Phase B to Phase A slowly with soft propeller mixing. Maintain, at all times, a creamy appearance. 5. Homogenize.

Hair Styling Gel

Ingredients:	%w/w
Cyclomethicone (and) Dimethiconol (and) Dimethicone (Abil OSW-12)	22.20
Phenyl Trimethicone (Abil AV-20)	3.00
Cyclomethicone (Abil B 8839)	3.00
Propylene Glycol	42.00
Hexylene Glycol	4.55
Water	22.00
Glycerin	1.00
Preservatives	Q.S.
Polyacrylamide (and) C13-14 Isoparaffin (and) Laureth-7	2.25

Procedure:
  Add the ingredients in order, mixing well between additions. When all ingredients are combined, continue mixing until viscosity increases. Final product is a translucent, colorless gel.
Formula BAM-5-43

SOURCE: Goldschmidt Chemical Corp.: Suggested Formulations

Hair Gloss to Control Gel Concentrate

Ingredients:	%w/w
Methacryloyl Ethyl Betaine/Methacrylates Copolymer	1.50
Water	20.00
Dimethicone (Abil 350)	4.00
Phenyl Trimethicone (Abil AV-20)	4.00
Cyclomethicone (and) Dimethiconol (and) Dimethicone	
(Abil OSW-12)	25.00
Propylene Glycol*	35.00
Hexylene Glycol	7.50
Preservatives	Q.S.
Fragrance	Q.S.
Polyacrylamide (and) C13-14 Isoperaffin (and) Laureth-7	3.00

Procedure:
   Blend the ingredients together in the order given with mixing. When all ingredients are combined, continue mixing until clear. Viscosity will develop after mixing.
   *Ethanol can be substituted for part of the propylene glycol if a lighter product is needed.

Iridescent Hair Gel Concentrate

Ingredients:	%w/w
Cyclomethicone (and) Dimethiconol (and) Dimethicone	
(Abil OSW-12)	27.27
Phenyl Trimethicone (Abil AV-20)	4.55
Propylene Glycol	40.93
Hexylene Glycol	4.55
Water	20.45
Preservatives	Q.S.
Polyacrylamide (and) C13-14 Isoparaffin (and) Laureth-7	2.25

Procedure:
   Add the ingredients in order, mixing well between additions. When all ingredients are combined, continue mixing until clear and iridescent.
Formulation BAM-3-5

SOURCE: Goldschmidt Chemical Corp.: Suggested Formulations

Hair Pomade

High luster hair pomade based on natural Simchin and Shebu. Pationic SSL and Ritapan DL provide moisturization for healthy looking hair.

Ingredients:	%W/W	Function
1. Mineral Oil - 90 wt.	5.00	Luster
2. Simchin Refined (Jojoba Oil)	2.00	Emollient
3. Petrolatum	85.70	Hold
4. Shebu (Shea Butter)	2.00	Luster
5. Lanolin USP (Lanolin)	2.00	Hold
6. Ritapan DL (Panthenol DL)	0.50	Health
7. Pationic SSL (Sodium Stearoyl Lactylate)	2.40	Moisture
8. Fragrance	0.20	Odor
9. Glydant	0.20	Preservative

Compounding Procedure:
Blend items 1-7 at 70C. Stir until uniform. Cool to 40C and add remaining items. Fill into containers.
Ref. No. 118-173

Hair Pomade

Quality hair pomade formulated with Simchin and Shebu for luster. Pationic ISL provides moisturization.

Ingredients:	%W/W	Function
1. Ritawax (Lanolin Alcohol)	1.50	Feel
2. Mineral Oil	5.00	Spread
3. Simchin Refined (Jojoba Oil)	2.00	Luster
4. Shebu (Shea Butter)	1.00	Luster
5. Petrolatum	84.60	Spread
6. Ritalan (Lanolin Oil)	3.00	Luster
7. Pationic ISL (Sodium Isostearoyl Lactylate)	2.50	Moisture
8. Glydant	0.20	Preservative
9. Fragrance - Jergens Type R-30318	0.20	Odor

Compounding Procedure:
Blend items 1-7 and heat until clear. Cool with mixing until 50C. Add perfume and preservative. Fill hot while mixing.
Ref. No. 118-178

SOURCE: R.I.T.A. Corp.: Suggested Formulations

Hair Relaxer - Glycerine

Glycerine based hair relaxer which moisturizes hair. Emulsion stability based on Ritachol 2000.

Ingredients:	%W/W	Function
1. Ritachol 2000 (R.I.T.A. Blend)	15.00	Emulsifier
2. Mineral Oil 70 Wt.	17.00	Spread
3. Ritachol (Mineral Oil and Lanolin Alcohol)	2.00	Feel
4. Petrolatum	17.50	Viscosity
5. Ritahydrox (Hydroxylated Lanolin)	0.50	Stability
6. Propylparaben	0.13	Preservative
7. Rita SA (Stearyl Alcohol)	2.00	Emulsifier
8. Glycerine	6.00	Moisture
9. Distilled/Deionized Water	29.67	--------
10. Methylparaben	0.20	Preservative
11. Sodium Hydroxide (25% Solution)	10.00	Relaxing

Compounding Procedure:
   Combine items 1-7 in Kitchen Aid mixer bowl and heat to 80C. In separate beaker heat items 8-10 to 85C. Add this water mixture to oil mixture and mix at low speed for 1 minute. Add cold water to cooling pan and cool to 65C. Add Sodium Hydroxide solution and mix. Then add ice to cooling pan and cool to 25C.
Note: Inversion will occur during mixing.
   Ref. No. 118-168

Hair Relaxer - Glycol

Ingredients:	%W/W	Function
1. Ritachol 2000 (R.I.T.A. Blend)	15.00	Emulsifier
2. Mineral Oil NF	17.00	Emollient
3. Ritachol (Mineral Oil and Lanolin Alcohol)	2.00	Spread
4. Petrolatum	17.50	Spread
5. Ritahydrox (Hydroxylated Lanolin)	0.50	Stability
6. Propylparaben	0.13	Preservative
7. Rita SA (Stearyl Alcohol)	2.00	Emulsifier
8. Propylene Glycol	6.00	Humectant
9. Distilled/Deionized Water	29.67	--------
10. Methylparaben	0.20	Preservative
11. Sodium Hydroxide (25% Solution)	10.00	Active

Compounding Procedure:
   Weigh and melt items 1-7 at 80C in Kitchen Aid mixer bowl. In separate beaker heat items 8-10 to 85C. Add water phase to oil phase and mix for one minute. Add cooling water to mixer cooling pan and cool to 65C. Add Sodium Hydroxide, then add ice to cooling pan, and then cool to 25C with constant agitation. Note inversion during cooling process.
   Ref. No. 118-169

SOURCE: R.I.T.A. Corp.: Suggested Formulations

Hair Sculpting Cream

This cream gives shine to the hair without being greasy and makes hair easy to style and manage. The Geahlene gives viscosity to the cream and helps hold hair in place. Inclusion of wheat protein adds body and conditioning effects.

Ingredient/Trade Name:	Weight%
A Deionized Water	62.86
Soluble Wheat Protein/Tritisol	1.00
Sodium Chloride	0.80
Panthenol/DL-Panthenol	0.10
Diazolidinyl Urea/Germall II	0.30
PVP/PVP K-30	1.00
B Phenoxyethanol/Emeressence 1160	0.70
Methylparaben	0.20
Propylene Glycol	5.00
C Polyglycerol-4 Isostearate (and) Cetyl Dimethicone Copolyol (and) Hexyl Laurate/Abil WE-09	5.00
Mineral Oil (and) Hydrogenated Butylene/Ethylene/ Styrene Copolymer (and) Hydrogenated Ethylene/ Propylene/Styrene Copolymer//Geahlene 750	20.00
Phenyl Trimethicone/Dow Corning 556	3.00
Fragrance	0.04

Procedure:
Mix part A ingredients until uniform. Heat part B to 60C with gentle stirring until clear. Add part B to part A with mixing. In a separate container, premix part C. Slowly add parts A and B to part C in small increments with high shear mixing until uniform.
SOURCE: Penreco: Suggested Formulation

Hair Conditioner

Part A:	
Deionized Water	90.60
Busan 1504	0.10
Panthenol	0.25
Aloe Vera 200X	0.05
Disodium EDTA	0.05
Cetearyl Alcohol & Cetearath-20	1.10
Stearyl Alcohol & Cetrimonium Bromide	5.00
Jojoba Oil	0.30
Part B:	
Acetamide MEA & Lactamide MEA	1.00
Hydrolyzed Wheat Protein & Wheat Oligosaccharides	1.25
Sodium PCA	0.30

Procedure:
Heat DI water to 75C. Add Busan 1504 preservative to water and mix until dissolved. Maintain temperature at 75C and add remaining Part A ingredients in order shown, ensuring that each is completely blended into batch before the addition of the next ingredient. Mix while cooling to 40C. Add Part B ingredients in order shown and mix until each is dissolved. Cool while mixing to 35C.
SOURCE: Buckman Laboratories, Inc.: Suggested Formulation

Hair Straightener

Formula:	%Wt.
Phase A:	
Deionized Water	56.00
Propylene Glycol	2.00
Phase B:	
Polawax	15.00
Petrolatum (Protopet 1S)	8.00
Hydrogenated Polyisobutene	10.00
Phase C:	
Sodium Hydroxide (25% Sol)	8.00
Phase D:	
Propylene Glycol (and) Diazolidinyl Urea (and)	
Methylparaben (and) Propylparaben	1.00

Procedure:
 Combine and heat water phase A and oil phase B separately to 75C. Add the oil phase to the water phase with rapid mixing. Cool to 40C before adding C and D. Cool to room temp. and package.

Conditioning Styling Gel

	Percent
Part A:	
Water, D.I.	q.s. to 100
Merquat 100	2.00
Part B:	
Blandol	10.00
Brij	21.00
Arlatone G	10.00
Propylene glycol	8.00
Glycerine	7.00
Part C:	
Fragrance	q.s.
Preservative and color	q.s.

Procedure:
 Mix Part A and B in separate vessels, heating each to 90C. Then, add Part B to Part A with moderate agitation. When the mixture is uniform begin to cool, continuing agitation. Add Part C at 53C and continue cooling. Pour at 50C.

SOURCE: Witco Corp.: Petrolatum Specialties Group: Suggested
 Formulations

High Petrolatum Mousse
Aerosol

Oil Phase:	%W/W
Protopet 1S Petrolatum	15.00
Crodamol PMP	5.00
Volpo 3	1.20
Water Phase A:	
Deionized Water	69.30
Carbopol 941	0.13
10% Sodium Hydroxide Solution	0.52
Water Phase B:	
Glycerine	5.00
Volpo S-10	2.85
Germaben II	1.00

Procedure:
Water Phase A:
   Charge a vessel with water and dust in the carbomer. Start heating to 80C. When completely dispersed, add the sodium hydroxide solution and continue mixing.
Water Phase B:
   Combine components in a separate vessel and mix, heat to 80C. When uniform add to Water Phase A.
Oil Phase:
   Combine all components of the oil phase, mix and heat to 80C. When uniform add the combined water phases slowly while mixing. Continue mixing to room temperature.
Fill Ratio:
   Concentrate 4.0%, Propellant A-46 96%.

Hot Oil Hair Treatment Mousse

	Percent
Water	57.80
Disodium EDTA	0.20
Carnation Mineral Oil	10.00
Avocado Oil	5.00
Dimethicone	2.00
Cetearyl Alcohol (and) Ceteareth-20 (Promulgen D)	5.00
Jojoba Oil	1.00
Mink Oil	1.00
Castor Oil	5.00
Octyldodecyl Stearoyl Stearate (Ceraphyl 847)	10.00
Quaternium 26 (Ceraphyl 65)	2.00
Germaben II	1.00

Procedure:
   Combine all ingredients except the Germaben II and heat to 80C. Cool to 50C and add the Germaben II. Cool to room temperature and fill. 95% concentrate/5% Propellant A-46.
Application:
   Apply small amount to wet hair. Cover hair with hot towel for 15-20 minutes. Rinse well.

SOURCE: Witco Corp.: Petroleum Specialties Group: Suggested Formulations

Hot Oil Conditioner

Ingredients:	%w/w
Cyclomethicone (and) Dimethiconol (and) Dimethicone	
(Abil OSW 12)	25.0
Dimethicone (100 cst)	50.0
Cyclomethicone	22.0
Phenyl Trimethicone (Abil AV 20)	3.0
Fragrance	Q.S.

Procedure:
    Combine all ingredients in order with mixing.

Hot Oil Super Conditioner

Ingredients:	%w/w
Cyclomethicone (and) Dimethiconol (and) Dimethicone	
(Abil OSW 12)	25.0
Dimethicone (100 cst)	42.0
Cyclomethicone	30.0
Phenyl Trimethicone (Abil AV 20)	2.5
Quaternium-80 (Abil Quat 3474)	0.5
Fragrance	Q.S.

Procedure:
    Combine all ingredients in order with mixing.

Soft Glossing Spritz

Ingredients:	%w/w
Phenyl Trimethicone (Abil AV-20)	10.0
Cyclomethicone (Abil B 8839)	22.7
Dimethicone (Abil 1000)	2.5
Dimethicone (Abil 100)	14.5
Cyclomethicone (and) Dimethicone (and) Dimethicone	
(Abil OSW-12)	50.0
Benzophenone-3	0.3
Fragrance	Q.S.

Procedure:
    Combine all ingredients in order - mixing well.

    Caution: Traces of water will cause turbidity.

SOURCE: Goldschmidt Chemical Corp.: Suggested Formulations

Intensive Conditioning Pomade

High oil, moisturizing pomade for control of dry, overworked hair.

Ingredients:	%W/W	Function
1. Paraffin (59-60C)	10.70	Control
2. Petrolatum	48.97	Moisture
3. Mineral Oil	30.00	Moisture
4. Forlan L (R.I.T.A. Blend)	5.00	Shine
5. Ritachol (R.I.T.A. Blend)	3.00	Shine
6. Patlac IL (Isostearyl Lactate)	2.00	Repair
7. Fragrance #163-478	0.30	Odor
8. Preservative	0.03	Preservative

Compounding Procedure:
Mix all items except perfume at 65C. Cool to 40C and add perfume and preservative. Let set up in proper container.
Ref. No. 118-122

Light Oil Free Conditioner

Basic conditioner to provide good combing and conditioning to all hair types.

Ingredients:	%W/W	Function
1. Distilled/Deionized Water	93.67	-----------
2. Rita-STAC (Steartrimonium Chloride)	1.50	Conditioner
3. Ritapro 165 (R.I.T.A. Blend)	2.00	Stability
4. Rita CA (Cetyl Alcohol)	2.00	Re-fatting
5. Ritawax ALA (R.I.T.A. Blend)	0.50	Shine
6. Fragrance - Salon #169-122	0.30	Odor
7. Kathon CG	0.03	Preservative

Compounding Procedure:
Heat water to 180F. In separate beaker heat items 2-5 to 165F. Add to water and mix until uniform. Cool to 120F and add perfume and preservative.
Ref. No. 118-115

SOURCE: R.I.T.A. Corp.: **Suggested Formulations**

Intensive Hair Conditioner

An intensive conditioner with nourishing oil to make hair more healthy.

Ingredients:	%W/W	Function
1. Distilled/Deionized Water	84.52	----------
2. Polyquta 3000 (Polyquaternium-10)	0.50	Anti-Static,Thickener
3. Methylparaben	0.15	Preservative
4. Rita-STAC (Steartrimonium Chloride)	1.50	Anti-Static, Conditioner
5. Rita-CTAC (Cetrimonium Chloride)	2.00	Anti-Static, Conditioner
6. Ritachol 2000 (R.I.T.A. Blend)	8.00	Emulsifier
7. Wheat Germ Oil	1.00	Nourishing Oil-Health
8. Tocopherol	0.20	--------
9. Propylparaben	0.10	Preservative
10.Kathon CG	0.03	Preservative
11.Distilled/Deionized Water	2.00	--------
12.Fragrance	q.s.	Odor

Compounding Procedure:
Disperse Polyquta 3000 in water. When dispersed, add Methyl-paraben. Heat to 60C. Add items 4 and 5 and heat to 70C. Heat items 6-9 to 70-75C and add with mixing. While mixing, cool to 45C and add items 10-12.
Ref. No. 116-168

Salon Strength Hair Conditioner

Balanced conditioning and repair are achieved through proper balance of Polyquta and emollient additives.

Ingredients:	%W/W	Function
1. Rita CA (Cetyl Alcohol)	2.00	Re-fatting
2. Mineral Oil 9NF	2.00	Shine
3. Rita SA (Stearyl Alcohol)	0.50	Re-fatting
4. Ritalan C (R.I.T.A. Blend)	1.00	Shine, Combing
5. Rita-CTAC (Cetrimonium Chloride)	2.00	Conditioning
6. Polyquta 3000 (Polyquaternium-10)	0.50	Combing
7. Distilled/Deionized Water	88.23	----------
8. Glycerine	2.00	Moisture
9. Sorbitol (70% Soln.)	0.50	Moisture
10.Supersat AWS-4 (PEG-20 Hydrogenated Lanolin)	1.00	Shine
11.Kathon CG	0.02	Preservative
12.Fragrance	0.25	Odor

Compounding Procedure:
Heat water to 70C. Add item 6 and mix until clear. Add items 8-10 with mixing and heat to 75-80C. Hear items 1-5 to 65C. Add to water mixture. Cool to 50C and add preservative and fragrance.
Ref. No. 119-25
SOURCE: R.I.T.A. Corp.: Suggested Formulations

Leave-On Scalp & Hair Conditioner

A dab of leave-on scalp and hair conditioner massaged into the hair and scalp, will provide conditioning and manageability to the hair while providing moisturization of the scalp. Phospholipid GLA will help maintain a healthy scalp and minimize itch and scaling caused by excessive dryness. It is also a delivery system for gamma linolenic acid, an essential fatty acid for normalizing skin.

Ingredients:	Wt.%
Water	90.2
Hydroxyethyl Cellulose	0.5
Glycol Distearate	1.5
Cetyl Alcohol	1.8
Monaquat SL-5	5.5
Phospholipid GLA	0.3
Preservative	q.s.

Procedure:
Charge water, slowly add hydroxyethyl cellulose with good agitation. Add remaining ingredients plus preservative and heat to 65-70C. Stir cool to 40C and add fragrance, color, etc. package.

    Appearance: White pearled lotion
    Viscosity: 2500 cP
    pH: 6.5
Formula F-653

Extra Hold Conditioning Mousse

	% By Weight
I. Water	37.35
Amphomer	3.75
Aminomethyl Propanol	0.60
Dow Corning 929 Emulsion	0.40
Phospholipid EFA	0.60
SD3A Alcohol	10.00
II. Hydroxyethyl Cellulose	0.30
Water	37.00
III. Propellant	10.00

Procedure:
Prepare Part I and II separately. To prepare Part II, carefully sprinkle hydroxyethyl cellulose into water with good agitation. Heat may be applied to help solubilization. Blend Part II to I and then aerosolize.
Formula F-554

SOURCE: Mona Industries, Inc.: Suggested Formulations

Light Oil Free Conditioner

Light oil free with a touch of foaming to cleanly condition hair to the ends.

Ingredients:	%W/W	Function
1. Grilloten LSE-87K (Sucrose Cocoate)	2.00	Conditioning, Mild
2. Grilloten PSE 141G (Sucrose Stearate)	4.00	Cleaning, Mild
3. Rita Cetearyl 70/30 (Cetearyl Alcohol)	2.50	Re-fatting
4. PEG-20 Cetearyl Alcohol	0.50	Combing
5. Propylene Glycol	1.00	Humectant
6. Polyquta 400 (Polyquaternium-10)	1.50	Combing, Style
7. Glycerine	3.00	Moisture
8. Distilled/Deionized Water	85.17	--------
9. Fragrance-Pert Type 189-724	0.30	Odor
10.Kathon CG	0.03	Preservative
11.Citric Acid (25% Soln.)	q.s.	pH Control

Compounding Procedure:
Heat 1-4 to 70C. Separately disperse item 6 in water. Then
add items 5 and 7. Heat to 80-85C. Then add items 1-4 to mixture.
Cool to 40C and add perfume and preservative. Adjust pH with
Citric Acid to pH 5.0.
Ref. No. 118-121

Aloe Vera Hair Conditioner (Light)

A light strength conditioner specially formulated for
normal hair. Adds sheen to hair.

Ingredients:	%W/W	Function
1. Distilled/Deionized Water	94.47	----
2. Ritaloe 200M (Aloe Vera Gel)	0.25	Shine
3. Rita CA (Cetyl Alcohol)	2.00	Emulsifier,Thickener
4. Rita SA (Stearyl Alcohol)	1.00	Emulsifier,Thickener
5. Rita-STAC (Steartrimonium Chloride)	1.50	Conditioner, Anti-static
6. Simchin Refined (Jojoba Oil)	0.75	Conditioner, Shine
7. Kathon CG	0.03	Preservative
8. Fragrance	q.s.	Odor

Compounding Procedure:
Heat items 1 and 2 to 70C. Heat items 3-6 to 65-70C and add
with stirring. Continue mixing until batch reaches 40-45C. Add
Kathon and Fragrance.
Ref. No. 116-174

SOURCE: R.I.T.A. Corp.: Suggested Formulations

Low Cost Relaxer

A lower cost relaxer with reduced oil phase materials.
Laneto-100 provides scalp protection from irritation.

Ingredients:	%W/W	Function
1. Ritachol 2000 (R.I.T.A. Blend)	15.00	Emulsifier
2. Rita SA (Stearyl Alcohol)	1.00	Emulsifier
3. Ritoleth-10 (Oleth-10)	2.00	Emulsifier
4. Petrolatum	8.00	Viscosity
5. Mineral Oil 70 wt.	10.00	Spread
6. Laneto-100 (PEG-75 Lanolin)	0.50	Feel
7. Propylene Glycol	2.00	Moisture
8. Distilled/Deionized Water	52.50	--------
9. Sodium Hydroxide (25% Soln.)	9.00	Relaxer

Compounding Procedure:
    Combine items 1-5 and heat to 60C. In a separate container,
heat items 6-8 and heat to 60C. Add to oil phase and mix at low
speed. Cool to 45C and add Sodium Hydroxide solution. Continue
mixing and fill to jars at 30C.
    Ref. No. 120-26

Extra Stable Relaxer

Very cost efficient, stable relaxer formulation which is
easy to process.

Ingredients:	% W/W	Function
1. Ritachol 5000 (R.I.T.A. Blend)	10.00	Emulsifier
2. Mineral Oil 70 Wt.	16.00	Spread
3. Ritachol (Mineral Oil and Lanolin Alcohol)	2.00	Feel
4. Petrolatum	17.00	Texture
5. Ritahydrox (Hydroxylated Lanolin)	0.60	Stability
6. Rita SA (Stearyl Alcohol)	2.00	Emulsifier
7. Ritox 52 (PEG-40 Stearate)	1.50	Stability
8. Glycerine	6.00	Moisture
9. Distilled/Deionized Water	34.70	--------
10.Methylparaben	0.20	Preservative
11.Sodium Hydroxide (20% Solution)	10.00	Relaxer

Compounding Procedure:
    Combine items 1-7 in orbital mixer at 80C. Combine items 8-10
at 80C and add to oil mixture. Cool to 65C. Add Sodium Hydroxide
solution with agitation. Cool in ice bath to 25C. Homogenize if
necessary to further improve stability.
    Ref. No. 117-21A

SOURCE: R.I.T.A. Corp.: Suggested Formulations

Medium Hold, 80% VOC Pump Hair Spray

Diaformer Z-A provides a medium hold in this clear, water white spray. It also reduces static and improves manageability when the hair is combed. Its clear, flexible film gives a natural looking hold while offering excellent curl retention.

Ingredients:	%W/W
Phase A:	
Diaformer Z-A	5.00
Dow Corning 190	0.20
Monamide 716	0.10
SD 40 Alcohol	75.00
Fragrance	0.10
Deionized Water	Q.S.

Procedure:
    Add Diaformer Z-A to alcohol, mixing well. Add water with mixing. Add remaining ingredients one at a time with mixing.

Properties:
    Appearance: Crystal clear, water white liquid

Formulation CHF-12

55% VOC Pump Hair Spray

This 55% VOC hair spray incorporates Diaformer Z-400. This deodorized polymer is well suited to lightly fragranced products. It provides a medium hold in this crystal clear, water white spray. Velsan P8-3 acts a a non-greasy superfatting agent.

Ingredients:	%w/w
Phase A:	
Diaformer Z-400	4.00
SD 40 Alcohol	40.00
Velsan P8-3	0.50
Fragrance	0.05
Deionized Water	Q.S.

Procedure:
    Add Diaformer Z-400 to alcohol, mixing well. Add water with mixing. Add remaining ingredients one at a time with mixing.

Properties:
    Appearance: Crystal clear, water white liquid

Formulation CHF-14

SOURCE: Sandoz Chemicals Corp.: Suggested Formulations

Medium Strength Hair Conditioner

Conditioning designed to provide nourishing and combing benefits while leaving a clean feel in the wet mode due to Dimethyl Quat.

Ingredients:	%W/W	Function
1. Rita-SBC (Stearalkonium Chloride)	1.14	Conditioner
2. Rita-CTAC (Cetrimonium Chloride)	0.48	Conditioner
3. Rita-CA (Cetyl Alcohol)	0.85	Re-fatting
4. Citric Acid @ 100%	0.09	pH Control
5. Dimethyl Stearamine (Adogen MA-108)	0.13	Conditioner
6. Distilled/Deionized Water	94.81	---------
7. Promois WK-HQ (Hydroxypropyl trimonium Hydrolyzed Keratin)	0.05	Body
8. Ritaloe 200M (Aloe Vera Gel)	0.05	Body
9. Ritapan DL (dl-Panthenol)	0.05	Body
10.Propylene Glycol	0.50	Clean Feel
11.Hydroxyethyl Cellulose (Natrosol 250 HHR)	1.00	Viscosity
12.Kathon CG	0.03	Preservative
13.Dow Corning 2-7224	0.40	Combing
14.Dow Corning 344 Fluid	0.17	Combing
15.Fragrance	0.25	Odor

Compounding Procedure:
To 15% of total water at 65C add separately items 1-4. Heat to 70C, add item 5 and mix 20 min. Add another 10% of the water at room temperature and continue mixing 20 min. To remaining 75% of water add items 7,8 and 9 at room temperature. Add item 10 and slowly sift in item 11. Agitate 15 min. while heating batch to 35C. When batch thickens, increase agitation for up to 60 min. until uniform. Add items 12,13 and 14, continue mixing. Then add quaternary/fatty alcohol phase (items 1-6) and fragrance. Mix until uniform.
Ref. No. 116-181B

Medium Hair Conditioner

Humectant conditioner to maximize combing from silicone and body from glycerine.

Ingredients:	%W/W	Function
1. Rita-STAC (Steartrimonium Chloride)	2.40	Conditioner
2. Rita GMS (Glyceryl Stearate)	1.00	Control
3. Rita CA (Cetyl Alcohol)	5.00	Coating
4. Ritaceti (Cetyl Esters)	2.00	Combing
5. Ritacet-20 (Ceteareth-20)	0.50	Emulsifier
6. Glycerine	1.00	Humectancy
7. Distilled/Deionized Water	85.77	--------
8. Dow Corning 344 Fluid	2.00	Combing, Shine
9. Fragrance - Finesse Type	0.30	Odor
10.Kathon CG	0.03	Preservative

Compounding Procedure:
Heat items 1-5 to 70C. In separate beaker heat items 6 and 7 to 75C. Add 1-5 to water. Mix and cool to 50C. Add silicone and mix. Add perfume and preservative.
Ref. No. 118-117

SOURCE: R.I.T.A. Corp.: Suggested Formulations

Moisture Balance Conditioner w/Cropeptide W

Although this easy-to-make conditioner contains few ingredients, it delivers big conditioning benefits.

Ingredients: %
Part A:
Incroquat CR Conc. (Cetearyl Alcohol (and) PEG-40 Castor
  Oil (and) Stearalkonium Chloride)   6.00

Part B:
Deionized Water   92.70

Part C:
Cropeptide W (Hydrolyzed Wheat Protein (and) Wheat
  Oligosaccarides)   1.00
Methyl paraben   0.20
Propyl paraben   0.10

Procedure:
    Heat Part A and Part B separately to 70-75C with mixing.
At 70-75C, add Part B to Part A with mixing and cool to 40C.
Add ingredients of Part C individually with mixing and cool
to desired fill temperature.
    pH: 4.0+-0.5   Viscosity: 2,000+-10% (@25C)
    N.A.T.C. Approved

SOURCE: Croda Inc.: Formulation HP-158

"Leave-On" Conditioner Concentrate

1. Incorporate fragrance. Take an exact weight of Concentrate and
   add exact weight of fragrance (depending on level after dilution. Blend together with thorough mixing.

   Note: Concentrate contains only enough preservative to protect
         the concentrate. Please add additional preservative to
         product diluted solution.

2. Dilute the above concentrated blend with 7 parts of deionized
   water for a firm set, or with 9 parts deionized water for
   regular setting product.

3. Apply finished product to shampooed and towel dried hair.
   Don't rinse. Style hair.

SOURCE: McIntyre Group Ltd.: Suggested Formulation

Natural Lipid Conditioner for Professional Salon

	Weight,%
Mackernium SDC-85 (Stearalkonium Chloride)	1.5
Mackalene NDC (Oleamidopropyl Dimethylamine (Oleamidopropyl Dimethylamine Lactate (and) Palmit-amidopropyl Dimethylamine Lactate (and) Palmitol-eamidopropyl Dimethylamine Lactate)	1.0
Mackpro NLP (Quaternium-79 Hydrolyzed Animal Protein) (Natural Lipid Protein)	2.0
Mackol 1618 (Cetearyl Alcohol)	1.8
Steareth-2	1.8
Mackstat DM (DMDM Hydantoin)	qs
Water, Fragrance, Dye qs to	100.0

Procedure:
1. Add first five components to water and heat to 70 degrees C.
2. Cool to 45 degrees C. and add remaining components.
3. Cool and fill.

Mild Opaque Conditioner

	Weight,%
Mackalene 326 (Stearamidopropyl Morpholine Lactate)	8.0
Cetyl Alcohol	1.8
Phosphoric Acid	0.6
Sodium Chloride	0.3
Mackstat DM (DMDM Hydantoin)	qs
Water, Dye, Fragrance, qs to	100.0

Procedure:
1. Add first four components to water and heat to 70 degrees C.
2. With stirring, cool and add dye, preservative and perfume at 40 degrees C.

Mild Pearl Conditioner

	Weight%
Mackalene 326 (Stearamidopropyl Morpholine Lactate)	7.0
PEG 400 Distearate	0.5
Sodium Chloride	0.5
Mackstat DM (DMDM Hydantoin)	qs
Water, dye, fragrance, qs to	100.0

Procedure:
1. Add the first three components to water and heat to 65 degrees C.
2. With continuous stirring, cool to 40 degrees C. and add dye, preservative and fragrance.

SOURCE: McIntyre Group Ltd.: Personal Care Formulary

Non-Aerosol Hair Spray with Protein

A non-aerosol hair spray based on protein (Promois WK-HQ) and Ritapan DL. Extra sheen is derived from Ritalan AWS.

Ingredients:	%W/W	Function
1. SD Alcohol - 40	66.70	Solubility
2. Ethyl Ester of PVM/MA-Copolymer	7.00	Hold
3. AMP-95	0.20	Neutralization
4. Dimethicone Copolyol	0.50	Sheen
5. Ritalan AWS (PPG-12-PEG-65 Lanolin Oil)	0.20	Sheen
6. Fragrance	0.10	Odor
7. Promois WK-HQ (Protein)	0.20	Repair
8. Ritapan DL	0.10	Moisture
9. Distilled/Deionized Water	25.00	--------

Compounding Procedure:
Combine item 2 to alcohol and mix until clear. Add AMP and mix well. Add items 4-6 in order with good agitation. Pre-mix items 7-9 and mix until uniform.

SOURCE: R.I.T.A. Corp.: Ref. No. 120-24

Hair Rinse for Stressed Hair

Component:	%
I. Emulgade 1000 Ni	4,0
Eutanol G	2,0
Cocopherol 1250	2,0
II. Dehyquart A	4,0
Water, demin./Preservation	88,0

pH-Value: approx. 4
Viscosity mPas, approx.: 10.000

Preparation in the Laboratory:
Melt phase I at 80-85C. Heat phase II to 80-85C and stir into Phase I. Stir at this temperature for 5 minutes. Cool down to 40C while stirring. Add preservative and,if necessary, heat sensitive additives. Cool down to 30C while stirring. Adjust the pH value e.g. with citric acid.

SOURCE: Henkel KGaA: Formulation No. 90/322/7

Pearly Lotion Conditioner

	Weight,%
Mackalene 316 (Stearamidopropyl Dimethylamine Lactate)	7.0
PEG 400 Distearate	0.5
Sodium Sulfate	0.5
Propylene Glycol	2.0
Mackstat DM (DMDM Hydantoin)	qs
Water, Dye, Fragrance qs to	100.0

Procedure:
1. Add first four components to water and heat to 65 degrees C.
2. With mild agitation cool to 45 degrees C. and add remaining components.
3. Cool and fill.

Protein Lotion Conditioner

	Weight,%
Mackine 301 (Stearamidopropyl Dimethylamine)	1.5
Mackol 16 (Cetyl Alcohol)	2.5
Lactic Acid 88%	0.7
Mackpro NLP (Quaterniun-79 Hydrolyzed Animal Protein) (Natural Lipid Protein)	1.5
Sodium Chloride	0.5
Mackstat DM (DMDM Hydantoin)	qs
Water, Fragrance, Dye qs to	100.0

Procedure:
1. Dissolve sodium chloride in water.
2. Add first four components and heat to 70 degrees C.
3. Blend until homogenous.
4. Cool to 45 degrees C. and add remaining components.
5. Cool and fill.

Spray Leave-On Conditioner

	Weight,%
Mackpro NLP (Quaternium-79 Hydrolyzed Animal Protein) (Natural Lipid Protein)	1.0
Mackalene 426 (Isostearamidopropyl Morpholine Lactate)	3.0
Mackstat DM (DMDM Hydantoin)	qs
Water, Dye, Fragrance qs to	100.0

Procedures:
1. Add components to water.
2. Heat to 40 degrees C. and blend until clear.

SOURCE: McIntyre Group Ltd.: Personal Care Formulary

Pump Gel Type Spritz

A gel setting lotion which can be dispensed from a pump.
Contains setting agents, anti-stats and Ritapan DL for repair
and luster.

Ingredients:	%W/W	Function
1. Distilled/Deionized Water	88.87	---------
2. Propylene Glycol	5.00	Humectant
3. Ritapan DL (Panthenol DL)	1.00	Repair
4. Ritaloe 200M (Aloe Gel)	1.00	Repair
5. Polyquta 3000 (Polyquaternium-10)	2.00	Set, Anti-Stat
6. Ritabate 20 (Polysorbate 20)	1.00	Stability
7. Hydroxyethylcellulose	1.00	Gelling
8. Rita CTAC (Cetrimonium Chloride)	0.50	Conditioning
9. Fragrance	0.20	Odor
10.Kathon CG	0.03	Preservative

Compounding Procedure:
Combine items 3 and 4 with the water and heat to 60C. Mix
until clear. Disperse items 5 and 7 into batch. Mix until
uniform (about 30 minutes). Then add remaining items in order
and mix well.
Ref. No. 118-180

Moisturizing Hair Styling Gel

Ingredient	%W/W	Function
1. Distilled/Deionized Water	58.20	---------
2. Acritamer 940 (Carbomer 940)	0.60	Gelling
3. Glycerine	10.00	Conditioning
4. Ritapan DL (Panthenol dl)	0.20	Moisturization
5. Ethanol @ 95%	10.00	Clarity
6. Laneto 50 (PEG-75 Lanolin)	5.00	Shine
7. Ritaphenone 3 (Benzophenone-3)	0.10	UV Absorber
8. Tetrasodium EDTA	0.10	Clarity
9. PPG-12-Buteth-16	5.00	Set
10.PVP	1.50	Set
11.Fragrance	0.10	Odor
12.DMDM Hydantoin	0.20	Preservative
13.Triethanolamine @ 99%	0.60	pH
14.Distilled/Deionized Water	5.40	--------
15.Ritalastin EL 30	3.00	Moisture

Compounding Procedure:
Combine items 1 + 3 and mix. Add item 2 slowly with agitation
until uniform. Separately combine items 4-12 and item 15. Mix
until uniform. Add to Acritamer mixture. Premix 13 & 14 for 10%
TEA solution. Add to mixture and agitate during thickening.
Ref. No. 118-142

SOURCE: R.I.T.A. Corp.: Suggested Formulations

Pump Spray Hair Detangler

Healthy shine detangling spray based on cationics.

Ingredients:	%W/W	Function
1. Rita-CTAC (Cetrimonium Chloride)	2.00	Cationic
2. Distilled/Deionized Water	88.50	--------
3. Dimethicone Copolyol (Abil B 88183)	1.00	Slip
4. Supersat AWS-4 (PEG-20 Hydrogenated Lanolin)	1.00	Feel
5. Glycerine @ 99%	4.00	Moisture
6. Propylene Glycol	2.00	Moisture
7. Ritapan DL (Panthenol DL)	1.00	Health
8. Simchin Refined (Jojoba Oil)	0.15	Shine
9. Glydant	0.20	Preservative
10.Fragrance - Pert Plus Type #189-724	0.15	Odor

Compounding Procedure:
    Heat water to 87C and add item 4. Mix until clear. Add
remaining ingredients except preservative and perfume. Cool
to 46C and add items 9 and 10.
    Ref. No. 118-171

Hot Oil Treatment

Hot oil treatment for deep penetration of the hair shaft.
Rejuvenates hair to its healthy state.

Ingredients:	%W/W	Function
1. Mineral Oil 65/70	89.00	Moisture, Shine
2. Pationic ISL (Na Isostearoyl Lactylate)	5.00	Substantive
3. Lauryl Myristyl Alcohol	2.00	Coupling
4. Phenoxyethanol	1.00	UV Protection
5. Shebu Refined (Shea Butter)	0.50	Emollient
6. Lanolin-X-tra Deodorized (Lanolin)	2.00	Repair
7. Fragrance-Floral #169-120	0.30	Odor
8. Glydant	0.20	Preservative

Compounding Procedure:
    Combine items 1-6 and heat to 165F. Mix and cool to 120F.
Add perfume and preservative.
    Ref. No. 118-119

SOURCE: R.I.T.A. Corp.: Suggested Formulations

Ringing Gel

Clear ringing gel with moisturization from Pationic ISL and gel structure from Supersat.

Ingredients:	W/W	Function
1. Distilled/Deionized Water	42.40	----------
2. Glycerine	15.00	Feel
3. Mineral Oil 70 sus.	12.00	Coat
4. Ritoleth 5 (Oleth-5)	10.00	Emulsifier
5. Supersat AWS 4 (Peg-20 Hydrogenated Lanolin)	17.00	Gel
6. Pationic ISL (Sodium Isostearoyl Lactylate)	1.00	Mildness
7. 2-Phenoxyethanol	2.00	Solubilizer
8. Germall II	0.40	Preservative
9. Fragrance	0.20	Odor

Compounding Procedure:
Heat items 1 and 2 to 165F. Heat items 3-7 to 165F. Add items 3-7 to items 1 and 2. Mix until uniform. Pour in jars at 135F.
Ref. No. 118-160

Micro Emulsion Ringing Gel

Shebu WS ringing micro emulsion gel. Ideal for hair dressing with mildness and shine.

Ingredients:	%W/W	Function
1. Distilled/Deionized Water	53.70	-----------
2. Glycerine	10.00	Feel
3. Mineral Oil, Light	12.00	Shine
4. Ritoleth 5 (Oleth-5)	10.00	Emulsifier
5. Supersat AWS 4 (PEG-20 Hydrogenated Lanolin)	12.00	Ring\Gel
6. Pationic ISL (Sodium Isostearoyl Lactylate)	1.00	Mildness
7. Shebu WS (PEG-50 Shea Butter)	1.00	Shine
8. Glydant (DMDM Hydantoin)	0.20	Preservative
9. Fragrance	0.10	Odor

Compounding Procedure:
Heat items 1,2 and 7 to 175F. Combine items 3,4,5 and 6 and heat to 175F. Add this to the first phase with agitation. (Note: process may need an add-back of 3-5% water). Cool to 130F or until thickening occurs. Add items 8 and 9. Add back water loss. Fill at this temperature into jars and cover.
Ref. No. 118-162

SOURCE: R.I.T.A. Corp.: **Suggested Formulations**

Salon Type Hair Conditioner

State of the art protein/panthenol based hair conditioner designed for achievement of salon type performance.

Ingredients:	%W/W	Function
1. Rita CA (Cetyl Alcohol)	0.85	Clean Feel
2. Rita-STAC (Steartrimonium Chloride)	0.44	Conditioner
3. Rita-CTAC (Cetrimonium Chloride)	0.48	Conditioner
4. Citric Acid @ 100%	0.09	pH Control
5. Dimethyl Stearamine (Adogen MA-108)	0.13	Conditioner
6. Distilled/Deionized Water	95.51	----------
7. Promois WK-HQ (Hydroxypropyl trimon-ium Hydrolyzed Keratin)	0.05	Body
8. Ritaloe 200M (Aloe Vera Gel)	0.05	Body
9. Ritapan DL (dl-Panthenol)	0.05	Body
10.Propylene Glycol	0.50	Clean Feel
11.Hydroxyethyl Cellulose (Natrosol 250 HHR)	1.00	Viscosity
12.Kathon CG	0.03	Preservative
13.Dow Corning 2-7224	0.40	Combing
14.Dow Corning 344 Fluid	0.17	Combing
15.Fragrance	0.25	Odor

Compounding Procedure:
To 15% of total water at 65C add separately items 1-4. Heat to 70C, add item 5 and mix 20 min. Add another 10% of the water at room temperature and continue mixing 20 min. To remaining 75% of water add items 7,8 and 9 at room temperature. Add item 10 and slowly sift in item 11. Agitate 15 min. while heating batch slightly to 35C. When batch thickens, increase agitation for up to 60 min. until uniform.
Ref. No. 116-181A

Salon Type Hair Conditioner

Salon type hair conditioning is achieved with the help of Shebu and an all-natural conditioning agent.

Ingredients:	%W/W	Function
1. Ritachol 1000 (R.I.T.A. Blend)	5.00	Emulsifier
2. Shebu Refined (Shea Butter)	0.50	Natural Shine
3. Ritoleth-10 (Oleth-10)	0.50	Application
4. Rita-CTAC (Cetrimonium Chloride)	2.00	Conditioning
5. Glycerine	3.00	Moisture
6. Distilled/Deionized Water	88.67	--------
7. Fragrance - Silkience Type #169-120	0.30	Odor
8. Kathon CG	0.03	Preservative

Compounding Procedure:
Combine items 1-4 and heat to 70C. In a separate beaker combine items 5 and 6 and heat to 75-80C. Add items 1-4 to water mixture and mix until uniform. Cool to 40C and add fragrance and preservative.
Ref. No. 118-120
SOURCE: R.I.T.A. Corp.: Suggested Formulations

Silicone Hair Gel

Silicone based setting gel. High glycerine appropriate for curl activation. Ritapan Dl for humectancy.

Ingredients:	%	Function
1. Distilled/Deionized Water	81.10	----------
2. Acritamer 940 (Carbomer 940)	0.80	Gelling
3. Glycerine	10.00	Curl
4. Propylene Glycol	5.00	Emollient
5. Dimethicone Copolyol	1.00	Shine
6. Ritapan DL (Panthenol dl)	0.80	Moisture
7. Laneto 50 (PEG-75 Lanolin)	0.50	Shine
8. Fragrance	0.10	Odor
9. Glydant (DMDM Hydantoin)	0.20	Preservative
10.TEA @ 99%	0.80	pH

Compounding Procedure:
Disperse slowly item 2 into 90% of the water with agitation. Premix items 4,5,6 and 7 with agitation. Premix items 8 and 9 into item 3. Combine item 10 with remaining 10% of the water. Neutralize Acritamer solution. When complete add other premixes and mix until uniform.
Ref. No. 118-156

Shebu Hair Setting Gel

Water based setting gel with Shebu. The setting agent is plasticized with Supersat for softer hold.

Ingredients:	%W/W	Function
1. Distilled/Deionized Water	60.80	----------
2. Acritamer 940 (Carbomer 940)	0.70	Gelling
3. Distilled/Deionized Water	26.50	----------
4. Glycerine	4.00	Moisture
5. Shebu WS (PEG-50 Shea Butter)	3.00	Luster
6. Supersat AWS 4 (Peg 20-Hydrogenated Lanolin)	1.00	Plasticizer
7. PVP K-30	2.00	Set
8. Ritapan DL (Panthenol dl)	0.20	Moisture
9. TEA @ 50%	1.40	pH
10.Fragrance	0.20	Odor
11.DMDM Hydantoin	0.20	Preservative

Compounding Procedure:
Slowly disperse Acritamer into item 1. Agitate until fully hydrated. Combine items 3-9 until uniform (slight heating may be necessary). Add this to the Acritamer mixture, add items 10 and 11, and agitate until uniform.
Ref. No. 118-158

SOURCE: R.I.T.A. Corp.: Suggested Formulations

Stiff Holding Hairspray

Ingredients:	Parts by Weight
Amphomer	4.75
AMP	0.80
Crotein AD	0.20
Dow Corning 556	0.15
Citroflex 2	0.20
Octyl Dimethyl PABA	0.05
SDA 40	6.85
Propellant A-46	30.00

Preparation:
   Dissolve AMP in the anhydrous alcohol. While maintaining good agitation, slowly add Amphomer to the vortex. Continue mixing until solution is complete. Add balance of ingredients. When completely dissolved and homogeneous, filter and fill concentrate Charge propellant.

Precision Specifications:
   Actuator: 21-8173 0.020" MB   Body: 07-3415 0.062" x 0.020" VT
   Stem: 0.4-1230 0.020"   Cup: 12-8700
   Stem Gasket: 05-0310 Buna   Dip Tube: 09-3530 0.060"
   Spring: 06-6010 SS
Formula 5887-80-4

Aerosol Shaping Hairspray

Materials:	Parts/Weight
Versatyl-42	3.75
AMP-95	0.96
DC-193 Silicone	0.10
DC-556 Silicone	0.10
Glycerine	0.10
Citroflex-2	0.10
Monamid 716	0.20
Sunarome OMC	0.05
Fragrance	Q.S.
Ethanol, Anhydrous	64.64
Propellant A-46	30.00

Valve: Precision
                .018" stem
                .018 x .013" body
                .018" FT Actuator
Spray Rate: 0.56 g/sec
Preparation:
   Add alcohol to the tank. While maintaining good agitation, slowly add Versatyl-42 to the vortex. Add AMP-95 and continue mixing until solution is complete. Add remaining ingredients of the concentrate. When completely dissolved and homogeneous, filter and fill concentrate to the can. Charge propellant.
Formula 6258-07

SOURCE: National Starch and Chemical Co.: Suggested Formulations

Styling/Conditioning Mousse

Ingredients:	%W/W
A Forestall	1.40
Brij 721	0.50
Dow Corning 929 Emulsion	0.10
Water, deionized	78.00
B Gaffix VC-713	5.00
SD Alcohol 40	10.00
C Hydrocarbon Propellant A-46	5.00

Procedure:
Heat (A) to 60C with stirring until uniform. Cool to 40C.
Add (A) to (B) with stirring. Pack in suitable aerosol
containers and pressurize with (C) at room temperature.
Formula PC 8216

Clear Gel Grooming Agent & Conditioner

Ingredients:	%W/W
A Mineral oil, Marcol 70	14.10
Brij 97	15.50
Arlatone G	15.50
Atlas G-3570	0.10
Propylene glycol	8.60
Sorbo	6.90
B Water, deionized	39.30

Procedure:
Heat (A) to 90C and (B) to 95C. Add (B) to (A) with moderate
stirring. Remove from heat and continue to stir as the mixture
cools. When the mixture turns clear (about 65C) pour into tubes
or jars.

Hair Straightening Cream

Ingredients:	%W/W
A Glycerol monostearate	20.00
Brij 35	2.00
Stearyl alcohol	5.00
Mineral oil	5.00
B Sodium lauryl sulfate	1.00
Water	40.00
*Preservative	
C Sodium hydroxide	3.50
Water	23.50
D *Perfume	
*q.s. these ingredients.	

Procedure:
Heat (A) to 60C and (B) to 62C. Add (B) to (A) with constant
agitation until cool. Add (C). Mix thoroughly. Add (D).
Formula HC-17

SOURCE: ICI Surfactants: Suggested Formulations

Styling Mousse
X22176-106 (Hydroalcoholic)

Concentrate:	%W/W
Distilled water	q.s.* to 100
Eastman AQ 55S polymer	8.0
Myvatex Texture Lite emulsifier	2.0
Monamid 150 ADD	1.0
Myvatex 60 emulsifier	0.3
SDA-40C alcohol	20.0
Fragrance	q.s.*
Citric acid	q.s.*

  *q.s.= quantity sufficient

Procedure:
1. Heat Eastman AQ 55S and water to 80-85C with mixing until
   the polymer is completely dispersed in the water.
2. Cool to room temperature.
3. Slowly add Myvatex Texture Lite with high-speed agitation.
   Care should be taken when mixing to avoid aeration.
4. When uniform, add Myvatex 60 and Monamid 150 ADD.
5. Slowly add SDA-40C alcohol.
6. Add fragrance.
7. Adjust pH to 6.5-7.0 with citric acid.
8. Aerosol final concentrate at 5.23 g/mL of A46 propellant
   (Aeropress).

Styling Mousse
X22176-107 (Alcohol-Free)

Concentrate:	%W/W
Distilled water	q.s.* to 100
Eastman AQ 55S polymer	8.0
Myvatex Texture Lite emulsifier	5.5
Monamid 150 ADD	1.0
Myvatex 60 emulsifier	0.3
Preservative	q.s.*
Fragrance	q.s.*
Citric acid	q.s.*

  *q.s.=quantity sufficient

Procedure:
1. Heat Eastman AQ 55S and water to 80-85C with mixing until the
   polymer is completely dispersed in the water.
2. Cool to room temperature and add preservative.
3. Slowly add Myvatex Texture Lite with high-speed agitation.
   Care should be taken when mixing to avoid aeration.
4. When uniform, add Myvatex 60 and Monamid 150 ADD.
5. Add fragrance.
6. Adjust pH to 6.5-7.0 with citric acid.
7. Aerosol final concentrate at 5.23 g/mL of A46 propellant
   (Aeropress).

SOURCE: Eastman Chemical Co.: Suggested Formulations

Water-Based Hair Spray
X21980-057(Pump)

Phase A:	%W/W
Distilled water	90.4
Eastman AQ 55S polymer	5.0

Phase B:	
Germall II diazolidinyl urea	0.3
Methylparaben	0.3

Phase C:	
Luviskol VA 73W PVP/VA copolymer--50% solids	4.0

Procedure:
1. Heat Phase A to 85C.
2. With mixing, hold at 80-85C for 15 minutes.
3. Cool to 60C, add Phase B, and mix until dissolved.
4. Cool to 40C and add Phase C.
5. Add water lost during heating.
6. Mix until uniform; filter and package.
   pH: 6.0+-1.0

Water-Based Hair Spray
X21980-058 (Aerosol)

Phase A:	%W/W
Distilled water	60.4
Eastman AQ 38S polymer	5.0

Phase B:	
Germall II diazolidinyl urea	0.3
Methylparaben	0.3

Phase C:	
Luviskol VA 73W PVP/VA copolymer--50% solids	4.0

Phase D:	
Dymel A dimethyl ether	30.0

Procedure:
1. Heat Phase A to 85C.
2. When mixing, hold at 80-85C for 15 minutes.
3. Cool to 60C, add Phase B, and mix until dissolved.
4. Cool to 40C and add Phase C.
5. Add water lost during heating.
6. Add phase D at room temperature.
7. Mix until uniform; filter and package.
8. Agitate aerosol container to ensure solution of propellant.
   pH: 6.0+-1.0

SOURCE: Eastman Chemical Co.: Suggested Formulations

Wheat Germ Foaming Conditioner

	Weight,%
Mackam 35 (Cocamidopropyl Betaine (Via Glyceride)	10.0
Mackalene 116 (Cocamidopropyl Dimethylamine Lactate)	8.0
Mackalene 716 (Wheat Germamidopropyl Dimethylamine Lactate)	1.0
Natrosol 250 HHR	0.7
Mackstat DM (DMDM Hydantoin)	qs
Water, Dye, Fragrance qs to	100.0

Procedure:
1. Completely hydrate Natrosol.
2. Add first three components and heat to 40 degrees C.
3. Blend until clear.
4. Add remaining components and cool.

Natural Lipid Styling Mousse

	Weight,%
PVP/VA E335	4.5
SDA Alcohol	21.5
Mackpro NLP (Quaternium-79 Hydrolyzed Animal Protein) (Natural Lipid Protein)	4.0
Deionized Water, Fragrance, Dye qs to	100.0

Procedure:
1. Combine components and blend until clear.
2. Pressurize with suitable propellant

Pump Type Hair Spray

	Weight,%
Resyn 26-1314	6.0
Mackpro NLP (Quaternium-79 Hydrolyzed Animal Protein) (Natural Lipid Protein)	1.0
Deionized Water	7.6
Ethanol, Fragrance qs to	100.0

Procedure:
1. Dissolve Resyn 26-1314 in alcohol.
2. Add remaining components and blend until clear.

SOURCE: McIntyre Group Ltd.: Personal Care Formulary

55% VOC Aerosol Hair Spray

Ingredients:	Parts by Weight
Lovocryl 47	8.00
AMP	1.38
Deionized Water	35.62
Anhydrous Ethanol, SDA-40	22.00
Fragrance	q.s.
Preservative	q.s.
DME	33.00

    Cloud Point: <-35C

Valve: Seaquist Valve NS-34
      Stem: 0.013"
      Gasket: Butyl 0.042" THK. Code: 501
      Cup: Regular, Epoxy top, Laminate Bottom, Dimpled,
         Code: 1610
      Spring: 0.020" SS
      Body: Capillary
      Vapor Tap: 0.013"
      Dip Tube: 0.040"
      Actuator: Excel 200 Misty 0.016" Misty

Preparation:
    Dissolve AMP in water and ethanol. While maintaining good agitation, slowly sift in Lovocryl 47. Mix until dissolved. Filter and fill, then charge with propellant.
Formula 7625:63B

80% VOC Non-Aerosol Hair Spray

    This 80% VOC non-aerosol hair spray has a natural feel and excellent drying time with a fine, misty spray.

Ingredients:	Parts by Weight
Lovocryl-47	5.00
NaOH	0.40
DC 193	0.30
dl-Panthenol	0.21
Citroflex-2	0.20
Deionized Water	13.89
Anhydrous Ethanol, SDA-40	80.00

Valve Specification: Calmar Mark II
                Actuator: WTS Head

Preparation:
    Dissolve NaOH in water and ethanol. While maintaining good agitation, slowly sift in Lovocryl-47. When solution is complete, add remaining ingredients and mix well until homogeneous. Filter and fill.
Formula 7577-65

SOURCE: National Starch and Chemical Co.: Suggested Formulas

80% VOC Hair Spray

Ingredients:	Parts by Weight
Amphomer	5.00
AMP	0.82
Citroflex-2	0.30
Monamid 716	0.20
Glycerine	0.10
dl Panthenol	0.10
Solulan-75	0.10
Neo-Heliopan AV	0.05
Anhydrous Ethanol, SDA-40	55.00
Propellant 152A	13.33
N-butane	25.00

Valve Specifications: Precision Valve Corp.
    Stem: 04-1215 .016"   Dip Tube: 09-2010
    Stem Gasket: 05-0350 Butyl Actuator Style: 21-8146 Kosmos
    Spring: 06-6010 SS   Orifice Size: .025" MB Concave
    Body: 07-7970 .016" LD   Mounting Cup: 32-7300-62 Flat, Epon
                                                T/B Dimpled Full Bond
Preparation:
    Dissolve AMP in ethanol. While maintaining good agitation,
slowly sift in Amphomer. When solution is complete, add remaining
ingredients and mix until homogeneous. Filter, fill and charge
with propellants.
Formula 7924:28

55% VOC Aerosol Hair Spray

Ingredients:	Parts by Weight
Lovocryl	5.00
AMP	0.89
Rhodorsil Oils 70041 VO .65	1.00
DI Water	38.11
Anhydrous Ethanol, SDA-40	22.00
Propellant DME	33.00

Valving Specifications:
    Seaquist NS 34
    Stem Orifice: 0.013"
    Gasket: Butyl 0.042" Thk.Code: 501
    Cup: Regular, Epoxy Top, Laminate Bottom, Dimpled, Code: 1610
    Spring: 0.020" SS
    Vapor Tap: 0.013"
    Tubing ID: 0.040"
    Actuator: Excell 200 Misty/0.106" Misty
Preparation:
    Dissolve AMP in ethanol and water. While maintaining good
agitation, slowly sift in Lovocryl. When solution is complete,
add Rhodorsil Oils 70041 VO .65 and mix until homogeneous.
Filter and fill-charge can with propellant.
Formula 7746-137
SOURCE: National Starch and Chemical Co.: Suggested Formulas

80% VOC Hair Spray

Resyn 28-2930	3.00
Amphomer LV-71	0.81
AMP	0.44
Citroflex-2	0.30
Monamid 716	0.20
Glycerine	0.10
Panthenol	0.10
Solulan-75	0.10
Neo-Heliopan AV	0.05
Anhydrous Ethanol, SDA-40	55.00
Propellant 152A	14.90
N-butane	25.00
Fragrance	q.s.

Cloud Point: <-35C

Valve: Precision Valve Corp.
       Stem: 04-1215   .016"
       Stem Gasket: 05-0350   Butyl
       Spring: 06-6010   SS
       Body: 07-7970   .016" LD
       Mounting Cup: 32-7300-62/Flat, Epon T/B Dimpled Full Bond
       Dip Tube: 09-2010
       Actuator Style: 21-8146 Kosmos
       Orifice Size: .025"   MB Concave
Formula 7318-93

High Humidity Hair Spray

Ingredients:	Parts by Weight
Resyn 28-2913	2.50
AMP	0.24
Crotein AD Anhydrous	0.20
DC-190	0.20
Fragrance	Q.S.
Anhydrous Ethanol, SDA-40	71.86
Propellant A-46	25.00

Valve: Precision
       Stem: 0.018" (#04-1220)
       Stem Gasket: Buna (#05-0310)
       Spring: SS (#06-6010)
       Body: 0.018" x 0.013" VT (#07-0131)
       Mounting Cup: Flat, Epon Top/Bottom, Dimpled (#12-7100)
       Actuator: 0.018" FT (#01-1836)
Preparation:
   Dissolve AMP in anhydrous ethanol, SDA-40. Slowly add Resyn
28-2913 to the solution while maintaining good agitation. Add
remaining ingredients and mix until homogeneous. Filter and
fill concentrate. Charge propellant.
    Formula 6472:103

SOURCE: National Starch and Chemical Corp.: Suggested Formulas

80% VOC Aerosol Shaper Hair Spray

Ingredients:	Parts by Weight
Amphomer LV-71	3.50
AMP Regular	0.60
Armeen DM 18D	0.25
Citroflex-2	0.20
Monamid 716	0.20
Solulan 75	0.10
Neo Heliopan AV	0.05
Varion C ADG LS	0.20
Deionized Water	14.90
Anhydrous Ethanol, SDA-40	50.00
N-butane	11.00
DME	19.00

Valve Specifications: Precision Valve
   Stem: 0.016" (04-1215)
   Stem Gasket: Butyl (05-0350)
   Spring: SS (06-6010)
   Body: 0.016" LD (07-7970)
   Mounting Cup: Flat, Epon T/B Dimpled Full Bond
   Actuator: Kosmos MB Concave 0.025" (21-8146)

Preparation:
   Dissolve AMP Regular and Armeen DM 18D in water and ethanol.
While maintaining good agitation, slowly sift in Amphomer LV-71.
When solution is complete, add remaining ingredients and mix
well until homogeneous. Filter and fill.
   Formula 8002:183

High Solids 80% VOC Hair Spray

Ingredients:	Parts by Weight
Lovocryl 47	10.00
AMP	1.73
Citroflex-2	0.20
Monamid 716	0.15
Glycerine	0.10
Panthenol	0.10
Deionized Water	7.72
Anhydrous Ethanol, SDA-40	50.00
N-butane	12.00
DME	18.00
Fragrance	q.s.

   Cloud Point: <-35C
Valve: Precision Valve Corp.
        Stem: 04-1215 0.016"   Gasket: 05-0350 Butyl
        Spring: 06-6010 SS   Body: 07-7970 0.016" LD
        Cup: 32-7300-62 Flat, Epon T/B, Dimpled, Full Bond
        Vapor Tap: None   Dip Tube: 09-2010 0.122"
        Actuator: 0.016" MB Concave Delta

Preparation:
   Dissolve AMP in water and ethanol. While maintaining good
agitation, slowly sift in Lovocryl 47. When dissolved, add
remaining ingredients and mix until homogeneous. Filter and
fill, then charge with propellants.
   Formula 7625:95

SOURCE: National Starch and Chemical Co.: Suggested Formulas

Section VII
Lotions

After Exercise Body Lotion

This refreshing lotion is an excellent carrier for fragrances
and will provide moisturizing along with a skin smoothing effect.

	% by Weight
Part I:	
Water	51.65
Carbomer 940	0.25
Triethanolamine (99%)	0.40
Part II:	
Water	30.00
Phospholipid SV	1.00
SD3A Alcohol	15.00
Part III:	
Phenyl Dimethicone	1.50
Titanium Dioxide (Cosmetic White)	0.20

Procedure:
    Slowly add Carbomer 940 to water with good agitation. After
the Carbomer 940 is dissolved, add triethanolamine. In a
separate container, mix water with Phospholipid SV and heat
to 65C with mixing. When dissolved, cool to 30-35C then add
SD3A alcohol. Blend Part II to I mixing well, and part III
as a slurry and blend together until a uniform white lotion
is obtained. Add color, fragrance and preservative as required
and package.

Therapeutic Humectant Lotion

	% by Weight
Part A:	
Phospholipid SV	1.7
Glycerin	15.0
TiO2	0.5
Water	72.6
Methyl Paraben	0.2
Part B:	
Steareth-2	3.3
Hexyl Laurate	3.0
C12-C15 Alcohol Benzoates	2.0
Dimethicone and Trimethylsiloxysilicate	1.5
Propyl Paraben	0.2

Procedure:
    Combine phases A and B separately with heating to 65C.
Homogenize B into A for a sufficient time to ensure good
emulsification. Stir cool to 45C, add fragrance, and package.
Comments:
    This smooth pourable lotion provides instant relief of dry,
chapped skin and provides a generous amount of emollients and
glycerin. The use of Phospholipid SV eliminates any greasiness
and leaves the skin with an elegant afterfeel.
SOURCE: Mona Industries, Inc.: Suggested Formulations

Alcoholic Milk Lotion

	Wt%
(O) Stearic Acid	2.0
Cetyl Alcohol	1.0
Isopropyl Myristate	2.0
Glyceryl Monostearate (Self Emulsifying Type)	1.1
Polyoxyethylene (20) Cetyl Ether	1.9
Methyl Paraben	q.s.
(W) Ajidew N-50	3.0
Propylene Glycol	5.0
Water	74.0
(E) Ethyl Alcohol	10.0

Procedure:
1. Heat (O) and (W) to 80C.
2. Add (W) to (O) slowly with agitation.
3. Cool to 50C with stirring and add (E).
4. Finish mixing at 35-40C.

pH: 5.1
Viscosity: 270 cps

Milk Lotion

	Wt%
(O) Liquid Paraffin (#70)	31.6
Paraffin Wax (mp 42-44C)	4.5
Cetyl Alcohol	4.5
Sorbitan Monostearate	1.8
Polyoxyethylene (20) Sorbitan Monooleate	2.8
Tocopherol Acetate	0.2
(W) Ajidew T-50	4.0
Water	50.5
Preservative	0.1

Procedure:
1. Heat (O) and (W) to 80C.
2. Add (W) to (O) with agitation.
3. Cool to 42C with stirring.

pH: 6.2
Viscosity: 25,000 cps

SOURCE: Ajinomoto USA, Inc.: Suggested Formulations

Alpha Hydroxy Acid Lotion

Resulting product is a highly absorbent lotion that can be
dispensed from a flexible tube or soft bottle. The alpha hydroxy
acid (AHA) contained in this formula is buffered lactic acid
with a use level of approximately 10%. The formulation utilizes
a cationic and nonionic emulsification system consisting of
Lexemul 561, Lexemul AR and Lexamine S-13. The cationic Lexamine S-13 enhances the substantivity of the formulation and
may act as an auxiliary AHA buffer. The product produces minimal whitening during rub-out.

	%w/w
A. Deionized Water	67.42
Glycerin USP (99.7%)	3.00
Methylparaben (Lexgard M)	0.20
2,4-Dichlorobenzyl Alcohol (Myacide SP)	0.20
Lactic Acid USP (88%)	0.60
B. Glyceryl Stearate (and) PEG-100 Stearate (Lexemul 561)	3.00
Glyceryl Stearate (and) Stearamidoethyl Diethylamine (Lexemul AR)	1.25
Stearamidopropyl Dimethylamine (Lexamine S-13)	0.50
Cetyl Alcohol (Adol 52)	0.50
Caprylic/Capric Triglyceride (Lexol GT-865)	8.00
Isobutyl Stearate (Lexol BS)	1.00
Propylparaben (Lexgard P)	0.10
C. Lactic Acid USP (88%)	5.90
Sodium Lactate (Purasal S/SP 60%)	8.33

Procedure:
1. Combine section "A". Begin heating to 75-80C.
2. Combine section "B". Heat to 80-85C with constant slow
   agitation.
3. When sections "A" and "B" are homogeneous and at the designated temperatures, slowly add "B" to "A". Mix with rapid
   agitation for 5 minutes then begin cooling.
4. Reduce mixing speed during cooling to prevent vortexing.
5. At 40-50C, add section "C" to sections "AB".
6. When the batch is homogeneous and at 35-40C adjust for water
   loss.
7. Mix to 30C.
8. Adjust final pH to 3.50-3.75 with Lactic Acid or Sodium
   Lactate, as required.

SOURCE: Inolex Chemical Co.: Formulation 398-156-1

Alpha Hydroxy Acid Lotion

Resulting product is a lotion that could be dispensed from
a flexible tube or flexible bottle. The alpha hydroxy acid (AHA)
contained in this formula is buffered lactic acid with a use
level of approximately 10%. This formulation utilizes a cationic
and nonionic emulsification system consisting of Lexemul 561
and Lexamine S-13. The cationic Lexamine S-13 enhances the
substantivity of the formulation and may act as an auxiliary
AHA buffer. The product produces minimal whitening during
rub-out.

	%w/w
A. Deionized Water	68.50
Glycerin USP (99.7%)	3.00
Hydroxyethylcellulose (Cellosize QP-4400-H)	1.00
B. Methylparaben (Lexgard M)	0.20
2,4-Dichlorobenzyl Alcohol (Myacide SP)	0.20
Lactic Acid USP (88%)	0.60
C. Glyceryl Stearate (and) PEG-100 Stearate (Lexemul 561)	2.00
Stearamidopropyl Dimethylamine (Lexamine S-13)	1.00
Cetyl Alcohol (Adol 52)	0.50
Caprylic/Capric Triglyceride (Lexol GT-865)	7.00
Isobutyl Stearate (Lexol BS)	2.00
Propylparaben (Lexgard P)	0.10
D. Lactic Acid (88%)	5.90
Sodium Lactate (Purasal S/SP 60%)	8.00

Procedure:
1. Combine section "A". Begin heating to 75-80C. Sift Hydrox-
   yethylcellulose into vortexing water.
2. When section "A" is homogeneous, add section "B" to section
   "A". Continue agitation. Continue heating to 75-80C.
3. Combine section "C". Heat to 80-85C with constant slow
   agitation.
4. When sections "AB" & "C" are homogeneous and at the designated temperatures, slowly add "C" to "AB". Mix with rapid
   agitation for 5 minutes then begin cooling.
5. Reduce mixing speed during cooling to prevent vortexing.
6. At 40-50C, add section "D" to sections "ABC".
7. When the batch is homogeneous and at 35-40C adjust for
   water loss.
8. Mix to 30C.
9. Adjust final pH to 3.50-3.75 with Lactic Acid or Sodium
   Lactate, as required.

SOURCE: Inolex Chemical Co.: Formula 398-157-2

Banana Hand Lotion

	Parts by Weight
Water	568.0gr
Carbomer 934	2.0gr
GMS-SE	4.0gr
Avocado Oil	16.0gr
Rosswax 573	4.0gr
Coconut Oil #76	16.0gr
Ross Jojoba Oil	4.0gr
TEA	4.0gr
Germaben II	6.0gr
Fragrance GK-17	q.s.

Procedure:
   Heat the water to 60C under agitation and slowly add the
Carbomer 934. When the water is fully mixed add the 573, GMS,
Avocado Oil, Coconut Oil and Jojoba Oil that have been heated
to 65C in separate kettle. As soon as the Oil Phase has been
mixed well, add the Germaben II and then the TEA under high
agitation, then the fragrance. Cool to 55C for filling.

O/W Lotion

Phase 1:	%
Ross Wax 63-0412	1.6
Ross Wax 1641	1.0
Mineral Wax #9	2.1
Ross Wax 63-0212	1.0
Amerchol L-101	5.2
Ross Jojoba Oil	2.1
GMS SE	2.1

Phase 2:	
Triethanolamine	1.0
Propylene Glycol	4.7
Water	78.2
Preservative Germaben II	1.0

| Novarome DE-47 Fragrance | q.s. |

Procedure:
   In separate kettles bring Phase (1) and (2) to 170F. When
temperature is reached add Phase (1) to (2) with agitation.
Cool to 120F and package.

SOURCE: Frank B. Ross Co., Inc.: Suggested Formulations

Behenyl Hand Lotion

This lotion features the combination of Incromine BB and
Crodacid B, that provide great stability (1 month at 50C). Crodamol PMP helps to formulate a non-oily lotion with smooth rub in.

Ingredients:	%
Part A:	
Incromine BB (Behenamidopropyl Dimethylamine)	3.0
Crodacid B (Behenic Acid)	2.5
Crodamol PMP (PPG-2 Myristyl Ether Propionate)	20.0
Dimethicone	3.0
Part B:	
Deionized Water	70.5
Part C:	
Propylene Glycol (and) Diazolidinyl Urea (and)	
Methyl Paraben (and) Propyl Paraben	1.0

Procedure:
   Combine ingredients of Part A with mixing and heat to 85-90C.
Heat Part B to 85-90C. Add Part B to Part A with mixing and cool
to 45C. Add Part C with mixing and cool to desired fill tempera-
ture.
   N.A.T.C. Approved
   Formula SC-215

Velvet 44-Behenyl Lotion

This lotion features a unique combination of emulsifiers
(Incromine BB and Crodacid B) that provide great stability
(3 months @ 50C) and leave a soft velvety feel on the skin.
Crodamol PMP helps reduce the oiliness of the mineral oil and
promotes a smooth rub in.

Ingredients:	%
Part A:	
Incromine BB (Behenamidopropyl Dimethylamine)	3.00
Crodacid B (Behenic Acid)	2.50
Mineral Oil (70ssu)	15.00
Crodamol PMP (PPG-2 Myristyl Ether Propionate)	10.00
Silicone Fluid (200 cps) DC 200	1.00
Part B:	
Deionized water	67.50
Part C:	
Germaben II	1.00

Procedure:
   Combine ingredients of Part A with mixing and heat to 85-90C.
Heat Part B to 85-90C. Add Part B to Part A with mixing and cool
to 40C. Add Part C with mixing and cool to desired fill tempera-
ture.
   N.A.T.C. Approved
   Formula SC-213
SOURCE: Croda Inc.: Suggested Formulations

Body Lotion

Ingredients:	%w/w
Phase A:	
Polyglyceryl-3 Methyl Glucose Distearate (Tego Care 450)	2.0
Caprylic/Capric Triglycerides (Tegosoft CT)	6.5
Octyl Stearate (Tegosoft OS)	5.7
Phase B:	
Glycerin	3.0
Water	81.4
Phase C:	
Mineral Oil	0.8
Carbomer 941	0.2
Phase D:	
Sodium Hydroxide (10% solution)	0.4
Phase E:	
Fragrance	Q.S.
Preservatives	Q.S.

Procedure:
1. Heat the ingredients of Phase A to 80C.
2. Heat the ingredients of Phase B to 80C.
3. Add A to B or B to A without stirring.
4. Stir.
5. Disperse Carbomer into the oil/ester add to A/B. Homogenize.
6. Cool to 35-40C with stirring.
7. Add Phase D/E. Stir.
8. Mix until viscosity is correct.

Emollient Lotion
O/W Cold Process

Ingredients:	%W/W
Phase A:	
PEG-25 Glyceryl Trioleate (Tagat TO)	0.5
Polyglyceryl-4 Isostearate (Isolan GI 34)	1.5
Mineral Oil	4.0
Isopropyl Palmitate (Tegosoft P)	3.0
Caprylic/Capric Triglycerides (Tegosoft CT)	2.0
Dimethicone (Abil 350)	0.2
Phase B:	
Water	83.6
Glycerin	3.0
Preservative	Q.S.
Phase C:	
Carbomer 934	0.3
Mineral Oil	1.2
Phase D:	
NaOH (10% Solution)	0.7

Procedure:
1. Homogenize A + B, oil particle size <1.0 um.
2. Add C and D, stir intensively for 30 minutes. Temperature of the phases A, B, C, D: are 20-25C

SOURCE: Goldschmidt Chemical Co.: Suggested Formulations

Body Lotion

Ingredients:	%W/W
A Isohexadecane	9.00
Mineral oil	5.00
Isostearic acid	3.00
Arlamol E	2.00
Cyclomethicone	1.00
B Xanthan gum	0.10
C Water	73.90
Sorbo, sorbitol solution	2.00
Arlatone 2121	4.00
D *Preservative	
E *Perfume	

*q.s. these ingredients.

Procedure:
    Heat (C) to 80C with occasional vigorous stirring. Add (B) to
(C) at 75C with vigorous stirring. Add (A) with vigorous stirring or homogenization. Add (D) and (E) upon cooling.

Moisturizing Hand Lotion

Ingredients:	%W/W
A Mineral oil	8.00
CR-15 alcohols benzoate	7.00
Jojoba oil	5.00
B Xanthan gum	0.10
C Water	70.90
Propylene glycol	4.00
Arlatone 2121	5.00
D *Preservative	
E *Perfume	

*q.s these ingredients.

Procedure:
    Heat (C) to 80C with occasional vigorous stirring. Add (B) to
(C) at 75C with vigorous stirring. Add (A) with vigorous stirring or homogenization. Add (D) and (E) upon cooling.

SOURCE: ICI Surfactants: Suggested Formulations

Concentrated Alpha Hydroxy Acid Lotion

Resulting product is a pourable lotion that could be dispensed from a glass or rigid plastic bottle. The formulation utilizes a buffered alpha hydroxy acid (AHA), Lactic Acid, at a use level of approximately 15%. The formula utilizes a cationic and nonionic emulsification system consisting of Lexquat AMG-IS and Lexemul 561. In addition to functioning as the primary emulsifier, Lexquat AMG-IS, also enhances skin substantivity. This formulation offers a high level of activity in a buffered form.

	%w/w
A. Deionized Water	52.30
Methylparaben (Lexgard M)	0.20
2,4-Dichlorobenzyl Alcohol (Myacide SP)	0.20
Propylene Glycol USP	2.70
Lactic Acid USP (88%)	0.50
B. Hydroxyethylcellulose (Cellosize QP-15,000H)	0.50
C. Isostearamidopropyl PG-Dimonium Chloride (Lexquat AMG-IS)	10.10
D. Glyceryl Stearate (Lexemul 515)	1.75
Glyceryl Stearate (and) PEG-100 Stearate (Lexemul 561)	3.40
Propylene Glycol Dicaprylate/Dicaprate (Lexol PG-865)	2.50
Cetyl Alcohol (Adol 52)	0.80
Jojoba Oil (Golden Jojoba Oil)	2.50
Myristyl Myristate (Ceraphyl 424)	0.85
Propylparaben (Lexgard P)	0.10
E. Lactic Acid USP (88%)	7.25
Sodium Lactate (Purasal S/SP 60%)	14.35

Procedure:
1. Combine section "A" heating to 80-85C. Use propeller mixer for agitation.
2. Sprinkle section "B" into section "A". Increase mixing speed to create a vortex during addition, then slow to highest speed which does not produce a vortex.
3. When sections "A" and "B" are homogeneous and at the designated temperatures add section "C" to sections "AB".
4. Combine section "D". Heat to 80-85C with continuous slow mixing.
5. When sections "ABC" and section "D" are homogeneous and at the designated temperatures, slowly pour section "D" into Sections "ABC". Increase mixing speed to a high speed. Maintain at 80-85C for 5 minutes, then begin cooling. Slow mixing speed to highest speed possible that will not cause vortexing.
6. Add section "E" to sections "ABCD" at 60-65C. Continue mixing and cooling.
7. At 40-45C adjust for water loss.
8. Mix batch until it reaches room temperature.
9. Adjust final pH to 3.60-3.80 with Lactic Acid or Sodium Lactate, as required.

Physical Properties:
Viscosity: 2400 cPs @ 24C (Brookfield RVT, TA @ 10 rpm)
pH: 3.7 @ 24C
SOURCE: Inolex Chemical Co.: Formulation 398-99-2

Deep Skin Cleansing Lotion

Materials:	% by Weight
1. Mineral Oil SUS 65/75	50.0
2. Amerchol L-101	3.0
3. Cerasynt PA	3.0
4. Tween 20	3.0
5. Stearic Acid T.P.	1.0
6. Propylparaben	0.1
7. Methylparaben	0.2
8. Water (deionized)	38.6
9. Borax U.S.P.	0.5
10. Bentone LT rheological additive	0.3
11. Perfume	0.3

Manufacturing Directions:
1. Part 1-In a stainless steel vessel, add items 1 through 7 and heat to 75C. Maintain this temperature and stir slowly until everything is in solution. Bring temperature to 60C.
2. Part 2-In a separate stainless steel vessel, add item 8 and heat to 60C. Add and dissolve item 9. Add item 10 and mix at medium speed (1200 rpm) for about 20 minutes, or until completely dispersed, while maintaining temperature at 60C.
3. Part 3-Add Part 2 to Part 1 slowly and mix together.
4. Cool lotion to 50C and add item 11. Stir slowly to room temperature.

Formula TS-105

Hand Lotion

Materials:	% by Weight
1. Water	87.59
2. Triethanolamine	0.42
3. Propylene Glycol	4.00
4. Methylparaben	0.10
5. Bentone LT rheological additive	0.80
6. Ceraphyl 424	3.00
7. Stearic Acid T.P.	0.79
8. Pluronic F-127	1.00
9. Glyceryl Monostearate	2.00
10. Propylparaben	0.10
11. Perfume	0.20

Manufacturing Directions:
1. Part 1-In a stainless steel jacketed kettle, add items 1 through 4 and heat to 60C. Using a homomixer at about 1400 rpm, add item 5 slowly to avoid lumps, and mix for 20 minutes or until homogeneous. If foam develops, add a minute amount of Antifoam 60. Heat Part 1 to 78C.
2. Part 2-In a separate vessel, add items 6 through 10 and heat to 78C. Mix until completely melted and homogeneous.
3. Mix Part 2 slowly in Part 1 at 78C using sweep blades.
4. Cool to 50C. Add item 10, mix and cool to room temperature.

Formula TS-200

SOURCE: Rheox, Inc.: **Suggested Formulations**

Deep Moisturizing Lotion

This after-bath lotion gives the benefits of potent skin conditioners while eliminating the tackiness associated with lanolin and petrolatum through the unique emolliency provided by Phospholipid EFA.

Part A:	% By Weight
Phospholipid EFA	4.00
Steareth-21	0.40
Water	82.00

Part B:	
Steareth-2	1.60
Anhydrous Lanolin	1.50
Petrolatum	3.00
Octyldodecyl Myristate	2.00
Cetearyl Alcohol	4.00
Dimethicone (100cS)	1.50

Combine ingredients in both phases separately and heat to 65C. Homogenize (B) into (A) with continued heating until sufficiently mixed. Stir-cool to 45-50C, then add fragrance, color, and preservative as needed before filling.

Hand and Body Lotion

A superior product designed for after-bath use on traditionally dry areas such as hands, elbows and heels. Phospholipid EFA is strongly substantive towards skin providing non-greasy moisturizing and a pleasant after feel.

Part A:	% by Weight
Phospholipid EFA	4.00
Water	83.00

Part B:	
Steareth-2	2.00
Light Mineral Oil	4.00
Cetearyl Alcohol	3.00
Octyldodecyl Myristate	2.50
Dimethicone (100cS)	1.50

Combine ingredients in both phases separately and heat to 65C. Homogenize (B) into (A) with continued heating until sufficiently mixed. Stir-cool to 45C. Add fragrance, color, and preservative as needed and fill.

SOURCE: Mona Industries, Inc.: PHOSPHOLIPID EFA: Suggested Formulations

Emollient Lotion

Illustrates the use of Veegum Pro magnesium aluminum silicate as an emulsion stabilizer and thickener. The formula is designed to serve as a guide for the development of new products or the improvement of existing ones.

A	Veegum Pro	1.5%
	Water	83.8
B	Triethanolamine	0.1
	Glycerin	3.5
C	Marcol 130	3.6
	Petrolatum	0.4
	Stearic acid XXX	1.6
	Cetyl alcohol	1.5
	Kessco Glycerol Monostearate SE	1.4
	Acetulan	2.0
	Dow Corning 200 Fluid	0.6
	Preservative	q.s.

Procedure:
Heat the water to 70 to 75C, then slowly add the Veegum PRO while agitating at maximum available shear. Mix until smooth. Add B to A with slow agitation until smooth. Maintain A/B at 70 to 75C, heat C to 75 to 80C. Add C to A/B and mix until cool.
Consistency: Medium viscosity lotion.
Suggested Packaging: Squeeze or pump bottle.
Comments: Veegum PRO effectively thickens and stabilizes
the lotion, even at elevated temperatures. This lotion is
absorbed rapidly, leaving the skin smooth and greaseless.

Hand and Body Lotion

Illustrates the use of Veegum PRO magnesium aluminum silicate as an emulsion stabilizer and thickener. The formula is designed to serve as a guide for the development of new products or the improvement of existing ones.

A	Veegum Pro	2.00%
	Water	70.75
	Glycerin	6.00
B	Marcal 130	10.00
	Petrolatum	4.00
	Arlacel 165	5.00
	Synchrowax AW1-C	1.25
C	Allantoin	1.00
	Preservative	q.s.

Procedure:
Heat the water to 70 to 75C, then slowly add the Veegum PRO while agitating at maximum available shear. Mix until smooth. Add glycerine and mix until uniform. Heat B to 75 to 80C. Add B to A and mix until cool. Add C and mix until uniform.
Consistency: Medium viscosity lotion.
Suggested Packaging: Squeeze or pump bottle.
Comments: Veegum PRO effectively thickens and stabilizes
the emulsion even at elevated temperatures. Glycerin helps
to rapidly hydrate dry skin and the selection of oils and waxes
produces a smooth and non-greasy feel. The allantoin provides
soothing relief for wounds, burns, and skin problems in general.
SOURCE: R.T. Vanderbilt Co., Inc.: Formula Nos. 417 & No. 420

European O/W Cleansing Lotion

Ingredients:	%W/W
A Mineral oil	15.00
Arlamol E	3.00
B Arlatone 2121	2.00
Propylene glycol	1.25
Glycerol	1.25
Water	72.50
C Carbomer 934 (3% solution)	5.00

D Sodium Hydroxide (10% solution) to pH of 6.5-7.0

E *Perfume

*q.s. these ingredients.

Procedure:
Disperse (C) in (B) at 75C with moderate stirring. Heat (A) and add at 75 with vigorous stirring. Homogenize during addition of (A). Stir vigorously until cool. Add (D) around 50C. Add (E) at below 35C.

Water/Oil Cleansing Lotion

Ingredients:	%W/W
A Mineral oil	15.00
Beeswax	0.50
Ceresin wax	0.50
Sorbo	27.00
Arlacel 186	3.00
Tween 80	0.50
B Water, deionized	53.50
*Preservative	

*q.s these ingredients.

Procedure:
Heat (A) to 70C. Heat (B) to 72C. Add (B) to (A) slowly with moderate stirring. Stir to 35C and add water lost due to evaporation.

SOURCE: ICI Surfactants: Suggested Formulations

Gentle Moisturizing Alpha Hydroxy Acid Lotion

Resulting product is a thixotropic lotion which can be dispensed from a flexible tube or soft plastic bottle. The product utilizes a blend of two alpha hydroxy acids (AHA's), and buffered citric acid at a combined use level of approximately 2.5%. This product could be used as a daily use moisturizing AHA product.

	%w/w
A. Deionized Water	76.70
Methylparaben (Lexgard M)	0.20
2,4-Dichlorobenzyl Alcohol (Myacide SP)	0.20
Hexylene Glycol	1.00
B. Hydroxyethylcellulose (Cellosize QP-15,000H)	1.00
C. Isostearamidopropyl PG-Dimonium Chloride	
(Lexquat AMG-IS)	2.00
D. Glyceryl Stearate (Lexemul 515)	3.00
Glyceryl Stearate (and) PEG-100 Stearate (Lexemul 561)	4.00
Propylene Glycol Dipelargonate	2.10
Isopropyl Palmitate (and) Isopropyl Myristate (and)	
Isopropyl Stearate (Lexol 3975)	1.00
Sesame Oil	2.00
Tocopherol (Copherol 1300)	0.10
Propylparaben (Lexgard P)	0.10
E. Deionized Water	4.00
Malic Acid (Granular)	1.00
Sodium Citrate USP-FCC	1.60

Procedure:
1. Combine section "A" heating to 70-75C. Use a propeller mixer for agitation.
2. Sprinkle section "B" into section "A". Increase mixing speed to create a vortex during addition, then slow to highest speed which does not produce a vortex.
3. When sections "A" & "B" are homogeneous and at the designated temperatures add section "C" to sections "AB".
4. Combine section "D". Heat to 75-80C with continuous slow mixing.
5. When sections "ABC" and section "D" are homogeneous and at the designated temperatures, slowly pour section "D" into sections "ABC". Increase mixing speed to a high speed. Maintain at 80-85C for 5 minutes, then begin cooling. Slow mixing speed to highest speed possible that will not cause vortexing.
6. Combine section "E" and mix until it becomes a clear solution.
7. Add section "E" to sections "ABCD" at 60-65C. Continue mixing and cooling.
8. At 40-45C adjust for water loss.
9. Mix batch until it reaches 30-35C.
10. Adjust final pH to 4.00-4.10 with a Malic Acid solution, as required.

Physical Properties:
Viscosity: 4,500 cPs @ 24C (Brookfield RVT, TA @ 10 rpm)
24 hour sample.
pH: 4.1 @ 24C

SOURCE: Inolex Chemical Co.: Formulation 398-87-3

Hand Lotion

Ingredients:	%W/W
A Isopropyl palmitate	2.00
Oleic acid	2.00
Stearic acid, triple pressed	2.00
Atmul 84S	3.20
Sorbo	2.00
Triethanolamine	1.00
Water, deionized and .1% preservative	83.30
B Alcohol, SDA 40	4.00
Fragrance	0.50

Procedure:
Heat (A) to 95C. Agitate until a milky emulsion forms and is cooled to 45C. Dissolve fragrance in the alcohol. Add (B) to (A). Stir to insure thorough agitation.

Hand Lotion

Ingredients:	%W/W
A Light mineral oil	4.00
Stearic acid	2.00
Acetylated lanolin	3.00
Arlacel 165	5.00
Cetyl alcohol	1.00
Silicone fluid, 200 cs	1.00
B Water	78.80
Sorbo, Sorbitol solution, USP	5.00
Methyl paraben	0.18
Propyl paraben	0.02

Procedure:
Heat (A) to 70C and (B) to 72C. Add (B) to (A) with agitation. Cool to set point and homogenize.
Formula SK-9

Hand Lotion

Ingredients:	%W/W
A Mineral oil	30.00
Brij 52	5.20
Brij 58	4.80
Triethanolamine	0.20
B Carboxy vinyl polymer, Carbopol 934	0.20
Water, deionized	59.60
*Preservative	
*q.s. these ingredients.	

Procedure:
Heat (A) to 70C. Heat (B) to 72C. Add (B) to (A) rapid agitation. Stir until set.
SOURCE: ICI Surfactants: Suggested Formulations

Hand Lotion

Ingredients:	%W/W
A Arlamol E	8.00
Arlasolve 200L	1.70
Brij 72	2.80
Stearyl alcohol	2.00
B Water, deionized	85.10
Carbopol 934	0.20
C Sodium hydroxide (10% W/W aqueous)	0.20
D *Preservative	
*Perfume	

*q.s. these ingredients.

Procedure:
    Disperse Carbomer in water and heat (B) to 70C. Heat (A) to 72C and add (B) to (A) with propeller agitation. Slowly add (C) and increase speed of the agitation as needed. Add (D) and replace water lost by evaporation.
Formula AE-9

Hand Lotion

Ingredients:	%W/W
A Stearyl alcohol	5.00
Brij 721S	1.50
B Water, deionized	92.80
Carbopol 934	0.20
C Sodium Hydroxide (10% W/W aqueous)	0.20
D Dowicil 200	0.10
E Fragrance	0.20

Procedure:
    Heat (A) to 70C and (B) to 75C. Add (B) to (A) slowly using propeller type agitation. Add (C). Add (D) when the temperature drops below 50C. Add (E) below 40C. Add water to compensate for evaporation. Homogenize. Adjust pH to 5.5-6.5

SOURCE: ICI Surfactants: Suggested Formulations

Hand Lotion

Ingredients:	%W/W
A Arlacel 165, acid stable g.m.s.	5.00
Cetyl alcohol	5.00
B Sorbo, 70% sorbitol solution U.S.P.	5.00
Water, deionized	85.00
*Preservative	
C *Mild acid to adjust pH	
*q.s. these ingredients.	

Procedure:
Heat (A) to 70C. Heat (B) to 72C. Add (B) to (A) with rapid
agitation. Cool to room temperature with stirring.

Protective Hand Lotion

Ingredients:	%W/W
A Silicone fluid, 50 cs	10.00
Brij 52	2.60
Brij 56	4.90
Triethanolamine	0.20
B Carboxy vinyl polymer, Carbopol 934	0.20
Water, deionized	82.10
*Preservative	
*q.s. these ingredients.	

Procedure:
Heat (A) to 70C. Heat (B) to 72C. Add (B) to (A) rapid
agitation. Stir until set.

Protective Hand Lotion

Ingredients:	%W/W
A Silicone fluid, 50 cs	10.00
Brij 52	4.00
Brij 58	3.60
Triethanolamine	0.20
B Carboxy vinyl polymer, Carbopol 934	0.20
Water, deionized	82.00
*Preservative	
*q.s. these ingredients.	

Procedure:
Heat (A) to 70C. Heat (B) to 72C. Add (B) to (A) rapid
agitation. Stir until set.

SOURCE: ICI Surfactants: Suggested Formulations

Hand Lotion

Materials:	% by Weight
Part A:	
1. Deionized Water	85.89
2. Triethanolamine	0.42
3. Glycerine	4.00
4. Methyl Paraben	0.10
5. Bentone EW rheological additive	1.50
Part B:	
6. Ceraphyl 424	3.00
7. Stearic Acid XXX	0.79
8. Isopropanol Lanolate Distilled	1.00
9. Pluronic F-127	1.00
10.Glyceryl Monostearate	2.00
11.Propyl Paraben	0.10
Part C :	
12.Fragrance	0.20

Manufacturing Directions:
1. Part A-In a stainless steel steam jacketed kettle, add item 1 to 4 and heat to 60C. Using a homomixer, add item 5 slowly to avoid lumps and mix for 20 minutes or until homogeneous. Heat to 80C.
2. Part B-In a seperate vessel, add items 6 to 11 and heat to 80C. Mix until completely melted and homogeneous.
3. Mix Part B slowly in Part A at 80C using sweep blades.
4. Cool to 50C. Add Part C, mix and cool to room temperature.
5. Viscosity of lotion after 24 hours should be 9100 cps.

Formula TS-246

Soft Hand Lotion

Materials:	% by Weight
1. Deionized Water	84.1
2. Bentone LT rheological additive	0.5
3. Glycerine	2.5
4. Triethanolamine	0.7
5. Methyl Paraben	0.1
6. Ceraphyl 424	1.5
7. Isocetyl Stearate	1.0
8. Acetol	1.7
9. Stearic Acid	4.0
10.Lexemul 55G	3.4
11.Volatile Silicone 7158	0.2
12.Propyl Paraben	0.1
13.Fragrance	0.2

Manufacturing Directions:
1. Using vigorous agitation, add the Bentone LT to the water. Mix for 20 minutes.
2. Add ingredients 3 through 5 to Step 1, heat to 80C.
3. In a separate vessel, combine ingredients 6,7,8,9,10 and 12. Heat to 80C.
4. Add the Oil Phase to the Water Phase.
5. Cool to 50C, then add ingredients 11 and 13.
6. Cool to 35C and package.

Formula TS-287

SOURCE: Rheox, Inc.: Suggested Formulations

Hand and Body Lotion

	%
Part 1:	
Water	78.6
Carbomer 934	0.2
Part 2:	
Modulan	1.6
IPP	3.8
Amerchol L-101	0.8
GMS SE	2.1
Rosswax 63-0412	4.0
IPM	4.0
Jojoba Oil	1.6
Part 3:	
Germaben IIE	1.0
Part 4:	
Fragrance	q.s.
Part 5:	
Triethanolamine	2.3

Procedure:
Part A:
   Disperse the Carbomer 934 in the water phase in a stainless steel kettle.
Part B:
   In a separate heated kettle, heat the oil phase until all ingredients are melted. When everything is melted add the Oil Phase to the Water Phase. When everything is blended add the preservative, the fragrance and the Triethanolamine with increased agitation. Cool to room temperature and package.

Jojoba Lotion

	%
Part A:	
Modulan	1.6
Amerchol L-101	0.8
Isopropyl Palmitate	5.0
Glyceryl Mono Stearate Pure	2.1
Rosswax 63-0412	4.0
Isopropyl Myristate	4.0
Ross Jojoba Oil	1.6
Part B:	
Water	74.4
Glycerine Pure Emery 916	4.2
Triethanolamine	2.3

Procedure:
   Heat Part (A) and Part (B) in separate vessels to 170F under agitation. When temperature is reached mix Part (A) to Part (B), and cool. Package in container at below 120F.

SOURCE: Frank B. Ross Co., Inc.: Suggested Formulations

High Petrolatum Lotion

High petrolatum for severely dry, chapped or chafed skin.

Oil Phase:	%W/W
White Protopet 1S	30.00
Crodamol PMP	10.00
Volpo 3	2.30

Water Phase A:	
Deionized Water	46.35
Carbopol 941	0.13
10% Sodium Hydroxide Solution	0.52

Water Phase B:	
Glycerin	5.00
Volpo S-10	5.70
Germaben II	qs

Procedure:
Water Phase A:
   Charge a vessel with water and dust in the carbomer. Start heating to 80C. When completely dispersed, add the sodium hydroxide solution and continue mixing.
Water Phase B:
   Combine components in a separate vessel and mix, heat to 80C. When uniform, add to water phase A.
Oil Phase:
   Combine all components of the oil phase, mix and heat to 80C. When uniform, add the combined water phases slowly while mixing. Continue mixing to room temperature.

Low Irritation Moisturizing Lotion

Disodium laurylsulfosuccinate	0.30%
Glycerin	5.00
Methylparaben	0.20
Cetyl alcohol	2.00
Carnation Mineral Oil	20.00
Methyl phenyl polysiloxane	2.00
Beeswax	0.20
Water	q.s. to 100.00

SOURCE: Witco Corp.: Petroleum Specialties Group: Suggested Formulations

Hydroalcoholic Mineral Oil Emulsion

Ingredients:	%W/W
A Light mineral oil	8.00
Stearyl alcohol	1.00
Brij 721, Steareth-21	2.28
Brij 72, Steareth-2	1.72
B Water, deionized	66.00
Carbomer 934	0.40
C Sodium Hydroxide Solution (10%)	0.40
D *Preservative	0.00
E Fragrance	0.20
F Ethanol, SDA-40, (190 Proof)	20.00

*q.s. these ingredients.

Procedure:
Heat (A) to 70C and (B) to 75C with good propeller agitation. Add (C) below 50C and mix thoroughly. Add the remaining phases below 35C. Replace water lost due to evaporation. Package.

Mineral Oil Lotion

Ingredients:	%W/W
A Mineral oil	10.00
Arlacel 165, acid stable g.m.s.	10.00
B Water, deionized	80.00
*Preservative	
C *Mild acid to adjust pH	

*q.s. these ingredients.

Procedure:
Heat (A) to 60C. Heat (B) to 62C. Add (B) to (A) with thorough but gentle stirring. Cool to 25C with gentle stirring.

SOURCE: ICI Surfactants: Suggested Formulations

Hydrogen Peroxide Lotion

Ingredients:	%W/W
A Stearic acid, triple pressed	10.00
Stearyl alcohol	0.50
Cetyl alcohol	1.00
Brij 721S	5.00
Silicone oil, 350 cs.	0.50
B Water, deionized	51.06
EDTA	0.20
Phenacetin	0.04
C Hydrogen peroxide, 27% dilution grade	22.20
D *Phosphoric acid, 10% C.P.	9.50

*q.s. these ingredients.

Procedure:
Heat (A) to 60C and (B) to 65C. Add (B) to (A) slowly with moderate agitation. Add (C) below 35C. Replace water lost by evaporation and adjust pH to 3.5-4.0 with dilute phosphoric acid (10% C.P.). Package in suitable container for possible evolution of oxygen.
Formula HC-14

Hydrogen Peroxide Lotion

Ingredients:	%W/W
A Cetyl alcohol	4.00
Brij 52	1.50
Brij 58	3.50
B Water, deionized	73.86
C Hydrogen peroxide, 35% dilution grade	17.14
D Phosphoric acid (10%C>P>) as required	

Procedure:
Heat (A) to 70C. Heat (B) to 72C. Add (B) to (A) slowly with thorough agitation. Cool to 45C with agitation. Compensate for lost water due to evaporation. Add (C). Adjust pH to 3.5-4.0 by adding (D).

SOURCE: ICI Surfactants: Suggested Formulations

Isopropyl Myristate Lotion

Ingredients:	%W/W
A Isopropyl myristate	10.00
Arlacel 60	1.60
Arlasolve 200L	3.33
B Water, deionized	84.07
Carbopol 934	0.40
C Sodium hydroxide (10% W/W aqueous)	0.40
D Dowicil 200	0.10
E Fragrance	0.10

*q.s. these ingredients.

Procedure:
Heat (A) to 65C and (B) to 60C. Add (A) to (B) slowly with moderate anchor type agitation and add (C). Add (D) at about 50C. Add (E) at 35C and add water to compensate for loss due to evaporation.
Formula PC 8186

Vanishing Lotion

Ingredients:	%W/W
A Arlamol ISML	4.00
Stearyl alcohol	1.00
Silicone oil (350 cs)	0.50
Arlamol E	1.00
Brij 700	2.00
Brij 72	2.00
B Water, deionized	79.05
Veegum (5% aqueous solution)	10.00
Carbopol 934	0.15
C Sodium hydroxide (10% aqueous)	0.15
D Germall II	0.10
E Herbal Fragrance SL 79-1224, PFW	0.05

Procedure:
Heat (A) to 70C and (B) to 72C. Add (B) to (A) with moderate agitation. Add (C). Add (D) below 50C. Add (E) at 35C and add water to compensate for loss due to evaporation.

SOURCE: ICI Surfactants: Suggested Formulations

Lanolin & Aloe Lotion

   Lotions have a wide range of applications in today's cosmetic market. Ideally they should leave a smooth luxurious afterfeel to the skin and help alleviate such problems as dry skin or chapping from wind and sunburn. SC-188 is an economical lotion which achieves both ends. Crodamol PTC, a non-oily ester which gives a "cushiony" feel at low levels of addition, also works with the lanolin and lanolin derived products to leave a satiny, non-greasy, protective film on the skin.

Ingredients:	%
Part A:	
Super Refined Babassu Oil (Babassu Oil)	2.00
Corona Pure New Lanolin (Lanolin)	2.00
Crodacol CS-50 (Cetearyl Alcohol)	1.50
Super Sterol Ester (C10-30 Cholesterol/Lanosterol Esters)	1.00
Crodamol PTC (Pentaerythrityl Tetracaprylate/Caprate)	0.80
Crodacol C-95 (Cetyl Alcohol)	0.50
Crodamol MM (Myristyl Myristate)	0.50
Incropol CS-20 (Ceteareth-20)	1.00
Mineral Oil (70ssu)	4.00
Stearic Acid XXX	0.50
Silicone L-45 (100cps)	0.30
Propyl Paraben	0.15
Part B:	
Deionized Water	81.55
Glycerin	3.00
Veragel 200 Powder	0.50
Methyl Paraben	0.30
Carbopol 941	0.10
Dowicil 200	0.10
PEG-15 Cocamine	0.20

Procedure:
   Dust the Carbopol 941 of Part B into the water with mixing. When completely dissolved, add the remaining ingredients of Part B and heat to 75-80C. Combine ingredients of Part A with mixing and heat to 75-80C. Add Part B to Part A with good mixing and cool to desired fill temperature. Adjust pH with 1.0% NaOH.

   pH: 6.0+-0.5
   Viscosity: 7,500+-10% (RVT Spindle #5, 20 rpm @ 25C)
   N.A.T.C. Approved

SOURCE: Croda Inc.: Formulation SC-188-2

Lotion

Ingredients:	%W/W
A Stearyl alcohol	8.00
Brij 721S	4.00
Silicone oil, 350 cs.	0.50
B Water, deionized	65.30
C Hydrogen peroxide, 27% dilution grade	22.20

Procedure:
Heat (A) to 60C and (B) to 65C. Add (B) to (A) slowly with moderate agitation. Add (C) below 35C. Replace water lost by evaporation and adjust pH to 3.5-4.0 with dilute phosphoric acid (10% C.P.). Package in suitable container for possible evolution of oxygen.
Formula PC-8202

Lotion

Ingredients:	%W/W
A Cetyl alcohol	3.00
Brij 721S	3.00
Silicone oil, 350 cs.	0.50
B Water, deionized	66.30
Sorbo	5.00
C Hydrogen peroxide, 27% dilution grade	22.20

Procedure:
Heat (A) to 60C and (B) to 65C. Add (B) to (A) slowly with moderate agitation. Add (C) below 35C. Replace water lost by evaporation and adjust pH to 3.5-4.0 with dilute phosphoric acid (10% C.P.). Package in suitable container for possible evolution of oxygen.
Formula PC-8203

Glossy Lotion

Ingredients:	%W/W
A Mineral oil, Carnation	8.00
Stearyl alcohol	1.00
Brij 721S	2.00
Brij 72	2.00
B Water, deionized	86.80
C Dowicil 200	0.10
D Fragrance	0.10

Procedure:
Heat (A) to 70C and (B) to 75C with good propeller agitation. Add (C) at 50C and mix thoroughly. Add (D) at 35C. Add water to compensate for loss due to evaporation.

SOURCE: ICI Surfactants: Suggested Formulations

Low Solids Almond Lotion II

	Parts by Weight
Water	568.0gr.
Carbomer 934	2.0gr.
Rosswax 573	4.0gr.
GMS SE	4.0gr.
Almond Oil-Lipoval A1M	16.0gr.
Coconut Oil #76	16.0gr.
Jojoba Oil	4.0gr.
TEA	4.0gr.
Preservative Germaben II	6.0gr.
Fragrance GG44	q.s.

Procedure:
Heat the water to 60C under agitation and slowly add the Carbomer 934. When the water is fully mixed add the 573, GMS, Almond Oil, Coconut Oil, and Jojoba Oil that have been heated to 65C in a separate kettle. As soon as they have been mixed well add the preservative, the fragrance and then the TEA under high agitation. Cool the batch to 55C, and package.

Peach Hand Lotion

	Parts by Weight
Water	568.0gr
Carbomer 934	2.0gr
Rosswax 573	4.0gr
GMS-SE	4.0gr
Almond Oil	16.0gr
Lipoval ALM	
Coconut Oil #76	16.0gr
Ross Jojoba Oil	4.0gr
TEA	4.0gr
Germaben II	6.0gr
Fragrance GK-16	q.s.

Procedure:
Heat the water to 60C under agitation and slowly add the Carbomer 934. When the water is fully mixed add the 573, GMS, Almond Oil, Coconut Oil, and Jojoba Oil that have been heated to 65C in a separate kettle. As soon as the Oil Phase has been mixed well, add the Germaben II, the Fragrance, and then the TEA under high agitation. Cool to 55C for filling.

SOURCE: Frank B. Ross Co., Inc.: Suggested Formulations

Milk Lotion

	Wt%
A. Liquid Petrolatum	4.7
Amiter LG-OD	2.2
Propylene Glycol Monostearate	0.4
POE (5) Hydrogenated Castor Oil (Emalex HC-5)*	1.3
POE (5) Glyceryl Monostearate (Emalex GM-5)*	2.8
Butyl Paraben	0.1
Amihope LL	3.0
B. Acylglutamate HS-11	0.3
1,3-Butylene Glycol	5.0
Methyl Paraben	0.2
C. Carboxyvinyl Polymer	0.2
Water	79.72
D. Sodium Hydroxide (NaOH)	0.08

*Nihon Emulsion Co.

Procedure:
1. Dissolve (A) at 80C.
2. Dissolve (C), and then neutralize with (D).
3. Add (B) to #2, and dissolve at 80C.
4. Add #3 slowly to (A) as mixing, and then cool to 30C.

Note: This milk lotion has light touch, and spreads well.

Milk Lotion

	Wt%
A Mineral Oil	5.00
Amiter LGOD	2.00
Propylene Glycol Stearate	0.50
PEG-5 Hydrogenated Castor Oil	1.50
PEG-5 Glyceryl Stearate	2.50
Butyl Paraben	0.10
Amihope LL	3.00
B Acylglutamate HS-11	0.30
Carbomer 941	0.20
Sodium Hydroxide	0.08
Butylene Glycol	5.00
Methyl Paraben	0.20
Water	79.62

Milk lotion with smooth touch and good spreadability.
Procedure:
Dissolve Carbomer 941 and sodium hydroxide in water first.
Add other ingredients of (B) to the solution and dissolve at
75-80 degrees C. Dissolve (A) ingredients at 80 degrees C and
add (A) to (B) with agitation. Cool down to room temperature
with agitation.
SOURCE: Ajinomoto USA, Inc.: Suggested Formulations

Milk Lotion

		Wt%
(O)	Liquid Paraffin (#70)	10.0
	Paraffin Wax (mp 54-56C)	5.0
	Cetyl Alcohol	3.0
	Isopropyl Palmitate	2.0
	Squalane	0.5
	Nikkol WCB	4.0
	Lanolin	0.5
	Polyoxyethylene (10) Lanolin Alcohol	1.0
	Polyethylene Glycol (6300) Monostearate	0.5
	Polyoxyethylene (5) Stearyl Ether	2.1
	Polyoxyethylene (25) Cetyl Ether	2.9
(W)	Ajidew N-50	3.0
	Polyethylene Glycol - 1500	3.0
	Water	62.5
	Preservative	q.s.

Procedure:
1. Heat (O) and (W) to 80C.
2. Add (W) to (O) slowly with agitation.
3. Cool to 35C with stirring.

pH: 5.3
Viscosity: 20,000 cps

Milk Lotion

		Wt%
(O)	Liquid Paraffin (#70)	10.45
	Beeswax	1.0
	Stearyl Alcohol	1.5
	Isopalmityl Alcohol	1.2
	Polyoxyethylene (10) Glyceryl Triisostearate	2.35
	Polyoxyethylene (15) Cetyl Ether	0.8
(W)	Ajidew N-50	3.0
	Water	79.7
	Preservative	q.s.

Procedure:
1. Heat (O) and (W) to 80C.
2. Add (W) to (O) slowly.
3. Cool to 40C with stirring.

pH: 5.2
Viscosity: 1,700 cps

SOURCE: Ajinomoto USA, Inc.: Suggested Formulations

Mineral Oil Lotion

Ingredients:	%W/W
A Mineral oil, Carnation brand	20.00
Brij 72	2.70
Arlasolve 200L	4.60
B Water, deionized	71.80
Carbopol 934	0.40
C Sodium hydroxide (10% W/W aqueous)	0.40
D Dowicil 200	1.00

Procedure:
Heat (A) to 70C and (B) to 72C. Add (B) to (A) with moderate
agitation. Add C. Add (D) below 50C. Add (E) at 35C and add
water to compensate for loss due to evaporation.
Formula PC-8184

Mineral Oil Lotion - O/W

Ingredients:	%W/W
A Mineral oil	30.00
Brij 52 (Ceteth-2)	5.20
Arlasolve 200 Liquid (Isoceteth-20)	6.60
B Water, deionized	52.80
Carbomer 934	0.20
Arlasolve DMI	5.00
C Sodium hydroxide-10% aqueous solution	0.20
D *Preservative	

*q.s. these ingredients.

Procedure:
Disperse Carbomer in the water and Arlasolve DMI with rapid
agitation. Heat (B) to 65 deg. C and (A) to 60 deg. C. Add (B)
to (A) with good agitation. Add (C). Add (D) below 50 deg. C.
Stir to 35 deg. C and replace evaporated water.

SOURCE: ICI Surfactants: Suggested Formulations

Mineral Oil Free Lotion Contains Ethanol (20%)

Ingredients:	%W/W
A Arlamol E	20.00
Brij 721S	2.00
Brij 72	4.00
Stearyl alcohol	4.00
B Water, deionized	48.80
Carbopol 934	0.40
C Sodium hydroxide (10% aqueous)	0.40
D Germall II	0.20
E Fragrance, H 45756 (P. Robertet)	0.20
F Ethanol, SDA-40 (190 proof)	20.00

Procedure:
    Heat (A) to 70C and (B) to 75C. Add (B) to (A) slowly using blade type agitation. Add (C). Add (D) when the temperature drops below 50C. Add (E) below 40C. Add water to compensate for evaporation. Homogenize. Adjust pH to 5.5-6.5.

Lotion

Ingredients:	%W/W
A Cetyl alcohol	1.00
Brij 721S	1.00
B Water, deionized	95.30
Sorbo, Sorbitol Solution USP	2.00
Carbopol 934	0.20
C Sodium hydroxide (10% W/W aqueous)	0.20
D Dowicil 200	0.10
E Fragrance	0.20

Procedure:
    Heat (A) to 70C and (B) to 75C. Add (B) to (A) slowly using propeller type agitation. Add (C). Add (D) when the temperature drops below 50C. Add (E) below 40C. Add water to compensate for evaporation. Homogenize. Adjust pH to 5.5-6.5

SOURCE: ICI Surfactants: Suggested Formulations

Moisturizing Lotion

Phase A:	Percent
Protopet 1S	2.00
PPG-15 Stearyl Ether (Arlamol E)	2.00
Stearyl Alcohol	1.00
Steareth-2 (Brij 72)	3.00
Steareth-20 (Brij 78)	1.00
Dimethicone	0.10

Phase B:	
Water	50.00
Carbomer 940 (Carbopol 940)	0.10

Phase C:	
Polyquaternium-19 (Arlatone PQ220)	10.00
Water	30.20
Triethanolamine	0.10

Phase D:	
Germaben II	0.50

Procedure:
   Disperse Carbopol 940 in water (B) and heat to 75C. Heat A
to 70C. Add A to B and mix using propeller-type agitation. Heat
C to 70C and add to AB with thorough agitation. Cool to 60C
with agitation and add D.

Pomade Lotion

Oil Phase:	Percent
Protopet 1S Petrolatum	30.00
Crodamol PMP	10.00
Volpo 3	2.30

Water Phase A:	
Deionized Water	46.35
Carbopol 941	0.13
10% Sodium Hydroxide Solution	0.52

Water Phase B:	
Glycerine	5.00
Volpo S-10	5.70
Germaben II	q.s.

Procedure:
   Water Phase A. Charge a vessel with water and dust in the
Carbomer. Start heating to 80C. When completely dispersed,
add the Sodium Hydroxide solution and continue mixing.
   Combine water Phase B in a separate vessel and mix and
heat to 80C. When uniform add to water Phase A.
   Combine the oil phase, mix and heat to 80C. When uniform
add the combined water phases slowly while mixing. Continue
mixing to room temperature.
**SOURCE: Witco Corp.: Petroleum Specialties Group: Suggested
         Formulations**

Neutralizing Lotion

Ingredients:	%W/W
A Forestall	1.40
Brij 35 SP	2.00
Water, deionized	92.30
*Stabilizer	
B Hydrogen peroxide, 35%	4.30
C *Phosphoric acid	

*q.s. these ingredients.

Procedure:
Mix (A) until uniform. Add (B) at 25 deg. C. Adjust pH to 4.5 to 5.0 with (C). Optional stabilizers may include sequestrants or antioxidants.
Formula HC-16

Neutralizing Lotion

Ingredients:	%W/W
A Sodium bromate	12.00
Brij 30	4.50
Polyglycol palmitic amide	4.50
Water	79.00
*Preservative	
B *Acetic acid, glacial	

*q.s. these ingredients.

Procedure:
Add the sodium bromate to water. Stir with heat until dissolved. Add the remainder of (A). Heat to 70-75C. Agitate continually until cooled to room temperature. Adjust pH to 6.5-7.0 with (B). Package.
Formula HC-18

Quick Dry Roll-On Lotion

Ingredients:	%W/W
A Cyclomethicone, DC 344 fluid	12.70
Brij 52	3.10
Brij 35SP	0.90
B Water, deionized	48.30
C Aluminum chlorhydroxide, 50% aqueous	35.00

Procedure:
Heat (A) to 70C and (B) to 72C. Add (B) to (A) with agitation. Cool to 35-40C with stirring and slowly add the remainder of ingredients.

SOURCE: ICI Surfactants: Suggested Formulations

O/W Broad Spectrum Protective Moisturizing Lotion with Vitamins*

Ingredients:	%w/w
A) Parsol MCX	2.00
Parsol 1789	1.00
Parsol 5000	0.60
Glyceryl Monostearate	2.00
Cetyl Alcohol Extra	2.00
Dermol 185	6.00
Ganex V-220	2.00
Prisorine 3505	2.00
Butylated Hydroxytoluene	0.05
Edeta BD	0.10
Phenonip	0.60
Amphisol A	1.50
B) Deionized Water	10.00
AMP 10% sol'n	3.97
C) Deionized Water	22.26
Propylene Glycol	5.00
Carbopol 981 1% sol'n	10.00
D) AMP 10% sol'n	1.02
E) Deionized Water	20.00
Parsol HS	0.60
AMP 10% sol'n	1.80
F) Vitamin E Acetate	2.00
G) Ropufa 25N-6 Oil	2.00
H) Panthenol	2.00
I) Ponceau Red SX 0.2% sol'n	0.25
J) Parflex 49915	0.40**

Procedure:
    Heat part A) to 85C while stirring. When homogeneous, add
parts B),C) and D) pre-heated to 75C, while mixing. Add Part E)
pre-heated to 75C, while mixing (be sure that the Parsol HS has
been completely dissolved, if traces remain, add a small quantity
of the neutralizing base until the solution is clear). Cool to
40C, add parts F), G), H), I) and J). Compensate for water loss
and continue stirring while cooling to ambient temperature.

Remark:
    *While not a sunscreen, this formulation scored an SPF 8 when
    tested according to the FDA/OTC method (range-finding assay
    on six subjects: IRI Ref. 582627).
    **Dermatologically tested perfume for sunscreens. The final
    pH value should be around 7.0 to prevent recrystallization
    of Parsol HS.

SOURCE: Givaudan-Roure: Formulation 08 COS 029

O/W Lotion

		Wt.%
1.	A-C Copolymer 580	2.0
2.	Mineral Oil (70 ss)	5.0
3.	Dow 556 Fluid	1.0
4.	Propylene Glycol Dipelargonate	10.0
5.	Amerchol 400	2.0
6.	Ethoxyol 24	1.0
7.	Arlacel 60	1.0
8.	Tween 60	2.0
9.	Propyl-P-Hydroxybenzoate	0.1
10.	Sorbitol (70%)	5.0
11.	Carbopol 941	0.5
12.	Methyl-P-Hydroxybenzoate	0.2
13.	Triethanolamine	0.75
14.	Water	69.45

Procedure:
    Disperse Carbopol in water. Weigh 1-9 and heat to 80-90C
with slow agitation. Add remaining ingredients to Carbopol/water
dispersion, except triethanolamine, and heat to 80-90C. Add the
water phase to the aqueous phase and shear in homomixer.
Continue shearing while cooling to 40C, then add triethanolamine,
mixing well. Cool to 30C, add perfume, de-aerate and package.
Ref: 5189-2-7

O/W Lotion

		Wt.%
1.	A-C Copolymer 580	2.0
2.	Distilled Isopropyl Lanolate	3.0
3.	Dow 556 Fluid	2.0
4.	Propylene Glycol Dipelargonate	13.0
5.	Ethoxyol 24	1.0
6.	Arlacel 60	1.0
7.	Tween 60	2.0
8.	Propyl-P-Hydroxybenzoate	0.1
9.	Sorbitol (70%)	5.0
10.	Carbopol 941	0.5
11.	Triethanolamine	0.75
12.	Methyl-P-Hydroxybenzoate	0.2
13.	Water	69.45

Procedure:
    Disperse Carbopol in water. Weigh 1-8 and heat to 80-90C
with slow agitation. Add remaining ingredients to Carbopol/water
dispersion, except triethanolamine, and heat to 80-90C. Add the
water phase to the aqueous phase and shear in homomixer. Continue
shearing while cooling to 40C, then add triethanolamine, mixing
well. Cool to 30C, add perfume, de-aerate and package.
Ref: 5189-2-8
SOURCE: Allied Signal Inc.: Suggested Formulations

O/W Lotion

	Wt.%
1. A-C Polyethylene 617	1.0
2. A-C Copolymer 540	1.0
3. Mineral Oil (70 ss)	5.0
4. Dow 556 Fluid	1.0
5. Propylene Glycol Dipelargonate	10.5
6. Hydroxyol	2.0
7. Ethoxyol 24	1.0
8. Arlacel 60	1.3
9. Tween 60	1.8
10.Propyl-P-Hydroxybenzoate	0.1
11.Sorbitol (70%)	5.0
12.Carbopol 941	0.25
13.Germall 115	0.4
14.Methyl-P-Hydroxybenzoate	0.2
15.Triethanolamine	0.75
16.Water	68.8

Procedure:

Disperse Carbopol in water. Weigh 1-10 and heat to 80-90C with slow agitation. Add remaining ingredients to Carbopol/ water dispersion, except triethanolamine, and heat to 80-90C. Add the wax phase to the aqueous phase and shear in homomixer. Continue shearing while cooling to 40C, then add triethanolamine, mixing well. Cool to 30C, add perfume, deaerate and package.
Ref: 5189-2-5

O/W Lotion

	Wt.%
1. A-C Copolymer 540	2.0
2. Mineral Oil, 70 ss	10.0
3. Glycerine	5.0
4. Isopropyl Palmitate	10.0
5. Anhydrous Lanolin	3.0
6. Arlacel 165	5.0
7. Propyl-P-Hydroxybenzoate	0.1
8. TEA	3.0
9. Tween 60	1.0
10.Methyl-P-Hydroxybenzoate	0.2
11.Germall 115	0.2
12.Water	60.5

Procedure:

Weigh 1-7 and heat to 85-95C. If higher solvating temperatures are used, solvation will be much faster. Hold wax blend at 85-95C. Weigh 9-12 and heat to 90-95C.

Place aqueous phase in mixing container, add wax phase, and shear in homomixer or equivalent. Cool to 40C while shearing. Add perfume and package.
Ref: 5011-38-9

SOURCE: Allied Signal Inc.: **Suggested Formulations**

O/W Lotion with α-Hydroxy Acids

Ingredients:	%w/w
Phase A:	
Ceteareth-15 (and) Glycerol Stearate (Tego Care 215)	6.0
Mineral Oil	4.0
Octyl Stearate (Tegosoft OS)	5.0
Caprylic/Capric Triglyceride (Tegosoft CT)	5.0
Stearyl Alcohol (Tego Alkanol 18)	5.0
Phase B:	
Glycerin	3.0
Water	62.0
Phase C:	
α-Hydroxy Acids (20%)*	10.0
Perfume, Preservatives	Q.S.

  *solution contains 10% citric acid and 10% malic acid. pH
  is adjusted to pH 3-3.5 with NaOH.

Procedure:
  Heat A and B to 65C and mix. Homogenize. Cool with stirring
to 40C. Add Phase C and cool to 30C or lower.

O/W Lotion
Cold/Cold Process

Ingredients:	%w/w
Phase A:	
PEG-25 Glyceryl Trioleate (Tagat TO)	2.0
Mineral Oil	5.5
Octyl Stearate (Tegosoft OS)	5.5
Caprylic/Capric Triglycerides (Tegosoft CT)	5.5
Phase B:	
Water	76.3
Glycerin	3.0
Preservatives	Q.S.
Phase C:	
Carbomer 934	0.3
Mineral Oil	1.2
Phase D:	
NaOH (10% solution)	0.7

Procedure:
1. Homogenize A + B, oil particle size <1.0 um.
2. Add C and D, stir intensively for 30 minutes. Temperature
   of the Phases A, B, C, D: are 20-25C.

SOURCE: Goldschmidt Chemical Corp.: Suggested Formulations

Oil-In-Water Body Care Lotion

Ingredients:	%W/W
A Arlatone 985	4.00
Brij 721	2.00
Arlamol HD	10.00
Almond oil	3.00
B G-2330	1.25
Propylene glycol	1.25
Alpantha	1.00
Water	74.00
C *Perfume	
*Preservative	
*q.s. these ingredients.	

Procedure:
   Heat (A) and (B) to 70C separately. Slowly add (B) to (A) while stirring vigorously. Homogenize around 40C. Remove from the homogenizer and allow to cool to 35C while stirring. Add (C) and cool to 30C. Package.

Oil/Water Foundation Lotion

Ingredients:	%W/W
Arlacel 165	6.00
Cetyl alcohol	0.50
Lanolin	0.50
Triethanolamine lauryl sulfate	1.00
Water, deionized	92.00

Procedure:
   Heat ingredients together to 90C. Use continuous vigorous agitation to form the emulsion. Continue agitation until temperature drops below 60 deg. C.

Oil/Water Foundation Lotion

Ingredients:	%W/W
A Synthetic spermaceti, Spermwax	5.00
Arlacel 165	12.00
B Glycerin	5.00
Water, deionized	78.00
*Preservative	
*q.s. these ingredients.	

Procedure:
   Heat (A) to 70C. Heat (B) to 72C. Add (B) to (A) slowly with moderate stirring. Stir to 35C and add water lost due to evaporation.
SOURCE: ICI Surfactants: Suggested Formulations

Oil/Water Emollient Lotion

Ingredients:	%W/W
A Octyl dimethyl PABA	7.00
Benzophenone-3	3.00
Arlamol E	7.00
Stearyl alcohol	2.50
Dimethicone	1.00
Arlasolve 200	3.10
Brij 72	3.90
B Water, deionized	72.10
Carbomer 934	0.20
C Sodium hydroxide (10% W/W aqueous)	0.20
D *Preservative	
*Fragrance	

*q.s. these ingredients.

Procedure:
   Disperse Carbomer in water and heat (B) to 60C. Heat (A) to 65C and add (B) to (A) with propeller agitation. Slowly add (C) and stir until uniform. Cool to 50C and add makeup water.

Oil/Water Emollient Lotion

Ingredients:	%W/W
A Octyl dimethyl PABA	5.00
Arlamol E	7.00
Stearyl alcohol	2.50
Dimethicone	1.00
Arlasolve 200	3.10
Brij 72	3.90
B Water, deionized	77.10
Carbomer 934	0.20
C Sodium hydroxide (10% W/W aqueous)	0.20
D *Preservative	
*Fragrance	

*q.s. these ingredients.

Procedure:
   Disperse Carbomer in water and heat (B) to 60C. Heat (A) to 65C and add (B) to (A) with propeller agitation. Slowly add (C) and stir until uniform. Cool to 50C and add makeup water.

SOURCE: ICI Surfactants: Suggested Formulations

Perfumed Body Lotion

This creamy body lotion rubs in easily and relieves dryness.
Skin is left feeling soft, smooth, and moisturized. Geahlene
enhances the richness and moisturizing properties of this
lotion.

Ingredient/Trade Name:	Weight%
A Deionized Water	67.20
Carbomer/Carbopol 940	0.20
Propylene Glycol	7.00
Methylparaben	0.20
Panthenol/DL-Panthenol	0.10
B Mineral Oil (and) Hydrogenated Butylene/Ethylene/	
Styrene Copolymer (and) Hydrogenated Ethylene/	
Propylene/Styrene Copolymer//Geahlene 750	7.00
Propylene Glycol Dicaprylate/Dicaprate//Myritol PC	7.00
Isostearyl Alcohol/Prisorine 3515	2.00
Propylparaben	0.10
Cetyl Alcohol/Lanette 16	2.00
Glyceryl Stearate (and) PEG-100 Stearate/Arlacel 165	2.50
Dimethicone/Dow Corning 200, 100 cSt	0.50
Potassium Cetyl Phosphate/Amphisol K	1.75
Tocopheryl Acetate/Vitamin E Acetate	0.10
C Triethanolamine, 99%	0.15
D Diazolidinyl Urea/Germall II	0.20
E Soy Lecithin/Sedermasome	1.00
Fragrance	1.00

Procedure:
Disperse Carbomer into rapidly agitated deionized water.
Add remaining part A ingredients. Heat to 75-80C and mix until
uniform and lump-free. Combine part B. Heat to 80C and mix
until all the solids are dissolved. Add part B to part A.
Mix for 30 minutes with good agitation. Add part C. Mix until
completely smooth and homogeneous. Cool to 50C. Add part D.
Cool to 40C. Add part E. Continue mixing and cooling to 30C.

SOURCE: Penreco: Suggested Formulation

Silk Protein Skin Lotion

	Weight,%
1. Mineral Oil	3.00
2. Mackester SP (Glycol Stearate Modified)	2.00
3. Emulsifying Wax N.F.	3.00
4. Glyceryl Stearate & PEG-100 Stearate	2.00
5. Polysorbate 80	0.66
6. Sorbitan Palmitate	0.60
7. Glycerin	2.00
8. Acetamide MEA 100%	1.00
9. Mackpro NSP (Oleyl/Palmityl/Palmitoleamidopropyl/ Silkhydroxypropyl Dimonium Chloride)	2.50
10.Mackstat DM (DMDM Hydantoin)	qs
11.Fragrance	qs
12.Deionized Water	qs

Procedure:
1. Melt 1,2,3,4,5,6,7,8, in a separate container to 75 degrees C.
2. In the mixing tank heat the water #12 to 78 degrees C. and
   add #9.
3. Start mixing and add the hot mixture of 1 thru 8 slowly with
   good agitation and mix well for 20 minutes.
4. Then start slow cooling with good mixing without aeration.
5. At 45 degrees C. add #10 and #11 and mix in.
6. Check pH and adjust if needed to 4.8-5.8.
7. Mix until cool.

Pearlescent Bath Lotion

	Weight,%
Sodium Lauryl Sulfate	40.0
Mackanate EL (Disodium Laureth Sulfosuccinate)	30.0
Mackam 35 HP (Cocamidopropyl Betaine)	5.0
Mackester SP (Glycol Stearate Modified)	1.5
Sodium Chloride	1.0
Mackstat DM (DMDM Hydantoin)	0.5
Water, Dye, Fragrance qs to	100.0

Procedure:
1. Add first four components to water and heat to 70C.
2. Blend until EGMS is completely dispersed.
3. Add sodium chloride and cool to 45C.
4. Add preservative, fragrance and dye.
5. Cool to room temperature and fill.

SOURCE: McIntyre Group Ltd.: Personal Care Formulary

Skin Care Body Lotion

This product imparts good skin feel and has a high quality appearance, though the production is cost effective. There are also barrier and moisturizing properties which act as a highly effective hydration system for all skin types. A formula of this type also help reduce transepidermal water loss.

Oil Phase:

Cetylstearyl Alcohol 1618 (P&G)	3.50%
NF White Beeswax (Koster Keunen)	1.50%
Light Mineral Oil (Witco)	4.00%
Isostearic Acid (Unichema)	1.00%
Amerchol L101 (Amerchol)	2.00%
Propyl Paraben (Sutton)	0.20%

Water Phase:

Water (Distilled)	76.85%
Glycerine (Unichema)	6.00%
Polyethylene Glycol 1450 (Union Carbide)	3.50%
Carboxymethyl Cellulose (CMC, Hercules)	0.30%
Carbopol 940 (BF Goodrich)	0.25%
Sodium Borate (Borax)	0.50%
Triethanolamine (Dow)	0.20%
Methyl Paraben (Sutton)	0.20%

Procedure:
Add to the water phase under agitation, in order; CMC until dissolved, then propyl paraben while mixing. Add carbopol, mix till homogeneous making sure there are no agglomerations. Then the remainder of the water phase components can be added, mix and heat to 75C. Add all the oil phase components, heat till 75C. and mix. Add slowly the oil phase to the water phase under agitation maintaining a temperature of 75C. When the oil phase is added, cool and pour into container.

Adaption of formula and its influence on the product:
By reducing the concentrations of mineral oil by 2.0%, polyethylene glycol 1000 by 0.5% and the addition of 2.5% Escalol 507 (Van Dyk) the cream will take on an SPF of 6-8. The product has the same appearance, skin feel and stability. Fragrance can also be added without affecting texture.

SOURCE: Koster Keunen, Inc.: Suggested Formulation

Skin Moisturizing Lotion

Phase A:	Percent
Polypentaerythrityl Tetralaurate (Miranol Ester PO-LM4)	4.00
Glyceryl Stearate (and) PEG 100 Stearate (Arlacel 165)	5.00
Carnation Mineral Oil	3.50
Isopropyl Myristate	2.00
Propylene Glycol Dipelargonate (Emerest 2388)	1.00
Beeswax	2.00
Stearic Acid	1.00
Stearyl Alcohol	0.50
Cyclomethicone (Dow Corning Fluid 344)	0.50

Phase B:	
Water	67.70
Carbomer 934, 3% solution (Carbopol 934)	7.50
Propylene Glycol	3.50

Phase C:	
Triethanolamine	0.80

Phase D:	
Germaben II-E	1.00

Procedure:
Heat Phase A & Phase B separately to 75C and add B to A with agitation. Then add Phase C. Cool to 40C and add Phase D.

Dry Skin Lotion

Phase A:	
Cetyl Alcohol	2.00%
Estol EHP 1543	2.00
Trivent NP-13	4.00
Carnation Mineral Oil	1.00
DC Silicone	1.00
Brij 58	1.00
Brij 20	1.00

Phase B:	
Deionized water	81.50
Carbopol 940	0.50
Dermacryl-79	1.00
Pricerine 9083	3.00
Triethanolamine (99%)	1.00

Phase C:	
Germaben II E	1.00

Phase D:	
Fragrance	q.s.

SOURCE: Witco Corp.: Petroleum Specialties Group: Suggested Formulations

Skin Softening Lotion

	%w/w
A. Glyceryl Stearate (and) PEG-100 Stearate (Lexemul 561)	3.00
Glyceryl Dilaurate (Lexemul GDL)	5.00
Cetearyl Alcohol (and) Ceteareth-20 (Lexemul CS-20)	3.00
Propylene Glycol Dicaprylate/Dicaprate (Lexol PG-865)	4.00
Mineral Oil	2.50
Propylparaben (Lexgard P)	0.10
B. Glycerin USP	3.00
Deionized Water	76.13
Methylparaben (Lexgard M)	0.25
C. Quaternium-76 Hydrolyzed Collagen (Lexein QX-3000)	3.00
D. 2-Bromo-2-Nitropropane-1,3-Diol (Lexgard Bronopol)	0.02

Procedure:
1. Heat phase "A" and phase "B" to 75C.
2. Add phase "A" to phase "B" with good agitation at 75C.
3. Continue to mix with good agitation and cool to 50C.
4. Add phase "C" to "AB" and mix till homogeneous.
5. When phase "ABC" is below 50C add phase "D".
6. Continues agitation and cool to 30C, then fill.

Formula 297-86

Skin Lotion

This moisturizing lotion applies effortlessly. The Lexorez 100 adds an elegant skin feel which also may act as a protectant aid.

	%w/w
A. Octyl Stearate (Lexol EHS)	7.50
Glyceryl Stearate (and) PEG-100 Stearate (Lexemul 561)	3.50
Glycerin/Diethylene Glycol/Adipate Crosspolymer (Lexorez 100)	2.00
Propylparaben (Lexgard P)	0.05
B. Carbomer (Carbopol 941)	0.30
Triethanolamine NF (99%)	0.20
Methylparaben (Lexgard M)	0.15
Deionized Water	86.30

Procedure:
1. Heat water to 70C and combine ingredients of Phase "B". Mix until all of the Carbomer is in solution.
2. In a separate vessel, melt components of phase "A" together. Heat to 70C.
3. Slowly add phase "A" to phase "B" with good agitation. Mix well until homogeneous.
4. Begin to cool. Continue agitation until mixture is at ambient temperature then fill.
5. Adjust for water loss.

Formula SK-114

SOURCE: Inolex Chemical Co.: Suggested Formulations

Soft & Silky Lotion

Part (A):	%
Rosswax 63-0412	1.6
Rosswax 1641	1.2
Rosswax 63-0212	1.0
GMS-SE	2.1
Ross Lotion Oil 2745	9.4

Part (B):	
Water	78.0
Propylene Glycol	4.7
Germaben II	1.0
Triethanolamine	1.0

Part (C):	
Fragrance	q.s.

Procedure:
    Heat Part (A) and Part (B) to 170F in separate steam jack-
eted kettles under agitation. When fully heated, add Part (A)
to Part (B) under agitation. Cool to 130F, Fragrance and
package.

Low Cost Low Solids Lotion

	Parts by Weight
Water	500.0gr.
Carbomer 934	4.0gr.
Rosswax 573	4.0gr.
GMS SE	4.0gr.
Jojoba Oil	3.0gr.
Dow Corning Silicone 344	6.0gr.
Triethanolamine	4.0gr.
Perfume	q.s.
Preservative Germaben II	q.s.

Procedure:
    Heat the water under agitation and slowly add the Carbomer
934. When fully mixed add the 573, GMS, Jojoba Oil and Silicone
have been blended in a separate kettle maintaining a temperature
of 140F. As soon as all the ingredients have been mixed well add
the preservatives, the Perfume, and add the TEA, under high agitation, cool to 120F and package.

SOURCE: Frank B. Ross Co., Inc.: Suggested Formulations

Strawberry Hand Lotion

	Parts by Weight
Water	568.0gr
Carbomer 934	2.0gr
GMS-SE	4.0gr
Apricot Oil Lipoval P	16.0gr
Rosswax 573	4.0gr
Coconut Oil #76	16.0gr
Ross Jojoba Oil	4.0gr
TEA	4.0gr
Germaben II	6.0gr
Fragrance DO-60	q.s.

Procedure:
   Heat the water to 60C under agitation and slowly add the
Carbomer 934. When the water is fully mixed, add the 573, GMS,
Apricot Oil, Coconut Oil and Jojoba Oil that have been heated
to 65C in a separate kettle. As soon as the oil phase has been
mixed well, add the Germaben II, the fragrance, and then the TEA
under high agitation. Cool to 55C for filling.

Apricot Hand Lotion

	Parts by Weight
Water	568.0gr
Carbomer 934	2.0gr
GMS-SE	4.0gr
Apricot Oil-Lipoval P	16.0gr
Rosswax 573	4.0gr
Coconut Oil #76	16.0gr
Ross Jojoba Oil	4.0gr
TEA	4.0gr
Germaben II	6.0gr
Fragrance GK-17	q.s.

Procedure:
   Heat the water to 60C under agitation and slowly add the
Carbomer 934. When the water is fully mixed, add the 573, GMS,
Apricot Oil, Coconut Oil and Jojoba Oil that have been heated
to 65C in a separate kettle. As soon as the Oil Phase has been
mixed well, add to the Water Phase with agitation. When fully
mixed, add the Germaben II and then the TEA under high agitation, then fragrance. Cool to 55C for filling.

SOURCE: Frank B. Ross Co., Inc.: Suggested Formulations

Super Moisturizing Lotion

This silky-feeling emulsion is stabilized and thickened using a synergistic combination of Veegum Magnesium Aluminum Silicate and Rhodigel Xanthan Gum. It also contains the sodium salt of pyrrolidone carboxylic acid as a natural moisturizing factor along with the well known humectant, glycerine. This lotion spreads easily and is quickly absorbed leaving the skin moist and supple.

Ingredient:	% by Wt.
A Veegum	1.0
Rhodigel	0.5
Deionized Water	74.5
B Sodium PCA (2)	3.0
Glycerine	5.0
C Hydrogenated Polyisobutene (3)	4.0
Mineral Oil (and) Lanolin Alcohol (4)	3.0
Cetyl Alcohol	2.0
Isopropyl Myristate	2.0
Sorbitan Palmitate	1.2
Polysorbate 40	3.8
D Citric Acid to pH 5.5	q.s.
Preservative, Dye, Fragrance	q.s.

(2) Ajidew N-50
(3) Polysynlane
(4) Amerchol L-101

Preparation:
Dry blend Veegum and Rhodigel and add to the water, mixing with maximum available shear until smooth and uniform. Add B ingredients and mix until dissolved. Mix C ingredients and heat to 50C until a uniform clear mixture is obtained. Add C to (A+B) with high speed mixing. Avoid incorporating air. Cool with continuous stirring to 30C and add D.

Consistency: Medium Viscosity Lotion (Viscosity: 1900-2400 cps)

Suggested Packaging: Plastic squeeze bottle or pump.

Comments:
This prototype formula is designed to serve as a guide for the development of new products or improvement of exixting ones.

SOURCE: R.T. Vanderbilt Co., Inc.: Formulation No. 437

TRF Facial Lotion

Phase A:	Percent
Cetearyl Alcohol (and) Ceteareth-20	0.80
Sorbitan Stearate (Arlacel 60)	0.50
Stearic Acid, triple pressed	0.50
Glyceryl Stearate (Emerest 2400)	1.00
Cetearyl Alcohol	1.40
Cetyl Acetate (and) Acetylated Lanolin Alcohol	0.50
C12-15 Alcohols Benzoate (Finsolv TN)	0.40
PPG-15 Stearyl Ether (Arlamol E)	0.40
Dimethicone	0.20
Carnation Mineral Oil	3.00

Phase B:	
Carbomer 941 (2% Disp.) (Carbopol 941)	7.50
Magnesium Aluminum Silicate (Veegum)	0.30
Potassium Hydroxide (pellets)	0.15
Tetrasodium EDTA	0.10
Glycerin	3.00
Water	78.55

Phase C:	
Tissue Respiratory Factors (Biodynes TRF)	0.70
Germaben II	0.50
Fragrance	0.50

Procedure:
    Melt Phase A to 75C. Heat water to 70C and disperse the
Veegum. Add the Carbopol slurry and the glycerin and EDTA.
Add to Phase A and add the potassium hydroxide. Cool to 50C
and add Phase C.

Facial Moisturizing Lotion (W/O)

Ingredients:	%Wt.
Carnation Mineral Oil	40.00
Aldo MCT	2.00
Synthetic Beeswax	10.00
Lanolin	1.00
Polyaldo 10-6-0	2.50
Deionized Water	43.90
Sodium Borate	0.60
Glydant Plus	q.s.

**SOURCE: Witco Corp.: Petroleum Specialties Group: Suggested
        Formulations**

Ultra Moisturizing Lotion

This creamy, oil-in-water emulsion is thickened and stabilized with a synergistic combination of Veegum Ultra and Carbomer. The well-known humectant glycerin performs the moisturizing function. Veegum Ultra also enhances the whiteness and brightness of the emulsion and helps adjust the pH to approximate that of the skin. The lotion spreads easily and is rapidly absorbed, leaving the skin moist and supple.

Ingredient:	% by Wt.
A Veegum Ultra (Magnesium Aluminum Silicate)	0.15
Carbomer 980 (2)	0.15
Deionized Water	73.70
B Glycerin	5.00
C Mineral Oil (and) Lanolin Alcohol (3)	4.00
Cetyl Alcohol	2.00
Isopropyl Palmitate	2.00
Hydrogenated Polyisobutene (4)	5.00
Isopropyl Myristate	3.00
Sorbitan Palmitate	1.20
Polysorbate 40	3.80
D Preservative, Fragrance	q.s.
Sodium Hydroxide, (10% Solution) to pH 6.0	q.s.

(2) Carbopol 980
(3) Amerchol L-101
(4) Polysynlane

Procedure:
Dry blend Veegum Ultra and Carbomer and add them slowly to the water while stirring with a propeller mixer at 700 rpm. Increase the mixer speed to 1500-1700 rpm and continue mixing for 30 minutes. Add B and mix 5 minutes. Mix C ingredients and heat to 50C. Heat (A and B) mixture to 50C. Add C to (A and B) and mix at 50C and 1500-1700 rpm for 10 minutes. Slow the mixer to 1000 rpm while cooling to 30C. Add D and mix until uniform.

Product Characteristics:
Viscosity: 2200-2800 cps
pH: 6.0+-0.2

Comments:
This prototype formula is designed to serve as a guide for the development of new products or the improvement of existing ones.

SOURCE: R.T. Vanderbilt Co., Inc.: Formula No. 450

Un-Buffered Alpha Hydroxy Acid Lotion

The resulting product is a thin lotion that would be compatible for use in a glass or rigid plastic bottle. The formulation contains Lactic Acid, as the alpha hydroxy acid (AHA), at a use level of approximately 4.5%.

	%w/w
A. Deionized Water	80.50
Methylparaben (Lexgard M)	0.20
2,4-Dichlorobenzyl Alcohol (Myacide SP)	0.20
B. Glyceryl Stearate (and) Stearamidoethyl Diethylamine (Lexemul AR)	1.50
Glyceryl Stearate (and) PEG-100 Stearate (Lexemul 561)	1.50
Glyceryl Stearate (Lexemul 515)	4.00
Caprylic/Capric Triglyceride (Lexol GT-865)	3.00
Stearamidopropyl Dimethylamine (Lexamine S-13)	2.00
Mineral Oil NF (Light)	2.00
Propylparaben (Lexgard P)	0.10
C. Lactic Acid (88%)	5.00

Procedure:
1. Combine section "A" heating to 75-80C. Use a propeller mixer for agitation.
2. Combine section "B" heating to 80-85C. Agitate slowly with a propeller mixer.
3. When sections "A" & "B" are homogeneous and at the designated temperatures slowly add "B" to "A", then begin cooling.
4. Reduce mixing speed during cooling to prevent vortexing.
5. At 55-60C add section "C" to sections "AB".
6. At 45C adjust for water loss.
7. Mix to room temperature.

Physical Properties:
Viscosity: 700 cPs @ 24C (Brookfield RVT, TA @ 10 rpm).
           24 hour sample
pH: 2.3 @ 24C

SOURCE: Inolex Chemical Co.: Formulation 398-30-3

W/O Lotion

		%
A.	Miglyol 840 Gel B	4.0
	Miglyol 840	7.5
	Miglyol 812	5.0
	Arlacel 481	3.0
	Arlacel 989	5.0
	Isopropyl myristate	5.0
	Vaseline	2.0
B.	Glycerol	5.0
	Magnesium sulphate	0.7
	Carbopol 934	0.2
	Water	to 100.0
C.	Perfume	qs
	Preservative	qs

Preparation:
   The constituents of A. are mixed and heated to 75-80C. Those of B. are mixed with a high speed stirrer, heated to the same temperature and emulsified, in small amounts, in A. with the high speed stirrer. C. is added at approx. 30C.

Body Lotion

		Wt%
A.	Miglyol 840	7.0
	Imwitor 960 flakes	4.0
	Marlophor T10 Na salt	2.0
	Cetyl alcohol	0.3
B.	Hostacerin gel 1%*	12.5
	Karion F	5.0
	Water	to 100.0
C.	Perfume	qs
	Preservative	qs

Preparation:
   The constituents of A. are mixed and heated to 75-80C. Those of B. are brought to the same temperature and gradually stirred into A. C. is incorporated at approx. 30C.

   *Preparation of the Hostacerin gel: Hostacerin PN73   1.0%
                                          Water   to 100.0%

   The Hostacerin is mixed with water until homogeneous and the mixture stirred until the gel is clear.

SOURCE: Huls America Inc.: Formulations for Cosmetics: Formulas

W/O Lotion with Eldew

Ingredients/Trade Name:	% by Weight
Part A:	
Di-(Cholesteryl, behenyl, octyldodecyl)	
N-Lauroyl-L-glutamic acid ester/Eldew CL-301	2.00
Cetearyl Octanoate	10.00
C12-C15 Alkyl Benzoate	5.00
Phenoxyethanol	0.60
Tocopheryl Acetate	0.05
Part B:	
Polyglyceryl-4 Isostearate (and) Cetyl Dimethicone	
Copolyol (and) Hexyl Laurate	5.00
Cetyl Dimethicone	3.00
Part C:	
Deionized Water	65.55
Sodium Chloride	0.80
Glycerin (99.5%)	5.00
Partially Deacetylated Chitin (1.0%)/Marine Dew	2.00
Part D:	
Methylparaben	0.20
Butylene Glycol	0.80

Procedure:
    Pre-melt part A at 50 degrees Centigrade. Add part B to part A. Pre-melt part D by heating to 50 degrees C. Add to part C. Slowly add Part C and D mixture to Parts A and B with high shear mixing.
        Appearance: White, smooth, shiny lotion   pH: 6.0-6.5
        Viscosity: 20,000-20,000 (RVT #6 @ 10rpm @ 25 degrees C)

Mild Lotion with Amihope

Ingredients:	% Weight
Phase A:	
Mineral Oil	5.00
Amiter LGOD	2.00
Propylene Glycol Stearate	0.50
PEG-5 Hydrogenated Castor Oil	1.50
PEG-5 Glyceryl Stearate	2.50
Butylparaben	0.10
Amihope-LL	3.00
Phase B:	
Acylglutamate HS-11	0.30
Carbomer 941	0.20
Sodium Hydroxide	0.08
Butylene Glycol	5.00
Methylparaben	0.20
Water	79.62

Procedure:
    Dissolve Carbomer 941 and Sodium Hydroxide in water first. Add all other ingredients of Phase B to the solution and dissolve at 75 to 80 degrees centigrade.
    Dissolve Phase A ingredients at 80 degrees centigrade. When both phases are at 80 degrees centigrade add A to B with agitation.
    Cool down to room temperature while continuing mixing.
    This lotion has a smooth feeling and good spreadability.
SOURCE: Ajinomoto USA, Inc.: Suggested Formulation

Section VIII
Shampoos

Acid Balanced Conditioning Shampoo

	Weight,%
TEA Lauryl Sulfate (40%)	35.0
Mackam 35HP (Cocamidopropyl Betaine)	10.0
Mackalane 426 (Isostearamidopropyl Morpholine Lactate)	6.0
Mackstat DM (DMDM Hydantoin)	qs
Water, Dye, Fragrance qs to	100.0

Procedure:
1. Add components to water and heat to 40 degrees C.
2. Blend until clear.
3. Adjust pH to 4.0 with citric acid.
4. Cool and fill.

All Purpose Shampoo

	Weight,%
Mackadet SBC-8 (Mild Blend)	20.0
Sodium Chloride	qs
Mackstat DM (DMDM Hydantoin)	qs
Water, Dye, Fragrance qs to	100.0

Procedure:
1. Add Mackadet SBC-8 to water and blend until clear.
2. Add Mackstat DM and adjust viscosity to 2000-3000 cps with sodium chloride.
3. Add dye, fragrance, and blend until clear.

Aloe Vera Gel Shampoo

	Weight,%
Aloe Vera Gel Liquid (1:1)	50.0
Water	14.5
Mackernium 007 (Polyquaternium 7)	3.0
Mackadet SBC-8 (Mild Blend)	32.0
Mackstat DM (DMDM Hydantoin)	qs
Fragrance, Dye qs to	100.0

Procedure:
1. Disperse Mackernium 007 in water and Aloe Vera Liquid.
2. Add Mackadet SBC-8 and heat to 45 degrees C.
3. Blend until homogeneous.
4. Adjust viscosity with sodium chloride.
5. Add remaining components and blend until clear.
6. Cool and fill.

SOURCE: McIntyre Group Ltd.: **Personal Care Formulary**

Aloe Vera Shampoo

Ingredients:	Percent
A. D.I. Water	64.84
Aloe Veragel 200 Powder	0.1
Sodium Chloride	1.3
Hydrolyzed animal protein	1.0
B. Sodium lauryl sulfate	26.0
Citric acid	0.40
Fragrance	0.15
D.M.D.M. Hydantoin	0.20
Germall 115 (preservative)	0.10
C. Richamide liquid	6.0

Procedure:
   Mix Phase "A" together. Mix Phase "B" together and add to
Phase "A". Blend together. Add Phase "C" and mix together.

Aloe Vera Premium-Type Shampoo with Protein

Ingredients:	Percent
Aloe Veragel 1:1	28.0
Cycloryl WAT	60.0
Cycloteric BET-C30	5.0
Peptein 2000	1.0
Cyclomide DC212S	4.0
NaCl	1.0
Citric Acid	q.s.
Perfume, Preservative, Color	q.s.

Procedure:
   Warm Aloe Veragel and WAT to 40C, and blend in ingredients
as listed. Adjust viscosity with NaCl and adjust pH with Citric
Acid.

SOURCE: Dr. Madis Laboratories Inc.: Suggested Formulations

Anti-Dandruff Lotion Shampoo

	Weight,%
Part A:	
Veegum	1.0
Methocel FYM	0.8
Water qs to	100.0
Part B:	
Sodium Olefin Sulfonate (40%)	35.0
Mackamide LLM (Lauramide DEA)	4.0
Mackamide S (Soyamide DEA)	1.0
Mackpro NLP (Quaternium-79 Hydrolyzed Animal Protein)	
(Natural Lipid Protein)	2.0
Part C:	
Zinc Omadine (48%)	4.0

Procedure:
1. Thoroughly disperse Veegum in water at 70 degrees C.
2. Then slowly add Methocel FYM and blend until homogeneous.
3. Add Part B to Part A and adjust pH to 6.5 with citric acid.
4. Add Zinc Omadine and blend until homogeneous.

Anti-Dandruff Shampoo Cream Type

	Weight,%
Sodium Lauryl Sulfate (30%)	61.8
Mackam 35HP (Cocamidopropyl Betaine)	10.0
Sodium Chloride	7.0
Triple Pressed Stearic Acid	5.0
Mackamide LLM (Lauramide DEA)	4.0
Propylene Glycol	4.0
Zinc Pyrithione (48%)	4.0
Mackamide PK (Palmkernelamide DEA)	2.0
Caustic Soda (50%)	1.6
Mackstat DM (DMDM Hydantoin)	qs
Water, Dye, Fragrance qs to	100.0

Procedure:
1. Heat stearic acid, Mackamide LLM, Mackamide PKM and propylene glycol to 70 degrees C.
2. Heat SLS, Mackam 35HP, Sodium Chloride, Caustic Soda and water to 70 degrees C.
3. Add oil to water and cool to 55 degrees C.
4. Slowly add Zinc Pyrithione.
5. Cool to 45 degrees C. and add remaining components.
6. Fill at 40 degrees C.

SOURCE: McIntyre Group Ltd.: Personal Care Formulary

Anti-Dandruff Shampoo

Anti-dandruff shampoo: with conditioning action coupled with natural suspension resulting in cosmetically elegant, highly efficacous anti-dandruff shampoo.

Ingredients:	%W/W	Function
1. Sodium Laureth-3 Sulfate (Standapol ES-3)	57.69	Lather, Cleaning
2. Pationic ISL/85*	3.00	Conditioning, Deposition
3. Rita EGDS (Glycol Distearate)	3.00	Opacity, Stability
4. Propylene Glycol	2.00	Stability
5. Zinc Omadine	2.00	Active
6. Ritavena-5 (Hydrolyzed Oat Flour)	4.50	Suspension,Clean Feel
7. Fragrance	0.25	Odor
8. Patlac LA (44%) (Lactic Acid)	0.40	pH Adjustment
9. Distilled/Deionized Water	27.16	----

* Sodium Isostearoyl Lactylate @ 85% in Propylene Glycol

Compounding Procedure:
Pre-mix items 1 to 4 at 165F. Add item 5 with high sheer mixing. Hydrate item 6 in 210F water. Cool to 165F and mix. Cool to 120F. Add perfume and adjust pH and viscosity.

Ref. No. 113-141C

Cleaning Shampoo Type

Clear viscous shampoo designed for oily hair types.

Ingredients:	%W/W	Function
1. Ammonium Lauryl Sulfate (Stepanol AM)	30.00	Cleaning, Lather
2. Ammonium Laureth Sulfate (Standapol EA-2)	25.00	Cleaning, Lather
3. Oleamidopropyl Betaine (Mirataine BET-0-30)	6.00	Mildness, Viscosity
4. Ritaloe 40X (Aloe Vera Gel)	0.50	No Static Fly-Away
5. Pationic ISL (Na Isostearoyl Lactylate)	2.50	Mildness,Conditioning
6. Kathon CG	0.30	Preservation
7. Fragrance - Squeeky Clean #165-047	0.03	Odor
8. Citric Acid (50% Soln.)	q.s.	pH Adjustment
9. Distilled/Deionized Wter	35.67	----

Compounding Procedure:
To the water add ingredients 1 to 5. Heat to 165F. Mix to uniformity. Cool to 120F and add preservative and perfume. Adjust pH if necessary to 5.8.
Ref. No. 118-69

SOURCE: R.I.T.A. Corp.: Suggested Formulations

Anti-Dandruff Shampoos

B 40/39:
```
Zetesol 856 T                                         25.0%
Amphotensid B 4                                        4.0%
Bio-sulphur CLR                                        1.0%
Water, perfume, sodium chloride,
 preservative                        q.s. to make 100.0%
```

B 41/24:
```
Zetesol 856T                                          20.0%
Perlglanzmittel GM 4175                                4.0%
Condipon                                               4.0%
Anti-dandruff-Usnat AO                                 2.0%
Perfume                                                0.3%
Water, sodium chloride, preservative  q.s. to make 100.0%
```

B 41/26:
```
Zetesol 2056                                          28.0%
Amphotensid B 4                                        8.0%
Perlglanzmittel GM 4175                                4.0%
Anti-dandruff-Usnat AO                                 2.0%
Perfume                                                0.5%
Water, preservative                   q.s. to make 100.0%
```

Shampoos for Greasy Hair

B 40/56:
```
Zetesol 856 T                                         30.0%
Mulsifan RT 275                                        2.0%
Purton SFD                                             1.0%
Water, perfume, preservative          q.s. to make 100.0%
```

B 41/27:
```
Zetesol 2056                                          28.0%
Amphotensid B 4                                        8.0%
Perlglanzmittel GM 4175                                4.0%
Mulsifan RT 275                                        1.0%
Perfume                                                0.5%
Water, preservative                   q.s. to make 100.0%
```

SOURCE: Zschimmer & Schwarz GmbH: Suggested Formulations

Antidandruff Shampoo

Materials:	% by Weight
1. Deionized Water	48.55
2. Bentone MA Rheological Additive	1.00
3. Methocel E 4M Premium	0.75
4. Zinc Omadine 48%	4.20
5. Super Amide GR	4.00
6. Standapol T	38.00
7. Triethanolamine	3.00
8. Color and Fragrance	0.50

Manufacturing Directions:
1. Heat Deionized Water to 70C, and with rapid stirring, add Bentone MA additive slowly and mix it for 20 minutes or until homogeneous. Maintain the temperature through step 4.
2. Add Methocel and mix it for 15 minutes with continuous stirring.
3. Add Zinc Omadine and mix for 5 minutes.
4. Reduce the stirrer speed and at low speed add Super Amide GR. Mix for 10 minutes.
5. Turn off the heat and while cooling, add Standapol T. Mix for 15 minutes at low speed.
6. Add Triethanolamine and color solution, stir slowly until mixed. Adjust the water loss and cool it to room temperature. Add fragrance and mix it.

SOURCE: Rheox, Inc.: Formulation TS-261

Protein Shampoo

Deionized Water	42.85
Busan 1504	0.10
Panthenol	0.25
Disodium EDTA	0.05
Citric Acid	0.20
Aloe Vera 200X	0.05
Sodium Laureth-2 Sulfate	42.00
Cocamide DEA	4.00
Cocamidopropyl Betaine	5.50
Hydrolyzed Collagen Protein-55%	5.00

Procedure:
  Add Busan 1504 preservative into water, and mix until dissolved. Follow with remaining ingredients in order shown, mixing well between each addition.

SOURCE: Buckman Laboratories Inc.: Suggested Formulation

Balsam & Protein Shampoo

Ingredients:	%W/W
A CP 1052 (shampoo base)	31.50
Water	67.20
Hydrolyzed Collagen (Hormel-Peptein 2000x1)	0.25
Balsam	0.25
B NaCl	0.80

Procedure:
Add water to shampoo base with moderate agitation. Add Protein and balsam with moderate agitation. Slowly add salt under faster stirring to build viscosity.
Formula CP 1060

Balsam & Protein Shampoo

Ingredients:	%W/W
A G-9600 (shampoo base)	31.50
Water	68.17
Hydrolyzed Collagen (Hormel-Peptein 2000x1)	0.30
Balsam	0.03
B *Preservative	

*q.s. these ingredients.

Procedure:
Add water to shampoo base with moderate agitation. Add Protein and balsam with moderate agitation.
Formula CP 1067

SOURCE: ICI Surfactants: Suggested Formulations

Cleaning Shampoo

Viscous, high foaming shampoo which maximizes cleaning by removing styling and residues from hair.

Ingredients:	%W/W	Function
1. Sodium (C14-C16) Olefin Sulfonate (Bioterge AS 40)	20.00	Cleaning, Lather
2. Sodium Lauryl Sulfate @ 25%	20.00	Cleaning, Lather
3. Ritamide C (Cocamide DEA)	3.00	Lather Density
4. Cocamidopropyl Betaine (Mackam 35 HP)	3.00	Mildness
5. Ritapeg 150 DS (PEG-150 Distearate)	1.00	Viscosity
6. Pationic ISL (Na Isostearyl Lactylate)	2.50	Mildness, Body
7. Kathon CG	0.03	Preservation
8. Fragrance--Nature Harvest #165-050	0.30	Odor
9. Citric Acid (50% Soln.)	q.s.	pH Adjustment
10.Sodium Chloride (25% Soln.)	q.s.	Viscosity Adjustment
11.Distilled Water	50.17	---------

Compounding Procedure:
To the water add ingredients 1 to 6 in order at 165F. Cool to 120F and add fragrance and preservative. Cool to 95F and adjust pH to 5.5 and viscosity as desired.

Ref. No. 118-18

Cleaning Shampoo

Neutralizing shampoo to thoroughly clean hair. Can be used after relaxer or hair straightener treatments.

Ingredients:	%W/W	Function
1. Alpha Olefin Sulfonate (Bioterge AS-40)	37.50	Cleaning, No Residue
2. Pationic ISL (Na Isostearoyl Lactylate)	3.00	Rinse Out, Combing
3. Ritamide C (Cocamide DEA)	3.00	Bubble Toughness
4. Fragrance-Nature Harvest 165-050	0.30	Odor
5. Citric Acid (50% Soln.)	q.s.	pH Adjustment
6. Distilled/Deionized Water	56.20	- --------------

Compounding Procedure:
To the water at 165F add 1 and 2. Cool to 120F. Pre-mix items 3 and 4 and add at 120F. Thoroughly mix and adjust pH to 5.5 with citric acid.

Ref. No. 118-51

SOURCE: R.I.T.A. Corp.: Suggested Formulations

Conditioning Shampoo

Ingredients:	%W/W
A Sodium laureth sulfate, 26% solution	15.50
Ammonium lauryl sulfate, 28% solution	15.00
Cocoamphocarboxypropionate, 38% sol.	12.50
Lauramide DEA	3.00
Tween 20	2.00
Water	50.60
B Forestall	1.40
C *Citric acid	

*q.s. these ingredients.

Procedure:
Mix (A) with gentle stirring and heat until homogeneous.
Add (B). Adjust pH to 5.0 to 5.5 with (C).

Conditioning Shampoo

Ingredients:	%W/W
A CP 1052 (shampoo base)	31.50
Water	66.84
Polyquaternium-11 (Gafquat 755)	1.32
B Hydroxypropyl methylcellulose (Methocel E4M)	0.34

*q.s. these ingredients.

Procedure:
Add water to shampoo base with moderate agitation. Add
polyquaternium with medium agitation. Heat (A) to 80C and
slowly add hydroxypropyl methylcellulose with faster stirring.
Stir with medium agitation until cooled to room temperature.
Formula PC 1057

Conditioning Shampoo

Ingredients:	%W/W
A CP 1052 (Shampoo base)	31.50
Water	67.40
Polyquaternium-11 (Gafquat 755)	1.00
B NaCl	0.10

Procedure:
Add water to shampoo base with moderate agitation. Add
Glycerin and Arlamol E with medium agitation. Heat (A) to 80C
and add glycol stearate with fast stirring. Stir with medium
agitation until cooled to room temperature and stir NaCl.
Formula CP 1062

SOURCE: ICI Surfactants: Suggested Formulations

Conditioning Shampoo

Ingredients:	%W/W
A Polyox WSR	0.02
Water	71.38
G-9600	25.00
Vitamin E	1.00
Protein	1.00
NaCl	0.60
Panthenol	1.00
B *Color	
*Preservative	
*Fragrance	

*q.s. these ingredients.

Procedure:
Add the Polyox WSR to the water until completely dissolved with heat if necessary. Add the remaining ingredients.

Conditioning Shampoo

Ingredients:	%W/W
A G-9600 (Shampoo base)	31.50
Water	68.01
Polyquaternium-11 (Gafquat 755)	0.30
B NaCl	0.19
*Preservative	

*q.s. these ingredients.

Procedure:
Add water to shampoo base with moderate agitation. Add poly-quaternium with medium agitation. Slowly add salt and stir with medium agitation.
Formula CP 1064

Professional Salon Type Shampoo

Ingredients:	%W/W
A CP 1052 (shampoo base)	31.50
Water	65.40
Roche-DL-Panthenol Cosmetic Grade	0.50
Calgon Merquat 550	2.50
B NaCl	0.10

Procedure:
Add water to shampoo base with moderate agitation. Add Panthenol and polyquaternium-7 with medium agitation. Slowly add salt under faster stirring to build viscosity.
Formula PC 1058

SOURCE: ICI Surfactants: Suggested Formulations

Conditioning Shampoo

Ingredients:	%
Part A:	
Deionized Water	49.05
SLES (3 mol)	40.00
Disodium EDTA	0.30
Part B:	
Incromide LR (Lauramide DEA)	5.50
Incromide CAC (Cocamide DEA Cocoyl Sarcosine)	1.00
Incromine Oxide BA (Babassuamidopropylamine Oxide)	2.00
Incrodet TD-7C (Trideceth-7 Carboxylic Acid)	0.15
Part C:	
Hydrotriticum 2000 (Hydrolyzed Whole Wheat Protein)	0.50
Crodacel QS (Stearyldimonium Hydroxypropyl Oxyethyl Cellulose)	0.50
Propylene Glycol (and) Diazolidinyl Urea (and) Methyl Paraben (and) Propyl Paraben*	1.00

Procedure:
   Combine ingredients of Part A with mixing and heat to 65-70C.
Add ingredients of Part B individually with mixing, and cool
batch to 45C. Add the ingredients of Part C individually with
mixing. Adjust the pH with a 10% citric acid solution.
Continue mixing and cool to desired fill temperature.
   pH: 5.5+-0.5
   Viscosity: 6,500 cps+-10%   *Germaben II
   N.A.T.C. Approved   Formula SH-71

Cold Mix Shampoo w/Crosultaine C-50

Ingredients:	%
Part A:	
Deionized Water	59.85
Incromide CA (Cocamide DEA)	2.00
Part B:	
Teals	25.00
Incromide BAD (Babassuamide DEA)	1.00
Incrosultaine C-50 (Cocamidopropyl Hydroxysultaine)	8.00
Croquat L (Lauryldimonium Hydroxypropyl Collagen)	0.25
Citric Acid (10% Soln)	2.90
Propylene Glycol (and) Diazolidinyl Urea (and) Methyl Paraben (and) Propyl Paraben*	1.00

Procedure:
   Combine ingredients of Part A with mixing. Add ingredients of
Part B in order with mixing. Adjust pH with a 10% citric acid.
   pH: 6.5+-0.5   *Germaben II
   N.A.T.C. Approved   Formula SH-78

SOURCE: Croda Inc.: Suggested Formulations

Conditioning Shampoo

Ingredients:	%W/W
Phase A:	
Water, Deionized	50.32
Sodium Laureth Sulfate (Standapol ES-1)	20.00
Sodium Lauryl Sulfate (Standapol WAQ-Special)	20.00
Lauramide DEA (Monamid 716)	3.00
Quaternium-22 (Ceraphyl 60)	2.00
Quaternium-26 (Ceraphyl 65)	2.00
Citric Acid (Fine Granular, 30% Aq. Solution)	1.40
Sodium Chloride	0.08
Phase B:	
Sodium Hydroxymethylglycinate (Suttocide A)	0.40
Phase C:	
Fragrance (Floral Spice-NP49)	0.50
Processed Ingredient:	
D & C Green No. 5 (0.1% Aq. Sol.)	0.19
FD & C Green No. 5 (0.1% Aq. Sol.)	0.11

Procedure:
1. Add ingredients in Phase A with mixing, being careful not
   to aerate batch, heat to 75C.
2. Cool Phase A to 40C. Add Phase B with slow mixing.
3. Add ingredients in Phase C with mixing until uniform.
4. Sweep to 25C.

SOURCE: ISP Van Dyk Inc.: Formulation #D9843-85-1

Everyday Hair Shampoo

	Ingredients:	% w/w
1	A) Genapol LRO liquid	35,00
2	B) Water demineralized	39,90
3	Phenonip	0,50
4	Euxyl K-200	0,30
5	C) Genapol AMG	8,00
6	Sericin	3,00
7	Sodium Chloride	3,00
8	Fragrance: Adriano 0/235970	0,30
9	Genagen CAB	10,00
10	D) Sodiumhydroxyd solution 30%	

Procedure:
   Dissolve item 3 and 4 in water (2). Dissolve phase B) in
item 1. Then add and dissolve items 5-9 one after another.
Adjust the pH with items 10 to pH 6-7.

SOURCE: Pentapharm Ltd.: Application No. F 011.A/05.94

Conditioning Shampoo

A gentle, pearlized, moderately viscous shampoo to thoroughly cleanse and condition hair. Helps maintain ideal moisture balance and leaves hair with added softness, manageability and body.

		% by Weight
Part A:	Deionized Water	50.28
	Polyquaternium-10 (Celquat SC-240)	00.25
	Panthenol (dl-Panthenol)	00.20
	Methylparaben	00.15
	Cocamidopropyl Betaine (Cycloteric BET C-30)	06.00
	Glycol Stearate (Lexemul EGMS)	01.00
	Sodium Laureth Sulfate (Standapol ES-2)	20.00
	TEA-Cocoyl Glutamate (Amisoft CT-12)	16.00
Part B:	Cocamide DEA (Cyclomide DC 212/S)	03.50
Part C:	Quaternium-75 (Finquat CT)	01.00
	Methylchloroisothiazolinone (and) Methylisothiazolinone (Kathon CG)	00.08
	Fragrance (Givaudan #PSC 10,435/6)	00.30
	Citric Acid	00.04
	Sodium Chloride	01.20

Compounding Procedure:
Disperse Celquat SC-240 in the water. Heat to 70C. Add the rest of Part A. Mix until completely uniform. Start cooling. At 60C, add Cyclomide DC 212/S. At 40C, add Part C in the given order.
    Color: Off-white (pearlized)
    pH: 5.50
    Viscosity (@ 25C): 5.000 cps (#4 spindle @ 10 rpm)

Gentle Shampoo

A clear, moderately viscous shampoo that provides rich, lubricous lather and softness to the hair. The pH-balanced formula is extremely gentle and provides effective cleansing without harsh stripping.

		% by Weight
Part A:	Deionized Water	45.55%
	Methylparaben	00.20
	Lauroamphoglycinate (and) Trideceth Sulfate (Miranol MHT)	30.00
	TEA-Cocoyl Glutamate (Amisoft CT-12)	20.00
	Sodium Cocoyl Glutamate (Amisoft CS-11)	01.00
	PEG-150 Distearate	01.50
	Sodium PCA (Ajidew N-50)	01.00
Part B:	Methylchloroisothiazaolinone (and) Methylisothiazolinone (Kathon CG)	00.05
	Fragrance (Noville #84504)	00.20
	Citric Acid	00.50

Compounding Procedure:
    Heat Part A to 65C until completely homogeneous. Cool to 40C. Add Part B.
    Color: Light straw - clear   pH: 5.65
    Viscosity: 3,750 cps (#4 spindle @ 10 rpm)

SOURCE: Ajinomoto USA, Inc.: **Suggested Formulations**

Conditioning Shampoo

	% by Weight
Water	45.35
Sodium Laureth Sulfate (2 Mole 26%) (Sipon ES2)	20.00
Sodium Lauryl Sulfate and Disodium Lauryl Sulfosuccinate (Monaterge 1164)	20.00
Trisodium Lauroampho PG Acetate Phosphate Chloride (Phosphoteric QL-38)	10.00
Dimethicone (Dow Corning 200 Fluid 200 CS)	2.50
Glycol Distearate (Kessco Ethylene Glycol Distearate, Armak)	1.00
Cocamide MEA (Monamid CMA)	1.00
Sodium Chloride	0.15

Procedure:
    Add ingredients in order listed with agitation. Heat to 70C.
Cool to 40C. Adjust pH to 5.5 to 6.0 with 50% citric acid. Add
fragrance, color and preservative as required.

Formulation Properties:
    Physical Appearance: White pearled lotion
    Viscosity @ 25C: 7,100 cps

    Formula F-578

Mild Conditioning Shampoo

	% by Weight
Water	34.0
Monateric CLV	8.5
Phospholipid EFA	1.5
Sodium Lauryl Sulfate (28%)	27.0
Sodium Laureth (2) Sulfate	29.0

Procedure:
    Blend in order listed, adjust pH down 5.5 to 6.0 with 50%
citric acid.

Typical Properties:
    Appearance: Clear Liquid
    Viscosity @ 25C: Approximately 17,000 cps.

    Formula F-644

SOURCE: Mona Industries, Inc.: Suggested Formulations

Cream Shampoo

	Weight,%
Mackanate LO-Special (Disodium Lauryl Sulfosuccinate)	88.0
Mackol 16 (Cetyl Alcohol)	2.0
Brij 52	2.0
Mackstat DM (DMDM Hydantoin)	qs
Water, Fragrance qs to	100.0

Solids, %: 40.0
pH (as is): 5.5
Appearance: Pearly Cream

Procedure:
1. Add Mackol 16, Brij 52 and water to Mackanate LO-Special and heat to 70 degrees C.
2. Blend until homogeneous.
3. Adjust pH to 5.5 to 6.0 with sodium hydroxide.
4. Cool to 50 degrees C. and add Mackstat DM and fragrance.
5. Adjust solid to 40.0+-1.0 at this point.
6. Cool and fill.

Highly Pearlescent Shampoo

	Weight,%
Sodium Lauryl Ether Sulfate 60%	20.0
Mackamide C (Cocamide DEA (1:1)	2.0
Mackester SP (Glycol Stearate Modified)	2.0
Stearic Acid	2.0
Magnesium Sulfate (7H2O)	6.0
Diethanolamine	0.67
Mackstat DM (DMDM Hydantoin)	qs
Deionized Water, Fragrance, Dye qs to	100.0

pH: 7.5-8.0
Viscosity (cps 25 degrees C): 1000-2500

Procedure:
1. Heat water to 75 degrees C. and add Magnesium Sulfate.
2. Dissolve completely then add other surfactants and DEA then add waxes.
3. Keep temperature at 70 degrees C. for 20 minutes start cooling slowly.
4. At 35 degrees C. add remainder of ingredients and cool while mixing to room temperature.
5. Adjust pH with DEA or Sulfuric Acid diluted solutions.

SOURCE: McIntyre Group Ltd.: Personal Care Formulary

Emollient Shampoo

Ingredients:	%W/W
A Arlamol E	0.50
Arlacel 186	0.50
B Tween 20	4.00
C Water, deionized	71.00
Tensagex EOC-670	6.00
D Sodium Cocoyl Sarcosinate	6.00
Lauryl Polyglucose	8.00
E PEG 6000 Distearate	4.00

Procedure:
Mix (A) and add (B). Mix until clear and add (C) slowly.
Mix (D) and with slight heat and add to (A,B,C). Heat to 65C
and add (E) while stirring.

Moisturizing Shampoo

Ingredients:	%W/W
A CP 1052 (shampoo base)	31.50
Water	66.50
Glycerine	0.50
Arlamol E, ICI (PPG-15 stearyl ether)	0.25
Jojoba Oil	0.25
B Glycol distearate (Stepan-Kessco Ethylene Glycol	
Distearate)	1.00

Procedure:
Add water to shampoo base with moderate agitation. Add Glycerin and Arlamol E with medium agitation. Heat (A) to 80C and add
glycol stearate with fast stirring. Stir with medium agitation
until cooled to room temperature.
Formula CP 1059

Shampoo Concentrate

Ingredients:	%W/W
Ammonium lauryl sulfate (Stepanol AM)	75.59
Water	0.30
Cocamidopropyl betaine (Tegobetaine L-7)	17.78
NaCl	0.47
Citric acid	0.53
Cocamide DEA (Monamid 150ADD)	5.33

Procedure:
Heat ingredients to about 45C. Stir each ingredient in one
at a time in order using moderate agitation.
Formula CP 1052

SOURCE: ICI Surfactants: Suggested Formulations

Ethnic Hair Care
Amphoteric Shampoo
(Cold Process)
Clear

Ingredients:	%W/W
Tetrasodium EDTA	0.1
Water	48.1
Sodium Lauryl Sulfate	20.0
Cocamidopropyl Betaine (Tego Betaine F)	25.0
Dimethicone Propyl PG Betaine (Abil B 9950)	0.5
Quaternium-80 (Abil Quat 3272)	0.3
Dimethicone Copolyol (Abil B 88183)	0.5
PEG-30 Glyceryl Laurate (Tagat L)	0.5
Citric Acid (25% Solution)	to pH 6
Fragrance	Q.S.
Cocamidopropyl Betaine (and) Glyceryl Laurate (Antil HS 60)	5.0
Sodium Chloride (25% Solution to adjust viscosity)	Q.S.

Procedure:
1. Mix ingredients in order.
2. Adjust viscosity with Sodium Chloride.

Ethnic Hair Care
Amphoteric Shampoo
(Cold Process)
Pearled

Ingredients:	%W/W
Tetrasodium EDTA	0.1
Water	45.1
Sodium Lauryl Sulfate	20.0
Cocamidopropyl Betaine (Tego Betaine F)	25.0
Dimethicone Propyl PG Betaine (Abil B 9950)	0.5
Quaternium-80 (Abil Quat 3272)	0.3
Dimethicone Copolyol (Abil B 88183)	0.5
PEG-30 Glyceryl Laurate (Tagat L)	0.5
Citric Acid (25% Solution)	to pH 6
Fragrance	Q.S.
Glycol Distearate (and) Steareth-4 (Tego Pearl N 100)	3.0
Cocamidopropyl Betaine (and) Glyceryl Laurate (Antil HS 60)	5.0
Sodium Chloride (25% Solution to adjust viscosity)	Q.S.

Procedure:
1. Mix ingredients in order.
2. Adjust viscosity with Sodium Chloride.

SOURCE: Goldschmidt Chemical Corp.: Suggested Formulations

Everyday Family Shampoo

Inexpensive shampoo utilizing Pationic ISL for mildness and conditioning shampoo designed to be used by all family members with all hair types.

Ingredients:	%W/W	Function
1. Sodium C14-C16 Olefin Sulfonate (Bioterge AS-40)	37.50	Cleaning
2. Pationic ISL (Na Isostearoyl Lactylate)	3.00	Mildness
3. Ritapeg 150 DS (PEG-150 Distearate)	3.50	Viscosity
4. Kathon CG	0.03	Preservative
5. Fragrance-Baby Powder #165-848	0.30	Odor
6. Sodium Chloride (25% Soln.)	q.s.	Viscosity
7. Sodium Hydroxide (20% Soln.)	q.s.	pH Adjustment
8. Distilled/Deionized Water	55.67	

Compounding Procedure:
Heat the water and item 1 to 165F. Heat the Pationic ISL and Ritapeg 150 DS in a separate vessel to 165F. Add to the water with good mixing. Cool to 120F and add fragrance and preservative. Add Sodium Chloride until salt curve response. Adjust pH to 6.5.

Ref. No. 118-37C

Everyday Family Shampoo

A mild everyday formula for all family members.

Ingredients:	%W/W	Function
1. Sodium Laureth-2 Sulfate (Standapol ES-2)	25.00	Cleaning
2. Ritamide C (Cocamide DEA)	8.00	Lather
3. Ritox 35 (Laureth-23)	3.00	Combing
4. Grillocam E-20 (Methyl Gluceth-20)	3.00	Mildness
5. Ritacet-20 (Ceteareth-20)	2.00	Body
6. Promois W-32 (Hydrolyzed Collagen)	0.20	Body
7. Citric Acid (50% Soln.)	q.s.	pH adjustment
8. Preservative - Kathon CG	0.03	Preservative
9. Fragrance - Nature Harvest	0.30	Odor
10.Distilled/Deionized Water	58.47	-------

Compounding Procedure:
Heat water to 165F. Combine items 3,4 and 5 and heat until clear. Add to the water with constant agitation. Add item 1 and mix. Cool to 120F. Add item 6. Pre-mix item 2 and fragrance. Add to mixture. Add preservative. Adjust pH to 5.5.

Ref. No. 118-85

SOURCE: R.I.T.A. Corp.: **Suggested Formulations**

Gel Shampoo

	% by Weight
Water	55.00
Sodium Chloride	1.00
Sodium Laureth (2) Sulfate (25%)	35.00
Monateric CAB	5.00
Phospholipid PTC	2.00
Monamid 1089	2.00

pH adjusted to 5.0-6.0

Procedure:
Add ingredients in the order listed making sure each one is completely dissolved before adding the next one. Add fragrance, coloring, or preservative as required.

Mild Conditioning Shampoo

	% by Weight
Water	34.0
Monateric CLV	8.5
Phospholipid EFA	1.5
Sodium Lauryl Sulfate (28%)	27.0
Sodium Laureth (2) Sulfate	29.0

Procedure:
Blend in order listed, adjust pH down 5.5 to 6.0 with 50% citric acid.

Typical Properties:
Appearance: Clear Liquid
Viscosity @ 25C: Approximately 17,000 cps.

Formula F-644

Conditioning Shampoo

	% by Weight
Water	31.0
Sodium Laureth (2) Sulfate (26%)	57.7
Monateric COAB	9.4
Phospholipid EFA	1.4
Sodium Chloride	0.5

Procedure:
Blend ingredients together. Adjust pH down to 5.0 with 50% citric acid. Add fragrances, color, and preservatives as required and package.

Formula F-621

SOURCE: Mona Industries, Inc.: Suggested Formulations

Hair Shampoo
Dilutable Viscous Shampoo (For Customer Dilution 1:4)

Rhodaterge DCA	80 parts
Water	18
Perfume, Preservative, Dye	2

Viscous Consumer Shampoo

Rhodaterge DCA	25.0 parts
Water	74.7
Perfume	0.3
Color, Preservative	Q.S.

Transparent Shampoo

	% by Weight
Water	64.5
Miracare ANK	33.0
Fragrance, Dye, Preservative	Q.S.
Sodium Chloride	2.5

Blending Procedure:
Charge water into mixing vessel and blend balance of ingredients in order listed.

Lotion Shampoo

	% by Weight
Water	67.5
Miracare ANK	25.0
Mirasheen 202	5.0
Fragrance, Dye, Preservative	Q.S.
Sodium Chloride	2.5

Blending Procedure:
Charge water into mixing vessel and blend balance of ingredients in order listed.

SOURCE: Rhone-Poulenc: Suggested Formulations

Hair Shampoo Clear

	Wt%
Ampholyt JB 130	17.8
Marlinat 242/28	13.9
Marlinat SL 3/40	11.4
Dionil OC/K	2.0
Lamepon S	2.0
Panthenol	0.2
Vitamin F	0.1
Perfume	qs
Colour	qs
Preservative	qs
NaCl	qs
Water	to 100.0

Preparation:
   The components are mixed together in sequence and stirred until homogeneous. The pH is adjusted to 5.5-6.5.

Hair Shampoo

	Wt%
Marlinat 242/28	29.0
Marlinat CM 105	14.0
Ampholyt JB 130	10.0
Dionil OC/K	3.0
Lamepon S	5.0
Panthenol	0.3
Hair Complex Aquosom	0.5
Camomile Special	2.0
Collagen CLR	0.5
Preservative	qs
NaCl	qs
Water	to 100.0

Preparation:
   The components are mixed together in sequence and stirred until homogeneous. The pH is adjusted to 5.5-6.5.

SOURCE: Huls America Inc.: Formulations for Cosmetics: Formulas

Hair Shampoos, Clear

B 40/5:

Zetesol NL	40.0%
Purton SFD	1.0%
Perfume	0.5%
Water, sodium chloride, preservative	q.s. to make 100.0%

B 40/30:

Zetesol 2056	30.0%
Purton SFD	1.0%
Water, perfume, sodium chloride, preservative	q.s. to make 100.0%

B40/31:

Sulfetal KT 400	37.0%
Amphotensid GB 2009	10.0%
Purton CFD	2.0%
Water, perfume, preservative	q.s. to make 100.0%

B40/50:

Extrakt 52	27.0%
Amphotensid B 4	12.0%
Purton SFD	0.5%
Extracts of plants, aqueous (10%)	2.0%
Water, perfume, preservative	q.s. to make 100.0%

Hair Shampoos, Pearlescent

B 41/1:

Zetesol 856 T	20.0%
Perlglanzmittel GM 4175	4.0%
Perfume	0.3%
Water, sodium chloride, preservative	q.s. to make 100.0%

B 41/3:

Sulfetal KT 400	25.0%
Perlglanzmittel GM 4175	8.0%
Amphotensid B 4	9.0%
Water, perfume, sodium chloride, preservative	q.s. to make 100.0%

B 41/25:

Zetesol 2056	28.0%
Amphotensid B 4	8.0%
Perlglanzmittel GM 4175	4.0%
Perfume	0.5%
Water, preservative	q.s. to make 100.0%

SOURCE: Zschimmer & Schwarz GmbH & Co.: Suggested Formulas

Hair Shampoo with Conditioner

	Wt%
Marlinat 242/28	35.0
Marlinat CM 105	10.0
Ampholyt JB 130	5.0
Marlamid DF 1218	2.0
Lamepon S	2.0
Marlazin KC 30/50	1.6
Panthenol	0.2
Vitamin F	0.1
Perfume	qs
Colour	qs
Preservative	qs
NaCl	qs
Water	to 100.0

Preparation:
   The components are mixed together in sequence and stirred until homogeneous. The pH is adjusted to 5.5-6.5.

Hair Shampoo with Conditioner for Everyday Use

	Wt%
Marlinat 242/28	46.0
Marlinat CM 105	7.0
Marlinat SL 3/40	3.8
Marlazin KC 30/50	1.6
Panthenol	0.2
Perfume	qs
Colour	qs
Preservative	qs
NaCl	qs
Water	to 100.0

Preparation:
   The components are mixed together in sequence and stirred until homogeneous. The pH is adjusted to 5.5-6.5.

SOURCE: Huls America Inc.: Formulations for Cosmetics: Formulas

Herbal Shampoo Pearlescent

	Wt%
Marlinat 242/28	30.0
Ampholyt JB 130	8.0
Marlamid PG 20	3.0
Hair Complex Aquosum	1.0
Stinging Nettle Special	0.5
Rosemary Special	0.3
Vitamin F	0.2
Horse-chestnut Special	1.3
Perfume	qs
Colour	qs
Preservative	qs
NaCl	qs
Water	to 100.0

Preparation:
The components are mixed together in sequence and stirred until homogeneous. The pH is adjusted to 5.5-6.5.

Highly Concentrated Hair Shampoo

	Wt%
A. Marlinat 242/28	54.0
Marlinat SL 3/40	25.0
Ampholyt JB 130	16.0
Softigen 767	3.0
Water	to 100.0
B. Perfume	qs
Preservative	qs
Colour	qs
NaCl	qs

Preparation:
The constituents of A. are added together in sequence, and stirred while warm until homogeneous. The ingredients of B. are added to A. at approx. 30C. The pH is adjusted to 5.5-6.6.

SOURCE: Huls America Inc.: Formulations for Cosmetics: Formulas

High Foaming 2 in 1 Shampoo

	Weight,%
Ammonium Lauryl Sulfate (28%)	65.0
Mackalene 426 (Isostearamidopropyl Morpholine Lactate)	6.0
Mackanate DC-30 (Disodium Dimethicone Copolyol Sulfo-succinate)	4.0
Ethylene Glycol Distearate	1.0
Mackamide PKM (Palmkernalamide MEA)	2.0
Mackernium 007 (Polyquaternium 7)	0.4
Mackstat DM (DMDM Hydantoin)	Q.S.
Water, Dye, Fragrance qs to	100.0

Procedure:
1. Combine the first five components and heat to 70 degrees C. with continuous mixing.
2. Dilute the Mackernium 007 in the remaining water and slowly add to the blend.
3. Blend until product is homogeneous and cool to 50 degrees C.
4. Add Mackstat DM, fragrance and dye.
5. Adjust pH with citric acid to 5.0-6.0 and cool.

Silicone Free 2:1 Shampoo

	Weight,%
Ammonium Lauryl Sulfate (30%)	40.0
Mackanate LA (Diammonium Lauryl Sulfosuccinate)	20.0
Mackalene 426 (Isostearamidopropyl Morpholine Lactate)	6.0
Mackamide CMA (Cocamide MEA)	2.0
Mackernium 007 (Polyquaternium 7)	1.2
Mackester EGDS (Glycol Distearate)	1.0
Sodium Chloride	0.8
Paragon (DMDM Hydantoin (and) Methyl Paraben)	q.s.
Water, Fragrance, Dye q.s to	100.0

Procedure:
1. Add the first seven components to water and heat to 70C.
2. Blend until completely homogeneous.
3. Cool to 50C. and add Paragon, fragrance and dye.
4. Adjust pH to 5.0-6.0 with citric acid.
5. Cool and fill.

SOURCE: McIntyre Group Ltd.: Personal Care Formulary

Light Conditioning Shampoo

Light conditioning shampoo based on hair substantivity of Pationic SSL (Sodium Stearoyl Lactylate). Promotes style retention.

Ingredients:	%W/W	Function
1. Sodium Lauryl Sulfate (Standapol WAQ-LC)	21.43	Cleaning
2. Sodium Laureth-2 Sulfate (Standapol ES-2)	21.43	Cleaning
3. Pationic SSL (Na Stearoyl Lactylate)	3.00	Style, Conditioning
4. Rita EGDS (Glycol Distearate)	3.50	Viscosity
5. Ritamide C (Cocamide DEA)	4.50	Lather Density
6. Fragrance - Pert Type #189-724	0.30	Odor
7. Kathon CG	0.03	Preservative
8. Citric Acid (50% Soln.)	q.s.	pH Adjustment
9. Sodium Chloride (25% Soln.)	q.s.	Viscosity Adjustment
10. Distilled/Deionized Water	45.81	----------

Compounding Procedure:
Heat items 3 and 4 until melted. Add to 165F water items 1 and 2. Blend in pre-mix with constant agitation. Cool to 140F. Heat item 5 and mix with perfume. Add these and preservative at 120F.
Ref. No. 118-98

Light Conditioning Shampoo

A light conditioning shampoo with reduced skin and eye irritation due to the effects of Grilloten (Sucrose Cocoate).

Ingredients:	%W/W	Function
1. Sodium Laureth-3 Sulfate (Standapol ES-3)	40.00	Lather
2. Cocamidopropyl Betaine (Mackam 35HP)	8.00	Mildness
3. Grilloten LSE-65K (Sucrose Cocoate)	2.80	Non-Irritation
4. Ritasynt IP (Glycol Stearate and others)	3.00	Conditioning
5. Polyquta-400 (Polyquaternium-10)	1.00	Combing, Style
6. Kathon CG	0.03	Preservative
7. Sodium Chloride (25% Soln.)	q.s.	Viscosity
8. Citric Acid (50% Soln.)	q.s.	pH Adjustment
9. Fragrance	0.30	Odor
10. Distilled/Deionized Water	44.87	-------------

Compounding Procedure:
To the water at 165F slowly add item 5. When solubilized add item 3, which has been melted. Then add 1, 2 and 4 with constant agitation at 165F. Cool to 120F and add perfume and preservative. Adjust pH to 5.5 and viscosity to desired level.
Ref. No. 118-2

SOURCE: R.I.T.A. Corp.: Suggested Formulations

Light Conditioning Shampoo

Light conditioning shampoo designed to leave the hair with a soft, conditioned feel from ethoxylated lanolin (Laneto 50).

Ingredients:	%W/W	Function
1. Sodium Laureth-2 Sulfate (Standapol ES-2)	50.00	Cleaning
2. Ritasynt IP (Glycol Stearate and Others)	4.50	Conditioning, Soft Feel
3. Pationic ISL (Na Isostearoyl Lactylate)	3.00	Mildness, Combing
4. Laneto 50 (PEG-75 Lanolin)	1.50	Soft Feel, Shine
5. Ritapeg 150 DS (PEG-150 Distearate)	0.50	Opacity
6. Ritapan DL (dl-Panthenol)	0.50	Conditioner, Body
7. Fragrance - Squeeky Clean 165-047	0.30	Odor
8. Kathon CG	0.03	Preservative
9. Sodium Hydroxide (20% Soln.)	q.s.	pH adjustment
10.Sodium Chloride (25% Soln.)	q.s.	Viscosity Control
11.Distilled Water	39.67	---------

Compounding Procedure:
To the water add 1-6, hear to 165F. Mix and cool to 120F. Add preservative and fragrance. Note pH should be near 6.0. Adjust with sodium hydroxide to 7.2-7.6 to maximize deposition of the Laneto 50 if desired.
Ref. No. 118-46

Light Conditioning Shampoo

A light conditioning shampoo optimized for permed hair. Contains moisturizers and repair agents to give hair new life.

Ingredients:	%W/W	Function
1. Sodium Laureth-2 Sulfate (Standapol ES-2)	53.60	Cleaning
2. Pationic ISL (Na Isostearoyl Lactylate)	2.25	Moisturization
3. Ritapeg 150 DS (PEG-150 Distearate)	0.75	Viscosity
4. Rita EGMS (Glycol Stearate)	2.00	Opacity
5. Ritalan DL (dl-Panthenol)	0.25	Body, Feel
6. Laneto 50 (PEG-75 Lanolin)	0.75	Feel
7. Propylene Glycol	2.50	Moisturization
8. Fragrance - Nature Harvest 165-050	0.30	Odor
9. Kathon CG	0.03	Preservative
10.Distilled/Deionized Water	37.57	-------

Compounding Procedure:
Heat items 2,3 and 4 until uniform. Separately heat water until 165F disperse item 5 and mix until uniform. Add item 1 followed by pre-mix. Add item 7 to batch. Pre-mix item 6 and perfume and add to batch which has been cooled to 125F. Mix thoroughly. Add other ingredients. pH should be near 5.5.
Ref. No. 118-94

SOURCE: R.I.T.A. Corp.: Suggested Formulations

Low pH, Protein Gel Shampoo

	Weight,%
Ammonium Lauryl Sulfate (30%)	35.0
Mackam 35HP (Cocamidopropyl Betaine)	12.0
Mackpro NLP (Quaternium-79 Hydrolyzed Animal Protein)	
(Natural Lipid Protein)	2.0
Mackamide LLM (Lauramide DEA)	2.0
Mackstat DM (DMDM Hydantoin)	qs
Lactic Acid	qs
Water, Dye, Fragrance qs to	100.0

Procedure:
1. Add first four components to water and heat to 60 degrees C.
2. Adjust pH to 5.0 with lactic acid.
3. Cool and add remaining components at 40 degrees C.

Mild Conditioning Shampoo

	Weight,%
Mackanate EL (Disodium Laureth Sulfosuccinate)	10.0
Mackam 35 (Cocamidopropyl Betaine (Via Glyceride)	25.0
Sodium Laureth Sulfate (60%)	10.0
Mackanate DC-30 (Disodium Dimethicone Copolyol	
Sulfosuccinate)	1.0
Mackamide C (Cocamide DEA (1:1)	2.0
Polysorbate 20	1.0
Mackstat DM (DMDM Hydantoin)	qs
Water, Dye, Fragrance qs to	100.0
pH: 5.5-6.7	
Viscosity (cps 25 degrees C.): 600-1200	

Procedure:
1. Add surfactants to water.
2. Start mixing at room temperature until all components are clearly dissolved.
3. Blend fragrance with Polysorbate and add to batch.
4. Adjust pH if necessary with citric acid.
5. Adjust viscosity with Sodium Chloride.

Economy Shampoo

	Weight,%
Mackadet SBC-8 (Mild Blend)	10.0
Sodium Chloride	qs
Mackstat DM (DMDM Hydantoin)	qs
Water, Dye, Fragrance qs to	100.0

Procedure:
1. Add Mackadet SBC-8 to water and blend until clear.
2. Add Mackstat DM and adjust viscosity to 3000-4000 cps with sodium chloride.
3. Add dye and fragrance and blend until clear.

SOURCE: McIntyre Group Ltd.: Personal Care Formulary

Mild Shampoo

Ultra mild shampoo for gentle cleaning of hair. Exceptionally easy to rinse out and leaves minimal residue.

Ingredients:	%W/W	Function
1. Sodium Laureth Sulfate (Standapol ES-1)	30.00	Cleaning
2. Cocamidopropyl Betaine (Mackam 35 HP)	20.00	Mildness
3. Ritabate 20 (Polysorbate 20)	5.00	Mildness
4. Ritapeg 150 DS (PEG-150 Distearate)	2.00	Light Conditioning
5. Preservative	q.s.	Preservative
6. Fragrance-Baby Powder #165-848	0.30	Odor
7. Sodium Chloride (25% Solution)	q.s.	Viscosity
8. Distilled/Deionized Water	42.70	

Compounding Procedure:
To water add items 1 and 4. Heat to 60C and mix until uniform. Add items 2 and 3 and continue mixing. Cool to 30C. Add fragrance and preservative. Adjust pH to 6.5. Adjust viscosity with Sodium Chloride solution.
Ref. No. 118-71

Mild Shampoo

Ingredients:	%W/W	Function
1. Sodium Laureth-2 Sulfate (Standapol ES-2)	46.50	Cleaning
2. Pationic ISL (Sodium Isostearoyl Lactylate)	2.25	Damage Repair
3. Ritapeg 150 DS (PEG-150 Distearate)	0.75	Viscosity, Manageability
4. Conditioning Agent*	q.s.	Conditioning
5. Fragrance-Nature Harvest 165-050	0.30	Odor
6. Kathon CG	0.03	Preservative
7. Sodium Hydroxide (20% Soln.)	q.s.	pH Adjustment
8. Sodium Choride (25% Soln.)	q.s.	Viscosity Control
9. Distilled/Deionized Water	50.17	-----

* Optional conditioning agent, such as jojoba oil, panthenol, protein, etc. could be added.

Compounding Procedure:
To the water add item 1 at 165F. Pre-mix items 2 and 3 and add at 165F. Cool to 120F and add perfume and preservative. Adjust pH to 6.0-6.5 and viscosity as desired.
Ref. No. 118-102

SOURCE: R.I.T.A. Corp.: Suggested Formulations

Moisture Balance Shampoo with Cropeptide W

This formula gives hair more body control in all humidities, an effect created by Cropeptide W. Especially recommended for dry, damaged hair, Cropeptide W can reduce its brittleness at low RH and its limpness at high RH. The shampoo is an easy-to-make and economical cold mix formula with very good cleansing properties.

Ingredients:	%
Part A:	
Incromide LR (Lauramide DEA)	5.00
Incronam 30 (Cocamidopropyl Betaine)	5.00
Sodium Pareth-25 Sulfate	15.00
Cropeptide W (Hydrolyzed Wheat Protein (and) Wheat Oligosaccharides)	1.00
Deionized Water	73.70
Part B:	
Methyl paraben	0.20
Propyl paraben	0.10

Procedure:

Combine ingredients of Part A with mixing until clear. Add the ingredients of Part B individually with good mixing. Adjust pH with a 10% citric solution.

pH: 5.0+-0.5
Viscosity: 7,500+-10% (RVT Spindle #4, 20 rpm, 25C)
N.A.T.C. Approved
Formula SH-82

Shampoo with Tritisol

A rich foaming shampoo with wheat derived Tritisol added for conditioning. Excellent wet combing characteristics are produced by this pearlescent product.

Ingredients:	%
Part A:	
SLES (3 mole)	16.50
Crosultaine C-50 (Cocamidopropyl Hydroxysultaine)	7.00
Deionized Water	61.50
Incromectant AMEA-100 (Acetamide MEA)	3.75
Part B:	
Incromide CAC (Cocamide DEA Cocoyl Sarcosine)	3.00
Incromide ALD (Almondamide DEA)	1.00
Incromate OLL (Olivamidopropyl Dimethylamine Lactate)	1.00
Part C:	
Tritisol (Soluble Wheat Protein)	2.25
Crodapearl Liquid (Sodim Laureth Sulfate (and) Hydroxyethyl Stearamide MIPA)	3.00
Germaben II	1.00

Procedure:

Combine ingredients of Part A with mixing and heat to 60-65C. Combine ingredients of Part B with mixing and heat to 80-85C. Slowly add B to A with mixing and cool to 40C. Add Part C with mixing. Adjust pH to 6.0 with a 10% citric acid solution. Cool to desired fill temperature.

pH: 6.00+-0.5   Viscosity: 2,500+-10% (@25C)
N.A.T.C. Approved   Formula SH-83-1
SOURCE: Croda Inc.: Suggested Formulations

Protein Shampoos

B 40/65:
Sulfetal KT 400   30.0%
Amphotensid B 4   10.0%
Purton SFD   1.5%
Hydrolastan   1.0%
Water, perfume, sodium chloride,
  preservative   q.s. to make 100.0%

B 40/148:
Part A:
Sulfetal CJOT 60   25.0%
Purton SFD   1.0%
Perfume   0.5%
Croquat L   1.0%
Part B:
Polymer JR 400   0.3%
Water, preservative   q.s. to make 100.0%

Vitamin Shampoos

B 40/29:
Zetesol 856 T   30.0%
Purton SFD   1.5%
Soluvit CLR   3.0%
Water, perfume, preservative   q.s. to make 100.0%

B 40/38:
Sulfetal KT 400   40.0%
Amphotensid B 4   5.0%
Purton CFD   1.5%
Vitamin F, water-soluble   1.0%
Water, perfume, sodium chloride,
  preservative   q.s. to make 100.0%

Shampoo for Dry Hair

B 40/131:
Zetesol 856 T   22.0%
Amphotensid B 4   8.0%
Oxypon 288   4.0%
Avocado oil   0.5%
Perfume   1.0%
Sodium chloride   approx. 2.0%
Water, preservative   q.s. to make 100.0%

SOURCE: Zschimmer & Schwarz GmbH & Co.: Suggested Formulations

Salon Conditioning Shampoo

Optimal conditioning level is achieved.

Ingredients:	%W/W	Function
1. Sodium Laureth Sulfate (Standapol ES-2)	54.35	Cleaning
2. Pationic ISL (Na Isostearoyl Lactylate)	3.00	Mild, Combing
3. Ritapeg 150 DS (PEG-150 Distearate)	1.00	Style,Viscosity
4. Ritapan DL (dl-Panthenol)	0.25	Body
5. Laneto 50 (PEG-75 Lanolin)	0.75	Combing
6. Polyquta 400 (Polyquaternium-10)	0.50	Combing,Style
7. Sodium Chloride (25% Soln.)	0.25	Viscosity
8. Kathon CG	0.03	Preservative
9. Fragrance - Nature Harvest 165-050	0.30	Odor
10.Distilled/Deionized Water	39.57	------

Compounding Procedure:
Disperse items 6 and 4 in order in water at 165F. Pre-dissolve item 3 in item 2. Add to 165F main batch and mix until uniform. Add items 5 and 1 in order and mix until uniform. Cool to 120F. Add preservative and perfume. Adjust viscosity with NaCl.
Ref. No. 118-28

Salon Conditioning Shampoo

A mild salon conditioning shampoo with extremely rich lather and aloe and panthenol to penetrate the hair shaft to produce healthy hair.

Ingredients:	%W/W	Function
1. Sodium Lauryl Sulfate (Standapol WAQ LC)	25.00	Cleaning
2. Cocamidopropyl Betaine (Mackam 35 HP)	10.00	Lather, Mild
3. Ritamide C (Cocamide DEA)	5.00	Lather, Density
4. Pationic ISL (Na Isostearoyl Lactylate)	1.75	Mildness
5. Ritaloe 200M (Aloe Vera Gel)	0.15	Conditioning
6. Ritapan DL (dl-Panthenol)	0.50	Health
7. Kathon CG	0.03	Preservative
8. Fragrance - Nature Harvest	0.30	Odor
9. Sodium Chloride (25% Soln.)	q.s.	Viscosity Control
10.Citric Acid (50% Soln.)	q.s.	pH Adjustment
11.Distilled/Deionized Water	57.27	-----

Compounding Procedure:
To the water add items 5 and 6, heat to 120F. Combine items 1, 2 and 4, add heat to 120F. Add to water mixture with uniform mixing. Pre-mix Ritamide C and perfume and add to batch. Add preservative. Adjust pH and viscosity as needed.
Ref. No. 118-63
SOURCE: R.I.T.A. Corp.: **Suggested Formulations**

Salon Style Shampoo

Salon conditioning is achieved through balanced blend of surfactants, polymers and proteins which give hair great combing and shine.

Ingredients:	%W/W	Function
1. Ammonium Lauryl Sulfate (Stepanol AM)	45.00	Cleaning
2. Cocamidopropyl Betaine (Mackam 35HP)	5.50	Mildness
3. Cocamidopropyl Amine Oxide (Mazox CAPA)	5.50	Lather
4. Ritamide C (Cocamide DEA)	1.00	Lather Density
5. Polyquta 400 (Polyquaternium-10)	0.10	Conditioning
6. Hydroxypropyl Methylcellulose (Methocel 40-100)	0.20	Body, Viscosity
7. Triethanolamine @ 99%	0.10	pH
8. Tetrasodium EDTA	0.20	Clarity
9. Methylparaben	0.20	Preservative
10. Germall 115	0.20	Preservative
11. Promois WK (Hydrolyzed Keratin)	0.40	Shine, Body
12. Fragrance-Nature Harvest #165-050	0.30	Odor
13. Citric Acid (50% Soln.)	q.s.	pH Control
14. Distilled/Deionized Water	41.30	-----

Compounding Procedure:
To the water add items 5 and 6 and neutralize with TEA. Add items 1,2,8,9,10 and 11 in order. Pre-mix items 3,4 and fragrance and add to mix. Mix until uniform. Adjust pH to 6.0 and viscosity.
Ref. No. 118-111

2 in 1 High Conditioning Shampoo

High conditioning 2 in 1 type shampoo with excellent wet combing/dry combing.

Ingredients:	%W/W	Function
1. Sodium Laureth-3-Sulfate (Standapol ES-3)	30.00	Cleaning
2. Ritamide C (Cocamide DEA)	3.75	Lather
3. Polyquta 400 (Polyquaternium-10)	1.00	Conditioning
4. Fragrance	0.30	Odor
5. Kathon CG	0.03	Preservative
6. Distilled/Deionized Water	64.92	----

Compounding Procedure:
To the water at 165F add items 3 and mix until clear. Add items 1 and 2 in order and mix until clear. Cool to 120F and add fragrance and preservative. Adjust pH with Citric Acid and viscosity with Sodium Chloride to desired ranges.
Ref. No. 115-205B

SOURCE: R.I.T.A. Corp.: Suggested Formulations

Shampoo for Permed Hair

	Weight,%
Mackanate OP (Disodium Oleamido MIPA Sulfosuccinate)	20.0
Mackanate CP (Disodium Cocamido MIPA Sulfosuccinate)	12.0
Sodium Laureth Sulfate (30%)	15.0
Mackamine WGO (Wheat Germamidopropylamine Oxide)	4.0
Mackalene 716 (Wheat Germamidopropyl Dimethylamine Lactate)	1.0
Mackstat DM (DMDM Hydantoin)	qs
Citric Acid to pH=6.0	
Sodium Chloride qs to 2000 cps	
Water, Dye, Fragrance qs to	100.0

Procedure:
1. Add surfactants to water and heat to 40 degrees C.
2. Blend until clear and adjust pH with citric acid.
3. Add remaining components and adjust viscosity with sodium chloride.

Pearlescent Shampoo Concentrate

	Weight,%
TEA Lauryl Sulfate	50.0
Mackamide LLM (Lauramide DEA)	30.0
Mackester SP (Glycol Stearate Modified)	5.0
Propylene Glycol	5.0
Sodium Chloride	1.0
Phosphoric Acid to pH = 7.5	
Mackstat DM (DMDM Hydantoin)	qs
Water, Fragrance, Dye qs to	100.0

Procedure:
1. Add first five components to water and heat to 70 degrees C.
2. Blend until homogeneous.
3. Cool to 40 degrees C. and add Mackstat DM, dye and fragrance.
Remarks: This product can be diluted one pint to a gallon with water. The viscosity can be controlled by regulating the propylene glycol.

Neutralizer Shampoo

	Weight,%
Mackanate OM (Disodium Oleamido MEA Sulfosuccinate)	30.0
Sodium Laureth Sulfate (30%)	20.0
Mackamine CAO (Cocamidopropylamine Oxide)	6.0
Mackamine WGO (Wheat Germamidopropylamine Oxide)	2.0
Mackstat DM (DMDM Hydantoin)	qs
Water, Dye, Fragrance qs to	100.0

    Solids, %: 19.5
    pH: 5.3
    Viscosity (cps, 25 deg C.): 1500
Procedure:
    Add surfactants to water and blend until clear. Adjust pH to 5.0-5.5 with citric acid. Add dye and fragrance.

SOURCE: McIntyre Group Ltd.: Personal Care Formulary

Sport Shampoo

		Wt%
A.	Marlinat DFN 30	35.0
	Marlinat CM 105	11.0
	Ampholyt JB 130	10.0
	Dionil OC/K	2.5
	Antil 141 liquid	qs
	Water	to 100.0
B.	Avocado Special	1.5
	Perfume	qs
	Colour	qs
	Preservative	qs

Preparation:
    The ingredients of A. are added together in sequence, warmed and stirred until homogeneous. The ingredients of B. are added to A. at approx. 30C. The pH is adjusted to 5.5-6.6.

Herbal Shampoo Against Greasy Hair

		Wt%
A.	Marlinat 242/28	28.0
	Marlinat CM 105	18.0
	Ampholyt JB 130	13.0
	Marlamid M 1218	3.0
	Water	to 100.0
B.	Avocado Special	1.0
	Camomile Special	0.5
	Perfume	qs
	Colour	qs
	Preservative	qs
	NaCl	qs

Preparation:
    The constituents of A. are added together in sequence and stirred while warm until homogeneous. The constituents of B. are added to A, at approx. 30C. The pH is adjusted to 5.5-6.6.

SOURCE: Huls America Inc.: Formulations for Cosmetics: Formulas

Two in One Shampoo

Ingredients:	%W/W
PC 9043, Silicone emulsion	10.00
Shampoo concentrate	90.00

Procedure:
Heat (A) to approx. 65C. Mix with very slowly with anchor type agitation.
Formula PC 9049

Two in One Shampoo

Ingredients:	%W/W
PC 9044, Silicone emulsion	10.00
Shampoo concentrate	90.00

Procedure:
Heat (A) to approx. 65C. Mix with very slowly with anchor type agitation.
Formula PC 9050

Two in One Shampoo

Ingredients:	%W/W
PC 9045, silicone emulsion	10.00
Shampoo concentrate	90.00

Procedure:
Heat (A) to approx. 65C. Mix with very slowly with anchor type agitation.
Formula PC 9051

Two in One Shampoo

Ingredients:	%W/W
PC 9046, silicone emulsion	10.00
Shampoo concentrate	90.00

Procedure:
Heat (A) to approx. 65C. Mix with very slowly with anchor type agitation.
Formula PC 9052

Two in One Shampoo

Ingredients:	%W/W
PC 9047, silicone emulsion	10.00
Shampoo concentrate	90.00

Procedure:
Heat (A) to approx. 65C. Mix with very slowly with anchor type agitation.
Formula PC 9053

Two in One Shampoo

Ingredients:	%W/W
PC 9048, silicone emulsion	10.00
Shampoo concentrate	90.00

Procedure:
Heat (A) to approx. 65C. Mix with very slowly with anchor type agitation.
Formula PC 9054

SOURCE: ICI Surfactants: Suggested Formulations

Two in One Shampoo

True 2 in 1 formula with exceptional wet and dry combing.

Ingredients:	%W/W	Function
1. Sodium Laureth-2-Sulfate (Standapol ES-2)	30.00	Cleaning
2. Cocamidopropyl Betaine (Mirataine BET C-30)	3.00	Lather
3. PEG-120 Methyl Glucose Dioleate	3.00	Mildness, Viscosity
4. Dow Corning 1401 Fluid	2.60	Combing
5. Ritamide C (Cocamide DEA)	4.00	Lather Density
6. Fragrance	0.20	Odor
7. Preservative - Kathon CG	0.03	Preservation
8. Citric Acid (50% Soln.)	q.s.	pH Adjustment
9. Distilled/Deionized Water	57.17	-------

Compounding Procedure:
    To the water at 60C add items 1,2 and 3. Premix items 4 and 5 and add to mix at 60C. Stir until uniform. Add fragrance, preservative and adjust pH to 6.0-6.5.
    Ref. No. 116-153

Two in One Shampoo

Unique combination of excellent combability, overall conditioning and mildness.

Ingredients:	%W/W	Function
1. Sodium Laureth-2-Sulfate (Standapol ES-2)	45.00	Cleaning, Lather
2. Grilloten LSE-65K (Sucrose Cocoate)	1.50	Mildness
3. Cocamidopropyl Betaine (Mirataine BET C-30)	10.00	Lather, Mildness
4. Dimethicone Copolyol (Abil B88183)	1.95	Combing
5. Quaternium 80 (Abil Quat 3272)	0.65	Conditioning
6. Fragrance - Squeeky Clean	0.30	Odor
7. Kathon CG	0.03	Preservative
8. Citric Acid (50% Soln.)	0.05	pH Adjustment
9. Sodium Chloride (25% Soln.)	q.s.	Viscosity Control
10.Distilled/Deionized Water	40.52	----

Compounding Procedure:
    To the water at 120F add items 4,5 and 3 in order. Pre-mix the items 1 and 2 while heating gently and add perfume and mix until uniform. Now mix both phases together. Adjust pH and viscosity.
    Ref. No. 119-4

SOURCE: R.I.T.A. Corp.: Suggested Formulations

2-In-1 Shampoo

Ingredients:	%W/W
A Dimethicone Copolyol Dow Corning Q2-5000	0.25
Dow Corning 345 fluid	0.25
Carbopol 1382	0.10
Water	66.13
B CP 1052 (shampoo base)	31.50
C Stepan-Kessco Ethylene glycol distearate	1.00
D NaCl	0.77

Procedure:
Blend ingredients in (A) together with moderate agitation at 35C. Add shampoo base to (A) with moderate agitation. Heat mixture to 80C and add glycol distearate with fast stirring. Allow to cool with medium agitation, stirring in salt at 50C or below.
Formula CP 1061

Two-in-One Shampoo

Ingredients:	%W/W
A Water	67.36
Dow Corning Q2-5200	0.10
Dow Corning 345 fluid	0.20
BF Goodrich Carbopol 1382	0.08
B Stepan-Kessco Ethylene glycol distearate	0.50
C G-9600 1053 (Shampoo base)	31.50
D NaCl	0.26
*Preservative	

*q.s these ingredients.

Procedure:
Blend ingredients in (A) together with moderate agitation at 35C. Add shampoo base to (A) with moderate agitation. Heat mixture to 80C and add glycol distearate with fast stirring. Allow to cool with medium agitation, stirring in salt at 50C or below.
Formula CP 1063

SOURCE: ICI Surfactants: Suggested Formulations

Three in One Shampoo

Resulting product is a clear rich shampoo which cleans; conditions and improves combability with Lexquat AMG-BEO; and styles with PVP/VA copolymer.

		%w/w
A.	Sodium Lauryl Sulfate (Standapol WAQ)	15.00
	TEA-Lauryl Sulfate (Standapol T)	10.00
	Cocamidopropyl Betaine (Lexaine C)	10.00
	Cocamide DEA (Standamid KD)	3.00
	Potassium Cocoyl-Hydrolyzed Collagen (Maypon 4C)	3.00
B.	Deionized Water	42.10
	Propylene Glycol USP	3.00
	Methylparaben (Lexgard M)	0.20
	Propylparaben (Lexgard P)	0.10
	Benzophenone-4 (Uvinul MS-40)	0.10
	Tetrasodium EDTA (Hamp-ene Na4)	0.10
C.	Behenamidopropyl PG Dimonium Chloride (Lexquat AMG-BEO)	7.00
D.	PVP/VA copolymer (Luviskol VA 73-W)	6.00
E.	Fragrance	0.20
	Color	QS
F.	Citric Acid NF	QS

Procedure:
1. Combine ingredients in phase "A". Heat to 75C with continuous slow oscillation. Do not entrap air.
2. Combine ingredients in phase "B". Heat to 75-80C. Mix rapidly to dissolve parabens.
3. When phases "A&B" are both homogenoeus and are 75C, add Phase "A" to "B". Adjust mixing speed to avoid foaming and air entrapment.
4. Combine phase "C" and mix until homogeneous.
5. Add phase "C" to "AB". Maintain 75C.
6. When phase "ABC" is homogenoeus add "D" to "ABC". Mix until homogeneous.
7. When phase "ABCD" is homogeneous begin cooling to ambient temperature.
8. At 30C add phase "E" to "ABCD". Continue mixing and cooling.
9. When batch reaches ambient temperature, adjust pH to 7 with phase "F".
10. Adjust for water loss.

SOURCE: Inolex Chemical Co.: Formulation SP-116

Three in One Shampoo

Resulting product is a rich clear shampoo which cleans; provides conditioning and improved combability with Lexquat AMG-BEO and quaternized chitosan; and styles with PVP/VA copolymer.

		%w/w
A.	Sodium Lauryl Sulfate (Standapol WAQ)	15.00
	TEA-Lauryl Sulfate (Standapol T)	10.00
	Cocamidopropyl Betaine (Lexaine C)	10.00
	Cocamide DEA (Standamid KD)	3.00
	Potassium Cocoyl-Hydrolyzed Collagen (Maypon 4C)	3.00
B.	Deionized Water	42.10
	Propylene Glycol USP	3.00
	Methylparaben (Lexgard M)	0.20
	Propylparaben (Lexgard P)	0.10
	Benzophenone-4 (Uvinul MS-40)	0.10
	Tetrasodium EDTA (Hamp-ene Na4)	0.10
C.	Behenamidopropyl PG Dimonium Chloride (Lexquat AMG-BEO)	7.00
	Polyquaternium-29 (Kytamer KC)	0.20
D.	PVP/VA co-polymer (Luviskol VA 73-W)	6.00
E.	Fragrance	0.20
	Color	QS
F.	Citric Acid NF	QS

Procedure:
1. Combine ingredients in phase "A". Heat to 75C with continuous slow oscillation. Do not entrap air.
2. Combine ingredients in phase "B". Heat to 75-80C. Mix rapidly to dissolve parabens.
3. When phases "A&B" are both homogeneous and are 75C, add Phase "A" to "B". Adjust mixing speed to avoid foaming and air entrapment.
4. Combine phase "C" and mix until homogeneous.
5. Add phase "C" to "AB". Maintain 75C.
6. When phase "ABC" is homogeneous add phase "D" to "ABC". Mix until homogeneous.
7. When phase "ABCD" is homogeneous begin cooling to ambient temperature.
8. At 30C add phase "E" to "ABCD". Continue mixing and cooling.
9. When batch reaches ambient temperature, adjust pH to 7 with phase "F".
10. Adjust for water loss.

SOURCE: Inolex Chemical Co.: Formulation SP-115

Wheat & Sesame Shampoo

Hydrotriticum WAA is the key ingredient in this shampoo, which acquires its unique quality from a blend of wheat derived materials. This shampoo has excellent foam with a rich, creamy lather and leaves the hair feeling refreshingly clean. Hydrotriticum WAA, a powerful humectant, helps to retain moisture that is vital to maintaining healthy hair. Incromectant AMEA-100 and Incromate SEL improve wet combing and detangling.

Ingredients: %

Part A:

Incromine Oxide WG (Wheat Germamidopropylamine Oxide)	3.00
Incronam WG-30 (Wheat Germamidopropyl Betaine)	3.00
Incromectant AMEA-100 (Acetamide MEA)	5.00
SLES (3 mol EO)	40.00
Disodium EDTA	0.10
Deionized Water	41.45

Part B:

Incromide WGD (Wheat Germamide DEA)	2.00
Incromate SEL (Sesamidopropyl Dimethylamine Lactate)	1.00
BHT	0.10

Part C:

Hydrotriticum WAA (Wheat Amino Acids)	3.00

Part D:

Methylparaben	0.25
Propylparaben	0.10
Propylene Glycol	1.00

Citric Acid (10% soln) to pH 5.4

Combine Part A and heat to 60C. Combine Part B and heat to 60C, mixing until homogeneous. Add Part B to Part A with mixing. Continue mixing and cool to 45C. Add Part C, followed by Part D, mixing until uniform. Adjust pH with citric acid solution.

N.A.T.C. Approved   Formula SH-84

Detangling Shampoo

Ingredients: %

Part A:

SLES (3 mol.)	20.0
Crosultaine C-50 (Cocamidopropyl Hydroxysultaine)	7.0
Deionized Water	58.0
Incromectant AMEA-100 (Acetamide MEA)	2.0

Part B:

Incromide CAC (Cocamide DEA Cocoyl Sarcosine)	3.0
Incromate ISML (Isostearamidopropyl Morpholine Lactate)	4.0

Part C:

Crodacel QM (Cocodimonium Hydroxypropy Cellulose)	2.0
Crodapearl Liquid (Sodium Laureth Sulfate (and) Hydroxyethyl Stearamide-MIPA)	3.0
Propylene Glycol (and) Diazolidinyl Urea (and) Methyl Paraben (and) Propyl Paraben*	1.0

Combine ingredients of Part A with mixing and heat to 60-65C. Combine ingredients of Part B with mixing and heat to 60-65C. Add Part B to Part A with mixing and cool to 40C. Add Part C with mixing and cool to desired fill temperature. Adjust pH with 10% citric acid solution.

Formula SH-92   *Germaben II

SOURCE: Croda Inc.: Suggested Formulas

Wheat Germ Conditioning Shampoo

	Weight,%
Mackanate OP (Disodium Oleamido MIPA Sulfosuccinate)	20.0
Sodium Laureth Sulfate (30%)	24.0
Mackanate WGD (Disodium Wheatgermamido PEG-2 Sulfosuccinate)	8.0
Mackam WGB (Wheat Germamidopropyl Betaine)	5.0
Citric Acid to pH = 5.5	
Sodium Chloride qs to viscosity = 20000 cps	
Mackstat DM (DMDM Hydantoin)	qs
Water, Dye, Fragrance qs to	100.0

Procedure:
1. Add surfactants to water and heat to 40 degrees C.
2. Adjust pH to 5.5.
3. Add remaining components and adjust viscosity with sodium chloride.

No-Sting 2:1 Shampoo

	Weight,%
Mackam 2C	35.0
Sodium Laureth-1 Sulfate	20.0
Mackanate DC-30	4.0
Mackernium 007	3.0
Mackester SP	2.0
Mackstat DM	q.s.
Citric Acid q.s. to pH	7.0-7.5
Water, Dye, Fragrance q.s. to	100.0

Procedure:
1. Add Mackam 2C, Sodium Laureth-1 Sulfate, Mackanate DC-30 and Mackester SP to water.
2. Heat to 70C to blend until homogenous.
3. Slowly add Mackernium 007.
4. Cool to 50C and add Mackstat DM.
5. Add fragrance, dye and adjust pH to 7.0-7.5.

Stripper Shampoo

	Weight,%
Dodecylbenzene Sulfonic Acid	21.5
Caustic Soda (50%)	5.4
Sodium Laureth Sulfate (60%)	4.0
Mackam 35 (Cocamidopropyl Betaine)	5.5
Sodium Xylene Sulfonate (40%)	8.0
Water, Dye, Fragrance qs to	100.0
Solids, %: 30+-1.0	
pH: 6.5-7.0	
Viscosity (cps, 25C): 250-350	
Cloud Point: 5C	

Procedure:
1. Add caustic soda to water and adjust pH to 7.0-8.0 with DDBSA.
2. Add remaining components and adjust pH to 6.5-7.0 with citric acid.
3. If necessary, lower viscosity with SXS, or raise viscosity with sodium chloride.

SOURCE: McIntyre Group Ltd.: Suggested Formulations

Section IX
Shaving Products

Aerosol Shaving Cream

Ingredients:	%W/W
A Stearic acid, triple pressed	6.30
Stripped coconut fatty acid	2.70
Sorbo	10.00
B Water, deionized	50.00
Allantoin	0.20
C Potassium hydroxide	1.70
Water	29.10
D *Perfume	

*q.s. these ingredients.

Procedure:
Heat (A) to 75C. Add (B) and reheat to 75C. Add (C) slowly, with agitation. Mix for 30 minutes, maintaining temperature at 75C. Cool slowly while agitating slowly. Add (D) at 35C. Prepare aerosol shaving cream cans and pressure fill each can with 3% to 4% isobutane-propane, 40 psig to 46 psig.

Nonionic Aerosol Foam

Ingredients:	%W/W
A Cetyl alcohol	4.30
Brij 721S	2.20
B Water, deionized	93.25
Sorbic acid	0.17
C Fougere Lavender, Fritzsche, D&O	0.08

-PRESSURIZE-

D Above concentrate (160.0g)
Difluoro ethane, Dymel 152A (8 mls)

Procedure:
Heat (A) to 70C and (B) to 75C. Add (B) to (A) slowly with agitation and add (C) at 35C. Adjust pH to 5.5 with dilute NaOH. Add water to compensate for loss due to evaporation. Continue agitation until viscosity is low enough to pour.

SOURCE: ICI Surfactants: Suggested Formulations

After Shave Balm

Dry Flo PC is a highly water resistant starch, provides a silky smooth elegant feel, and is talc free.

Ingredients:	%W/W
Phase A:	
Estol 1550	4.00
Cutina CP	2.50
Lonzest 143-S	3.00
Ivarlan 3230	1.50
Phase B:	
Deionized Water	61.60
Alcolec BS	1.00
Carbopol 941 (2% Aq. Soln.)	10.00
NaOH (25%)	0.40
Methylparaben	0.15
Propylparaben	0.15
Phase C:	
Glycerin (99.5%)	7.00
Dry-Flo PC	8.00
Phenoxyethanol	0.20
Phase D:	
Fragrance	0.50

Procedure:
Combine Phase B, heat to 80C. Combine Phase A, heat to 80C. At 80C add A to B, mix for 15 minutes. Cool to 40C. Phase C; slurry Dry-Flo PC into Glycerin, add to A/B at 40C. Add Phenoxyethanol, mix well. Add Phase D. Mix until uniform.
SOURCE: National Starch and Chemical Co.: Formula 8062-16

Shave Gel

The product will raise the hair off the face to make for a closer cut without irritation. The Hexanediol Behenyl Beeswax stabilizes the silicone oil to leave the skin silky and smooth.

Phase A:	
Hexanediol Behenyl Beeswax (Koster Keunen)	3.0%
Orange Wax (Koster Keunen)	2.0%
Permethyl 104A (Permethyl)	3.0%
Polybutene (Amoco)	4.0%
Isostearic Acid (Unichema)	0.6%
Palmitic Acid (Proctor & Gamble)	1.4%
Phase B:	
Purified Water	63.9%
Triethanolamine (Dow)	0.8%
Carbopol 940 2% dispersion (B.F. Goodrich)	12.0%
Aloe Vera Gel (Active Organics)	0.8%
Phase C:	
Silicone 345 (Dow)	6.7%
Allantoin (Sutton)	0.8%
Germaben II (Sutton)	1.0%

Heat and mix Phase A and add to a heated and mixed Phase B. Cool to 50C and add phase C, one at a time. Cool to 40C and pour into containers.

Fragrances, actives and other silicone oils can be incorporated into this type of formula, with only minimal changes in stability and performance.
SOURCE: Koster Keunen, Inc.: Suggested Formulation

After Shave Conditioner

This multi-phase emulsion (o/w/a) is very stable and has wonderful skin feel. Besides the previously mentioned advantages. Cera Bellina produces a product with good viscosity characteristics in a product of low solids content.

Oil Phase:

Light Mineral Oil, Carnation (Witco)	7.5%
Cera Bellina (Pg-3 Beeswax, Koster Keunen)	4.0%
Glycerol Monostearate (Henkel)	3.3%
Deodorized Orange Wax	2.0%
Vitamin E (Van Dyk)	0.5%
Propyl Paraben (Sutton)	0.1%

Water Phase:

Water (distilled)	64.0%
Butylene Glycol (Arco)	5.0%
Glycerol (Unichema)	1.0%
Methyl Paraben (Sutton)	0.3%
Triethanolamine (Dow)	1.0%

Alcohol Phase:

SDA-30 (Quantum)	10.0%
Benzocaine (National Starch)	1.0%
Carboxy Methyl Cellulose (Hercules)	0.3%

Procedure:

Combine components of the wax phase in a vessel, melt and mix maintaining a temperature of 75C. Heat the water phase to 75C in a separate vessel making sure the components are all dissolved. Slowly add the oil phase to the water phase under low shear. Cool to 35 to 40C and then add the alcohol phase under low shear. This will ensure that the friction will not increase the temperature which will evaporate the alcohol.

Adaptation of formula and its influence on the product:

Alterations in component concentrations can be achieved by the addition of the secondary emulsifying agents. This well allow for active ingredients to be added at the same time as maintaining the stability and rheological properties.

SOURCE: Koster Keunen, Inc.: Suggested Formulations

Alcohol Free After Shave Lotion

Ingredients:	% by Weight
Part I:	
Water	66.00
Phenyl Dimethicone	1.50
Carbomer 940	0.30
Triethanolamine (99%)	0.60
Part II:	
Water	30.00
Phospholipid SV	1.00
Menthol	0.10

Procedure:
  Slurry Carbomer 940 into Dimethicone then mix well with the water. After the Carbomer 940 is dissolved, add triethanolamine. In a separate container blend and heat the water, menthol and Phospholipid SV to 75-80C. When dissolved, blend Part II to Part I mixing well. Add color, fragrance and preservative as required and package.

Conditioning After Shave Toner

	% by Weight
Part I:	
Water	51.00
Carbomer 940 (Premix)	0.35
Phenyl Dimethicone (Premix)	2.00
Triethanolamine (99%)	0.60
Part II:	
Water	30.05
Phospholipid SV	1.00
SD3A Alcohol	15.00

Procedure:
Part I:
  Slurry Carbomer 940 into water by first mixing it with Phenyl Dimethicone. Allow to mix until homogenous. When Part II is ready, neutralize Part I with TEA. Add Part II and blend together until a smooth lotion is formed.
Part II:
  Mix water and Phospholipid SV. Heat to 65 with agitation until the Phospholipid SV is dissolved. Cool to 30-40C and add SD3A alcohol. Add to Part I.

SOURCE: Mona Industries, Inc.: Suggested Formulations

Brushless Shave Cream

This formulation features Vanseal NACS-30, sodium cocoyl-sarcosinate, Vanseal CS, cocoylsarcosine and potassium cocoate as high foaming yet mild surfactants. Sorbitol adds humectancy while PVP and talc provide lubricity. Stearic acid, propylene glycol stearate, and cetyl alcohol are included as thickeners and to provide pleasant after-feel.

Ingredient:	% by Wt.
A Vanseal NACS-30	15.00
Vanseal CS	2.50
Deionized Water	25.75
Potassium Cocoate (2)	35.00
Sorbitol, 70%	5.00
PVP (3)	0.75
B Talc (4)	5.00
C Stearic Acid	7.00
Propylene Glycol Stearate (5)	2.50
Cetyl Alcohol	1.50
D Preservative, Dye, Fragrance	q.s.

(2) Liquid Coconut Soap, 40%
(3) PVP K-30
(4) AGI Talc
(5) Cerasynt PA

Preparation:
Mix A ingredients together and heat to 55C with gentle stir-ring until clear. Add B to A with adequate agitation. Heat C to 60C. Add C to (A + B), mixing until uniform and homogeneous. Cool to 30C and add D.

Consistency: Flowable gel (Viscosity: 2500-3500 cps)

Suggested Packaging: Plastic bottle or pump.

Comments:
This prototype formula is designed to serve as a guide for the development of new products or improvement of existing ones.

SOURCE: R.T. Vanderbilt Co., Inc.: Formula No. 434

Emulsion for Shaving

Ingredients:	%W/W
A Cetearyl alcohol	3.00
Arlacel 186	1.00
Brij 721S	3.00
B Water, deionized	93.00
C *Dowicil 200	

*q.s. these ingredients.

Procedure:
Heat (A) to 70C and heat (B) to 72C. Add (B) to (A) and
stir. Cool to about 35C and add (C) and water to compensate
for the loss due to evaporation. Stir to room temperature
and package. Adjust pH to 5-6.
Formula CP 1082

Soap Free - Nonionic Shaving Emul

Ingredients:	%W/W
A Cetearyl alcohol	3.00
Arlacel 186	1.00
Brij 721S	3.00
B Water, deionized	83.00
Sorbo, 70% sorbitol solution	10.00
C *Dowicil 200	

*q.s. these ingredients.

Procedure:
Heat (A) to 70C and heat (B) to 72C. Add (B) to (A) and stir.
Cool to about 35C and add (C) and water to compensate for the
loss due to evaporation. Stir to room temperature and package.
Adjust pH to 5-6.
Formula CP 1091

SOURCE: ICI Surfactants: Suggested Formulations

Hemostatic After Shave

Ingredients:	% w/w
1 A) Irgasan DP 300	0,01
2 Crodamol DA	1,00
3 Ethanol 95%	50,00
4 B) Propylenglycol USP	2,00
5 Water demineralized	44.49
6 C) Fragrance: Courage 0/243101	0,50
7 D) Cephalipin	2,00

Procedure:
    Dissolve phase A).
    Mix phase B).
    Slowly add phase B) to phase A). Then add item (6).
    Finally add item (7).
Application No. I 001.A/4.91

Shaving Foam

Ingredients:	% w/w
1 Water demineralized	71,08
2 Glycerin	4,00
3 Triethanolamine 95%	3,21
4 Genapol LRO liquid	5,00
5 Euxyl K-200	0,30
6 Stearic Acid	8,30
7 Luvitol EHO	1,00
8 Nipasol M	0,20
9 BHT	0,01
10 Cocamide DEA	0,40
11 Fragrance	0,50
12 Cephalipin	3,00
13 Phytaluronate	3,00

Procedure:
    The filling data are: Concentrate   95%
                         Isobutane (at 3,2 psi)   5%
Application No. I 003.0/02.93

SOURCE: Pentapharm Ltd.: Suggested Formulations

Shaving Creme

		%
A.	Softisan 601	20.0
	Miglyol 812	12.0
	Marlowet TA 6	5.0
	Imwitor 900	3.0
B.	NaCl	5.0
	Ampholyt JB 130	5.0
	Locron L	5.0
	Water	to 100.0
C.	Perfume	qs
	Preservative	qs

Preparation:
    The constituents of A. are mixed together and heated to
75-80C. B. is heated to the same temperature and gradually
stirred into A. Perfume is introduced at approx. 30C.

Shaving Gel

		%
A.	Marlinat 242/28	50.0
	Ampholyt JB 130	10.0
	Marlamid DF 1218	5.0
	Antil 141 liquid	1.0
	Water	to 100.0
B.	Softigen 767	2.0
	Perfume	qs
	Sicomet Patentblau*	0.05
	Preservative	qs

*Sicomet Patentblau (Patent blue) 80E 131 1% in Softigen 767

Preparation:
    The constituents of A. are mixed together and warmed to
approx. 40C. The constituents of B. are mixed together and
added to A. Viscosity at 20C measured in a Brookfield visco-
meter: 1900 mPa.s.

SOURCE: Huls America Inc.: Formulations for Cosmetics:
        Suggested Formulations

Ultra Aerosol Shave Cream for Sensitive Skin

Veegum Ultra, magnesium aluminum silicate, is used in this emulsion formula to improve the stability of the luxurious lather produced by combining Vanseal NACS-30 (sodium cocoyl sarcosinate) with stearic and coconut acid soaps. Vanox PCX acts as an antioxidant in this formulation.

Ingredient:	% by Wt.
A Veegum Ultra (Magnesium Aluminum Silicate)	1.00
Deionized Water	75.80
B Glycerin	3.00
Triethanolamine	4.00
C Stearic Acid XXX	6.00
Coconut Acid (2)	1.30
Mineral Oil	2.50
Cetyl Alcohol	1.00
Vanox PCX (BHT)	0.20
D Vanseal NACS-30 (Sodium Cocoyl Sarcosinate)	5.00
Methylparaben	0.20
Fragrance	q.s.

(2) Emery 622

Procedure:
   Sift Veegum Ultra into the water while stirring with a propeller mixer at 700 rpm. Increase the mixer speed to 1700 rpm and mix for 30 minutes. Add the B ingredients in the order shown while mixing at 1700 rpm and heating to 75C. Mix the C ingredients and heat to 75C. Slowly add C to (A and B). Adjust the mixer speed to avoid air entrapment and excessive foaming. Mix for 10 minutes at 75C, then slow the mixer while cooling to 30C. Add the D ingredients in the order shown and mix each until uniform. Package in aerosol containers using 95.0 wt % of Formula No. 451 and 5.0 wt.% A-46 propellent.

Product Characteristics:
   Viscosity: 500-700 cps
   pH: 8.0+-0.2

Comments:
   This prototype formula is designed to serve as a guide for the development of new products or the improvement of existing ones.

SOURCE: R.T. Vanderbilt Co., Inc.: Formulation No. 451

Section X
Soaps and Hand Cleaners

Economy Heavy-Duty Hand Cleaners

	Wt.%
Rhodapon SB 8208/S	17.5
Alkamide LE	0.7
Rhodacal A-246/L	3.8
Triethanolamine	0.5
Oleic Acid	1.0
Glycerine	0.6
Alkamuls EGMS	0.9
Sodium Chloride	3-4
Fragrance, Dye, Preservative	Q.S.
Water	Q.S. to 100

	Wt.%
Rhodapon SB 8208/S	17.5
Alkamide DIN 295/S	0.7
Rhodacal A-246/L	3.8
Triethanolamine	0.5
Oleic Acid	1.0
Glycerine	0.5
Alkamuls EGMS	0.9
Sodium Chloride	2-3
Fragrance, Dye, Preservative	Q.S.
Water	Q.S. to 100

Procedure:
   Charge water into reactor and warm to 75-80C. With rapid but smooth agitation, blend in ingredients as listed. Cool to 30-35C and blend in fragrance, dye, preservative.

Typical Properties:
   Appearance @ 25C: Pearl, lotion-like liquid
   Viscosity @ 25C: 4,000-5,000 cps

   Formula HA-0097

SOURCE: Rhone-Poulenc: Suggested Formulations

Emulsion Cleaner

Ingredients:	% as Supplied
Water	88.1
Acusol 820 Stabilizer (30%)	1.7
Deodorized Kerosene	10.0
NaOH (50%)	0.2

Brookfield Viscosity cps, @ 0.5 rpm: 21,000
                                                  @ 12 rpm: 2,300
pH: 9.2

Mixing Instructions:
Add the ingredients in the listed order.
High-shear mixing is necessary to disperse the solvents
    (kerosene, oil)

Waterless Hand Cleaner

Ingredients:	% as Supplied
Water	47.1
Acusol 820 Stabilizer (30%)	1.7
NPE10	3.0
Deodorized Kerosene	38.0
Mineral Oil	10.0
NaoH (50%)	0.2

Brookfield Viscosity cps, @ 0.5 rpm: 4,000,000
pH: 7.8

Mixing Instructions:
Add the ingredients in the listed order.
High-shear mixing is necessary to disperse the solvents
    (kerosene, oil).

SOURCE: Rohm and Haas Co.: ACUSOL 820 Stabilizer: Suggested
            Formulations

Emulsion Cleaner and Waterless Hand Cleaner
Emulsion Cleaner

Ingredients:	% as supplied
Water	88.1
Acusol 820 Stabilizer (30%)	1.7
Deodorized Kerosene	10.0
NaOH (50%)	0.2

Brookfield Viscosity cps @ 0.5 rpm: 21,000
@ 12 rpm: 2,300
pH: 9.2

Waterless Hand Cleaner

Ingredients:	% as supplied
Water	47.1
Acusol 820 Stabilizer (30%)	1.7
NPE10	3.0
Deodorized Kerosene	38.0
Mineral Oil	10.0
NaOH (50%)	0.2

Brookfield Viscosity @ 0.5 rpm: 4,000,000
pH: 7.8

Mixing Instructions:
Add the ingredients in the listed order.
High-shear mixing is necessary to disperse the solvents (kerosene, oil)

SOURCE: Rohm and Haas Co.: Acusol 820 Stabilizer: Suggested Formulation

Fast Set Waterless Hand Cleaner with d-Limonene F-453

	% By Weight
d-Limonene*	40.0
Monamine 1255	12.0
Monafax 785	0.5
Glycerine	1.0
Mineral Oil	2.0
Water	44.5

Procedure:
  Combine in order listed, adding water last with good
agitation.

Formulation Properties:
  Physical Appearance: White ringing gel

  * TABS-D from Union Camp may be substituted for d-Limonene

Clear Gel d-Limonene Waterless Hand Cleaner F-454

	% By Weight
d-Limonene*	20.0
Glycerine	20.0
Monamine 1255	12.0
Monafax 785	0.5
Water	47.5

Procedure:
  Combine in order listed, adding water last with good
agitation.

Formulation Properties:
  Physical Appearance: Clear ringing gel

  * TABS-D (Union Camp) may be substituted for d-Limonene

SOURCE: Mona Industries, Inc.: Suggested Formulations

Hand Cleaners
Hand Cleaning Paste

B 22/2:

Sulfetal TC 50	4.0%
Purton SFD	1.0%
Carboxymethyl cellulose 100%	1.0%
Soft soap	12.0%
Quartz sand	60.0%
China clay	6.0%
Sodium tripolyphosphate	1.0%
Water, perfume, preservative	q.s. to make 100.0%

Hand Cleaning Cream, Free of Solvents

B 27/8:
Part A:

Mulsifan CPA	4.0%
Oxypon 288	1.0%
Paraffin oil DAB	25.0%
Oleic acid	5.0%
Triethanolamine	2.5%

Part B:

1,2-Propylene glycol	6.0%
Water	56.2%

Part C:

Perfume	0.2%
Preservative	0.1%

Hand Cleaning Cream with Solvent

B 27/2:
Part A:

Zusolat 1005/85	10.5%
Shellsol T	34.5%
Paraffin oil DAB	9.0%
Aerosil 200	1.0%
Perfume	0.1-0.2%

Part B:

Oxypon 2145	2.5%
Extrakt 52	15.0%
Euxyl K 400	0.2%
Water	27.1-27.2%

Manufacture:
Part A and Part B are to be mixed separately. Finally add
Part B to Part A whilst stirring.

SOURCE: Zschimmer & Schwarz GmbH & Co.: Suggested Formulations

Hand Cleaners from Concentrate
Economy

	Wt.%
Miracare ANL	30.0
Fragrance, Dye, Preservative	Q.S.
Sodium Chloride	2-3
Water	67.5

Pearlescent

	Wt.%
Miracare ANL	32.5
Mirasheen 202	7.5
Fragrance, Dye, Preservative	Q.S.
Sodium Chloride	0.5-1.5
Water	59.0

Premium

	Wt.%
Miracare ANL	40.0
Fragrance, Dye, Preservative	Q.S.
Sodium Chloride	0.5-1.5
Water	59.0

Lotion

	Wt.%
Miracare ANL	37.5
Opacifier E-305	0.5
Fragrance, Dye, Preservative	Q.S.
Sodium Chloride	1-2
Water	60.5

Procedure:
   Charge water into mixing vessel and slowly blend in Miracare ANL. Mix until completely uniform and then blend in remaining ingredients in order listed. Adjust formulation viscosity to desired level with addition of sodium chloride, as needed.

Physical Properties:
   Appearance: Clear, yellow, pearlized or opaque
   pH, as is: 7-8
   Viscosity: 2000-4000 cps

   Formula HA-0096

SOURCE: Rhone-Poulenc: Suggested Formulations

Industrial and Institutional Liquid Hand Soaps
Glucopon & AS

	Wt %
Water	61.0
Standapol WAQ-SP	27.0
Glucopon 425 CS	6.0
Velvetex BA-35	2.8
Sodium Chloride	3.2

Glucopon & LAS

	Wt %
Water	77.2
Sodium LAS (60%)	10.0
Glucopon 600 CSUP	8.0
Velvetex BA-35	2.8
Standamid SD	2.0

Glucopon & AES

	Wt %
Water	58.9
Standapol ES-2	30.0
Glucopon 600 CSUP	6.0
Standamid SD	1.0
Velvetex BA-35	2.8
Sodium Chloride	1.3

Glucopon & AOS

	Wt %
Water	67.0
Sodium AOS (40%)	20.0
Glucopon 600 CSUP	6.0
Standamid SD	1.0
Velvetex BA-35	2.8
Sodium Chloride	3.2

Viscosity of Products: 4000-5000 cPs

Procedure:
Add in the order listed, dissolving each ingredient completely using moderate agitation. Adjust pH to 6.5-7.0 with citric acid. Incorporate color and fragrance as needed.

SOURCE: Henkel Corp.: Suggested Formulations

Liquid Hand Soap Concentrate (Antimicrobial)

Components:	Wt %
Velvetex BA-35	11.7
Glycerine	1.0
Propylene Glycol	1.0
Triclosan*	1.0
Glucopon 625 CSUP	15.6
Sodium Chloride	2.0
Tetrasodium EDTA	1.0
Standapol WAQ-LC	66.7

Procedure:
    Add in the order listed, dissolving each ingredient completely using moderate agitation. Adjust pH to 6.5-7.0 with Citric Acid. Incorporate preservative; color and fragrance as needed.
    For use, concentrate should be diluted by the addition of 25 wt % concentrate to 71.5 wt % deionized water and 3.5 wt % sodium chloride solution (25%). The resulting clear liquid hand soap will have a viscosity of approximately 6,000 cps.
    * May also use P-Chloromela xylenol (PCMX)

Waterless Hand Cleaner (Gel)

Oil Phase:	Wt %
Deodorized Kerosene	40.0
Emersol 213	7.5
Trycol 5963	5.0
Standamid KD	1.0
Water Phase:	Wt %
Water	41.0
Triethanolamine	3.5
Glycerine	2.0

Waterless Hand Cleaner (Lotion)

Oil Phase:	Wt %
Deodorized Kerosene	40.0
Emersol 213	7.5
Glycerine or PEG Stearate	2.0
Water Phase:	
Water	40.0
Triethanolamine	3.0
Propylene Glycol	3.0
Glucopon 425 CS	4.5

Procedure:
    Mix the oil and water phases separately in the order listed, heating both to 60C. Slowly pour the oil into the water while stirring. Continue stirring while cooling until a homogeneous product is formed. Incorporate color and fragrance as needed.

SOURCE: Henkel Corp.: Suggested Formulations

Liquid Hand Soap Formulations

Miracare ANK is also an ideal base for preparation of emollient "Liquid Hand Soap" products. The rich creamy lather typical of formulations derived from Miracare ANK is extremely stable even in the presence of heavy soils.

Lotion Hand Soap
Requires No Heat

Water	61.0
Miracare ANK	30.0
Mirasheen 202	7.5
Fragrance, Dye, Preservative	Q.S.
Sodium Chloride	1.5

Blending Procedure:
   Charge water into mixing vessel and blend balance of ingredients in order listed.

Cream Soap
(Requires Heat)

Water	62.0
Miracare ANK	35.0
Alkamuls EGMS	1.5
Fragrance, Dye, Preservative	Q.S.
Sodium Chloride	1.5

Blending Procedure:
   Charge water and ANK into mixing vessel and heat to 75C.
Add and disperse Alkamuls EGMS. Cool with mixing to 40C and blend in remaining ingredients.

Liquid Hand Soap

	Parts by Weight
Rhodaterge SMC	8
Water	92
Perfume, Dye, Preservative	Q.S.

SOURCE: Rhone-Poulenc: Suggested Formulations

Liquid Soap

	%
Marlinat 242/28	32.0
Marlinat CM 105	18.0
Ampholyt JA 140	10.0
Dionil OC/K	3.0
Lamepon S	2.0
Camomile Special	0.5
Lime Blossom Special	0.5
Collagen CLR	0.8
Perfume	qs
Colour	qs
Preservative	qs
NaCl	qs
Water	to 100.0

Preparation:
    The components are mixed together in sequence and stirred until homogeneous. The pH is adjusted to 5.5-6.5.

Liquid Soap

	%
Marlinat 242/28	22.0
Lamepon S	10.0
Marlinat DFN 30	6.0
Marlamid PG 20	3.0
Marlamid DF 1218	1.0
Perfume	qs
Colour	qs
Preservative	qs
NaCl	qs
Water	to 100.0

Preparation:
    The components are mixed together in sequence and stirred until homogeneous. The pH is adjusted to 5.5-6.5.

SOURCE: Huls America Inc.: Formulations for Cosmetics:
        Suggested Formulations

Liquid Synthetic Soaps

B 21/65:

Zetesol NL	43.00%
Purton SFD	3.00%
Sodium sulfate	3.00%
Perfume	0.20%
Citric acid	0.05%
Water, preservative	q.s. to make 100.00%

B 21/64:

Extrakt 52	30.0%
Purton SFD	2.0%
Water, perfume, preservative, sodium chloride	q.s. to make 100.0%

B 21/61:

Zetesol 856 T	25.0%
Setacin 103 Spezial	10.0%
Perlglanzmittel GM 4055	5.0%
Purton SFD	1.0%
Water, preservative, perfume, sodium chloride, citric acid	q.s. to make 100.0%

B 21/92:

Zetesol 2056	25.0%
Setacin 103 spezial	30.0%
Perlglanzmittel GM 4175	10.0%
Oxypon 288	2.0%
Sodium chloride	2.0%
Water, preservative	q.s. to make 100.0%

B 21/87:

Zetesol 2056	20.0%
Amphotensid B 4	9.0%
Purton CFD	1.5%
Perfume	0.3%
Water, preservative	q.s. to make 100.0%

SOURCE: Zschimmer & Schwarz GmbH & Co.: Suggested Formulations

Liquid Soap with Aloe Vera Gel

This is a cleansing and conditioning liquid soap with Aloe
Vera.

	%w/w
Phase A:	
Ammonium lauryl sulfate, 28% active	35.70
Ammonium laureth sulfate, 26% active	19.20
Phase B:	
Citric acid, anhydrous	0.20
Phase C:	
Deionized water	20.45
Polyquaternium-10	0.75
Phase D:	
Lauramide DEA	3.00
Lauryl methyl gluceth-10 hydroxypropyl dimonium chloride	8.00
PEG-120 methyl glucose deoleate	2.00
Tetrasodium EDTA	0.20
Phase E:	
Aloe Vera Gel (Terry AG002)	10.00
Phase F:	
Suttocide A	0.30
Fragrance	qs

Procedure:
   Add Phase A ingredients to main mixing vessel, and combine
with propeller agitation. Heat to 60C. Dissolve Phase B
ingredients. In a premix container, disperse Phase C ingredients.
Heat to 60C to hydrate. When Phase A is uniform, add Phase C to
main vessel, agitating between each addition. Cool batch to 40C.
Slowly add Phases E and F with adequate mixing.
SOURCE: Terry Laboratories, Inc.: **Suggested Formulation**

Pearlescent Liquid Hand Soap

Ingredient:	Wt%
Water	54.70
DDBSA, Na salt (60%)	23.50
Add, mix well	
Lauryl ether sulfate, Ammon. salt (60%)	16.50
Add, mix well	
Lytron 295 opacifier	0.10
Add, mix well	
Euperlan PK-771 pearlizer	2.20
Add, mix well	
Tomah AO-728 Special	3.00
Add, mix well	

   pH: 7.85
   Recommended Dilutions: 1 oz./gal
   Applications: Hand soap or dishwash
   Notes: High foam, degreasing-lanolin can be added for feel
          Viscosity 360 cps

SOURCE: Tomah Laboratories, Inc.: **Formula PC-101**

Low Cost Clear Cleansing Bar

Ingredients:	%
Part A:	
Propylene Glycol	11.0
Glycerin	7.0
Crovol PK-70 (PEG-45 Palm Kernel Glycerides)	3.0
Incromectant AMEA-70 (Acetamide MEA)	9.3
Crodasinic LS30 (Sodium Lauroyl Sarcosinate)	6.7
Incromide CAC (Cocamide DEA Cocoyl Sarcosine)	7.0
Teals*	9.0
Deionized Water	6.0
Sodium Cumenesulfonate (45%)	2.0
TEA (99%)	2.0
Part B:	
Sucrose	9.0
Part C:	
Tallow Soap #7325	28.0

Procedure:
    Combine ingredients of Part A with mixing and heat to 80-90C.
Add Part B with mixing until uniform. Add Part C with mixing
while heating to 95-100C. Mixing until tallow is dissolved
(20-30 minutes). Avoid water loss by covering mixing vessel.
Skin off surface foam and pour into molds. Allow molded soap to
cool.
    N.A.T.C. Approved   *Standamide T
SOURCE: Croda Inc.: Formulation SC-252

Shower Soap

	Weight,%
Mackanate EL (Disodium Laureth Sulfosuccinate)	20.0
Mackanate OM (Disodium Oleamido MEA Sulfosuccinate)	15.0
Sodium Lauryl Sulfate	10.0
Mackamide LLM (Lauramide DEA)	6.0
Mackpearl LV (Pearl Agent)	3.0
Mackernium 007 (Polyquaternium 7)	2.5
Mackstat DM (DMDM Hydantoin)	qs
Citric Acid qs to pH 6.0	
Sodium Chloride qs to 10,000 cps	
Water, Dye, Fragrance qs to	100.0

Procedure:
1. Disperse Mackernium 007 in water.
2. Add remaining component and heat to 40 degrees C.
3. Adjust pH with citric acid.
4. Adjust viscosity with sodium chloride.
5. Cool and fill.

SOURCE: McIntyre Group Ltd.: Suggested Formulation

"Medicated" Skin Cleaner

	Wt.%
Rhodapon LT-6	25.0
Rhodapon 21LS	30.0
50% Sodium Hydroxide Aqueous Solution	1.S. to pH 8.0
PCMX	1.0
Water	23.4
Alkamide DL 203/S	2.5
Cycloteric BET C 30	7.5
Geropon SBFA 30	10.0
Alkamuls EGMS	0.6
Fragrance and Dye	Q.S.

Procedure:
1. Charge Rhodapon LT-6 and Rhodapon 21LS into mixing vessel. Warm system to 45-50C. With smooth agitation, adjust pH to 8.0 with sodium hydroxide solution (50% Aq.), as needed.
2. With rapid but smooth agitation, slowly blend in PCMX. Mix until system is completely uniform.
3. With smooth agitation, slowly blend in Alkamide DL 203/S and water. Heat system to 70-75C with smooth agitation.
4. Slowly blend in Cycloteric BET C 30, Geropon SBFA 30 and Alkamuls EGMS. Maintain 70-75C temperature until system is completely uniform. Cool to 40-45C and blend in compatible fragrance and dye.

Typical Properties After 24 Hours:
Appearance: Viscous, pearlescent liquid
pH (as is): 8.0-8.5
Viscosity @ 25C: 10,000-14,000 cps
Approx. Solids (%): 30.4
Active Ingredient: Parachlorometaxylenol (PCMX)
Formula HA-0100

Heavy-Duty Hand Cleaner

	Wt.%
Rhodapon SB 8208/S	30.0
Rhodacal A-246/L	6.3
Oleic Acid	1.6
Triethanolamine	0.9
Mirasheen 202	10.0
Sodium Chloride	1-3
Fragrance, Dye, Preservative	Q.S.
Water	Q.S to 100

Procedure:
Charge Rhodapon SB 8208/S, Rhodacal A-246/L, Triethanolamine, Mirasheen 202 and Water into reactor and mix until uniform. Slowly blend in oleic acid followed by fragrance, dye, preservative. Adjust viscosity by blending in sodium chloride. Continue agitation until system is uniform and then package.

Typical Properties:
Appearance: White, pearlized liquid
pH, as is: 7-8
Viscosity: 2000-3000 cps
Formula HA-0098

SOURCE: Rhone-Poulenc: Suggested Formulations

Mild Lotion Soap

	Wt.%
Rhodapon L-22	17.0
Rhodapex MA-360	3.0
Cycloteric BET-C-30	2.0
Alkamide DC 212/S	1.5
Geropon SBFA-30	5.0
Alkamuls EGMS	1.0
Citric Acid	Q.S. to pH 5-6
Sodium Chloride	0.5-1.5
Water	69.5
Fragrance, Dye, Preservative	Q.S.

Procedure:
Charge water into mixing vessel and slowly blend in Rhodapon L-22, Rhodapex MA-360, Cycloteric BET-C-30, Alkamide DC 212/S, and Rhodacal SBFA-30. Mix until completely uniform. Heat systems to 65-70C. Disperse Alkamuls EGMS into heated system and mix until uniform. Cool to 40-45C and blend in citric acid, fragrance, dye and preservative. Add sodium chloride, as needed, to adjust formulation to desired range (3,000-6,000 cps at 25C).

Typical Properties:
Appearance at 25C: Pearlescent liquid
Viscosity at 25C: 3,000-6,000 cps (as desired)
pH (as is): 5-6
% Solids: 13.5-14.5

Formula HA-0103

Liquid Hand Soap

	Wt.%
Mirataine CB	10.0
Rhodapon LSB	30.0
Cyclomide DC-212/S	2.0
Water	58.0

Procedure:
Add the water. With mixing, add the remaining materials and mix until uniform.

Physical Properties:
Appearance: Clear, very light yellow solution
pH: 8.9-9.0
Viscosity: 600-700
Specific Gravity: 1.04

Formula HA-0071

SOURCE: Rhone-Poulenc: Suggested Formulations

Premium Emollient Hand Soap

	Wt.%
Rhodacal A-246/L	23.0
Cycloteric BET-C 30	4.5
Alkamide LE	1.5
Alkamul EGMS	2.0
Disodium EDTA	0.1
Glycerine	0.2
Citric Acid	Q.S. to pH 6.2-6.7
Sodium Chloride	Q.S. to 3,000-5,000 cps
Fragrance, Dye, Preservative	Q.S.
Water	Q.S. to 100

Procedure:
    Combine Rhodacal A-246/L and Water and with moderate agit-
ation, warm to 70-75C. Blend in Alkamide LE, Alkamul EGMS,
Disodium EDTA and Glycerine. Once system is uniform, cool to
35C with smooth agitation and blend in fragrance, dye, preservative. Adjust pH to 6.2-6.7 with citric acid and add sodium
chloride as needed to viscosity to 3,000-5,000 cps (25C).

Typical Properties:
    Appearance: Pearlized liquid
    pH, as is: 6.2-6.7
    Viscosity: 3000-5000 cps

    Formula HA-0095

Hand Cleaner

	Wt.%
Miranol C2M Conc. NP	16.0
Mirataine CB	5.0
Rhodapon LSB	16.0
Water	63.0

    Adjust pH to 7.5 with citric acid.

Procedure:
    Add the water. With mixing, add the remaining materials and
mix until uniform.

Physical Properties:
    Appearance: Slightly hazy gel
    pH: 7.5
    Viscosity: 1500-2000
    Specific Gravity: 1.0408

    Formula HA-0070

SOURCE: Rhone-Poulenc: Suggested Formulations

All quantities indicated in the following formulations are expressed as parts by weight of the finished soap. These weights are on a percentage basis so that quantities necessary to produce any size batch may be conveniently calculated.

6) Triethanolamine Oleic-Coconut Shampoo (40% Real Soap)

Emery 621 Coconut Fatty Acid (acid value 263)	12.7
Emery 213 Low Titer Oleic Acid (acid value 203)	12.7
Triethanolamine (avg. mol. wt. 140)	14.7
Carbitol	7.0
Water	to 100.0

Procedure:
Blend the oleic and coconut fatty acids. Dissolve the triethanolamine in the required amount of water and proceed as for the coconut shampoo, formula 4, with the exception that heating is not essential. Longer standing is required without heat, but the ability to make soaps without heating may be valuable in particular situations.
Properties and Variations:
This soap product is a clear liquid of amber to red color, very mild and nonirritating. The Carbitol gives the soap a pleasing odor and contributes to the cleansing power of the shampoo.
Triethanolamine soaps cannot be checked for neutrality by the usual methods, but even if an excess of triethanolamine is present, the soap will be only mildly alkaline.

7) Coconut-Oleic Hand Soap (15% Real Soap)

Emery 621 Coconut Fatty Acid (acid value 263)	6.5
Emersol 213 Low Titer Oleic Acid (acid value 203)	6.5
Caustic potash (100%)	3.0
Water	to 100.0

Procedure:
Prepare as for formulation 4:
Dissolve the caustic in the necessary amount of water, heating to 49-54C (120-130F), then add the fatty acid in a slow steady stream. Agitate while mixing the fatty acid and caustic until saponification is complete. The reaction mixture should be heated to 66-71C (150-160F) during the final stages of saponification.
Check the neutrality of the soap and adjust as necessary. Perfume as desired. Allow the soap to stand and settle for several days at near freezing temperatures, if possible, and decant or filter the soap.
Properties and Variations:
Distilled pine oil or other perfume may be added. This is a typical formula for soaps used in liquid soap dispensers, and many such products contain fluorescein, a dye that imparts a greenish-yellow fluorescence.
SOURCE: Henkel Corp./Emery Group: Fatty Acids and Their
    Water Soluble Soaps: Suggested Formulations

Water Free Hand Cleaner

Ingredients:	%W/W
Isostearic acid	20.00
Mineral oil	60.00
Brij 30	15.00
Tween 81	5.00

Procedure:
    Mix well at room temperature.

Waterless Hand Cleaner

Ingredients:	%W/W
A Isoparaffinic hydrocarbon	30.00
Lanolin	1.00
Cetyl alcohol	3.00
Paraffin wax	2.00
Arlacel 60	1.00
Tween 60	4.00
B Glycerin	5.00
Water, deionized	54.00
*Preservative	
*q.s. these ingredients.	

Procedure:
    Heat (A) to 70C. Heat (B) to 72C. Add (B) to (A) slowly with
moderate stirring. Stir to 35C and add water lost due to
evaporation.

Waterless Hand Cleaner

Ingredients:	%W/W
A Magnesium aluminum silicate	2.50
Water, deionized	30.00
B Arlacel 60	2.00
Tween 60	8.00
Deodorized kerosene	35.00
C Methylcellulose, 4000 cps	0.50
Water, deionized	22.00
*Preservative	
*q.s. these ingredients.	

Procedure:
    A-Add MAS to H2O slowly, agitating continually until smooth.
(About 1 hour). Heat (A) to 62C. B-Heat to 60C. Add (A) to (B).
Stir until cool. C-Heat 1/2 of H2O to 90C and add methylcellulose
slowly with agitation until dispersed. Add (C) to (A&B) with
agitation.
SOURCE: ICI Surfactants: Suggested Formulations

Waterless Hand Cleaner

		Wt%
A.	Montmorillonite (1)	3.0
	Water	39.5
B.	Potassium Hydroxide	1.5
	Water	10
C.	Oleic Acid	6.0
	Deodorized Kerosene	25.0
	Polysorbate 80, Acetylated Lanolin Alcohol,	
	Cetyl Acetate (2)	3.0
	Petrolatum	2.0
D.	Acuscrub 50	10.0
	Preservatives	q.s.

Procedure:

Slowly add the Montmorillonite to the water while agitating at maximum available shear. Dissolve the potassium hydroxide in the water. Add B to A with medium shear. Heat A/B to 60C. Combine C and heat to 65C. Cool to 50C, then slowly add Acuscrub 50 until cooled to 30C. Add preservatives and mix until smooth and uniform.

(1) Mineral Colloid BP   (2) Solulan 98

Waterless Hand Cleaner

	Wt%
A. Oil Phase:	
Petroleum Distillates (1)	30.0
Oleic Acid	6.0
C12-15 Pareth-3 (2)	4.0
Lanolin	0.5
B. Water Phase:	
Water	44.5
Triethanolamine	3.0
Glycerin	2.0
C. Acuscrub 50	10.0

Procedure:

Heat both oil and water phases separately to 55C. Add Acuscrub 50 slowly to the oil phase with moderate agitation. Slowly pour the water phase into the oil phase with constant stirring until a smooth and uniform gel is formed.

(1) Shell Sol 71   (2) Neodol 25-3

SOURCE: Allied Signal Inc.: Suggested Formulations

Waterless Hand Cleaner No. 390

	Van Gel B	1.7
A	CMC 7MF	0.3
	Water	44.0
B	Potassium hydroxide	1.0
	Water	3.0
	Oleic acid	10.0
C	Carnation White Mineral Oil	10.0
	Deodorized kerosene	30.0
	Preservative	q.s.

Procedure:
Blend Van Gel B and CMC. Slowly add to the water while
agitating at maximum available shear. Continue mixing until
smooth. Add B to A and mix until uniform. Add C to A/B and
mix until emulsion is smooth and uniform. Add desired preservative.
Consistency: Medium viscosity cream.
Suggested Packaging: Pump or wide mouth container.
Comments:
Van Gel B/CMC thickens this cold process formula to a medium
viscosity cream while stabilizing the emulsion from bleed and
separation. Potassium oleate allows good cleansing action at a
pH of 8. The mineral oil minimizes defatting of the skin caused
by the deodorized kerosene.

Liquid Waterless Hand Cleaner No. 243

	Veegum	2.0
A	Water	73.0
	Glycerin	4.0
B	Tergitol NP-10	3.0
	2-Amino-2-methyl-1-propanol	0.5
	Deodorized kerosene	10.0
	Oleic acid	1.5
C	Arlacel 186	5.0
	Clearlan	1.0
	Preservative	q.s.

Procedure:
Slowly add Veegum to the water while agitating at maximum
available shear. Continue mixing until smooth. Add B to A and
heat to 50C. Heat C to 55C. Add C to A/B. Mix until uniform.
Add desired preservative.
Consistency: Pourable liquid
Suggested Packaging: Pump dispenser
Comments:
Veegum stabilizes and thickens this liquid emulsion. In this
formulation, glycerin serves as a humectant while the lanolin
helps prevent defatting of the skin.
SOURCE: R.T. Vanderbilt Co.: Suggested Formulations

Section XI
Sun Care Products

Absorbing Sunscreen

Formula:	%Wt.
Phase A:	
Veegum Ultra	1.50
Deionized Water	70.50
Glycerin	5.50
Phase B:	
PEG-150 Distearate	3.00
Dioctyl Malate	2.00
Carnation Mineral Oil	4.00
Cetyl Alcohol	0.50
Benzophenone-3	3.00
Octyl Dimethyl PABA	7.00
Steareth-2	0.90
Steareth-20	2.10
Phase C:	
Preservative, Fragrance	q.s.

Synergistic Sunscreen

4-tert-Butyl-4-methoxydi-benzoylmethane	2.00%
Amyl N,N-dimethyl-p-aminobenzoate	3.00
2-Hydroxy-4-methoxy-benzophenone	2.00
Cetyl alcohol	3.00
Stearic acid	3.00
Protopet 1S Petrolatum	3.00
Olive Oil	3.00
Squalane	5.00
Propylene glycol	3.00
Potassium hydroxide	0.20
Talc	5.00
Trisodium EDTA	0.02
Water	q.s to 100.00

SOURCE: Witco Corp.: Petroleum Specialties Group: Suggested
    Formulations

Absorbing Sunscreen

Formula:
Phase A:

L-Tyrosine	0.20%
Phenylalanine	0.10
Aspartic acid	0.10
Escin	0.05
Copper gluconate	0.01
Hydrolyzed collagen	1.50
Carrot oleoresin	0.15
St. John's Wort extract	1.00
Dihydroxyacetone	0.50
Cetyl alcohol	1.00
Preservative	q.s.
Emulsifiers, thickening agents	4.00
Carnation Mineral Oil	15.00
Lanolin	3.00
Stearin	3.00
Triethanolamine (85%)	1.00
Water	q.s. up to 100.00

Water Resistant SP4 Sunscreen Composition

Water	67.00%
Glycerin	5.00
Polyglyceryl-8 oleate	3.00
Padimate O	4.00
Cyclomethicone	10.00
Carnation Mineral Oil	5.00
Lanolin alcohol	5.00
Propylene glycol (and) diazolidinyl urea (and) methyl paraben (and) propyl paraben (Germaben II)	0.50
Perfume oil	0.50

Non-Alcoholic Sunscreen

Octyl dimethyl PABA	7.00%
Benzophenone-3	3.00
Carnation Mineral Oil	10.00
Oleth-2	6.25
Isoceteth-20	18.75
Water	40.00
Propylene glycol	4.00
Sorbitol	11.00

SOURCE: Witco Corp.: Petroleum Specialties Group: Suggested Formulations

After Sun Lotion

Part I:	Parts by Weight
Water	500.0
Carbomer 934	2.0

Part II:	
Rosswax 2540	6.0
Rosswax 1824	15.0
Coconut Oil #76	25.0
GMS SE	6.0
Ross Jojoba Oil	4.0

Part III:	
Aloe Vera Liquid	10.0

Part IV:	
Germaben II	6.0

Part V:	
Fragrance	q.s.

Part VI:	
Triethanolamine	4.5

Procedure:
   Heat the water in a steam jacketed kettle and add the Carbomer 934 with agitation. In a separate jacketed kettle heat Part II until clear. Next add Part III, then Part IV, then Part V, fragrance and finally add Part VI. Cool to 130F and package.

Solar Tanning Oil Mousse

Part (A):	%
Ross Base Oil 2539	62.3
Escalol 507	5.0
Arlacel 60	3.0
Tween 60	4.0

Part (B):	
Water	24.7
Germaben II	1.0
Fragrance	q.s.

Procedure:
   Heat Part (A) and Part (B) in separate stainless steel vessels under gentle agitation to 170F. When temperature is reached and both are clear, add Part (B) to Part (A), cool to 120F, Fragrance and package.
Aerosol Fill: 90% of above concentrate
                10% of A-46 Propellent
Note: Pack in Epon lined cans with Precision Valve Systems.

SOURCE: Frank B. Ross Co., Inc.: Suggested Formulations

Aloe After Sun Lotion (40% Aloe)

Ingredients:	Percent
A. Water	74.00
Glycerin	3.0
Triethanolamine	1.0
Germaben II	0.5
B. Stearic Acid	8.0
Light mineral oil	5.0
Finesolv TN	2.0
Cetyl Alcohol	1.0
Silicon Fluid 225	0.5
Cocoa Butter	2.0
Isopropyl Lanolate	2.0
C. Aloe Veragel Liquid Concentrate 1:40	1.0
D. Fragrance	q.s.

Procedure:
Heat phases to 80C. At 80C add oil phase to water phase.
Mix and cool to 55C. Add Aloe concentrate to batch at 55C.
Add fragrance at 45C.

Suntan Lotion

Ingredients:	Percent
A. Lexemul 561	4.00
Stearic acid	2.00
Cetyl alcohol	1.00
Mineral oil	4.00
Sesame oil	1.00
Laneth 10 Acetate	0.50
Ethyl Dihydroxypropyl PABA	3.00
Mink oil	0.10
Polysorbate 20	0.20
Lexgard P	0.05
B. Propylene Glycol	4.00
Carbomer 934	0.20
Triethanolamine	qs to pH 7.2-7.5
Aloe Vera Liquid 1:1	10.00
Lexgard M	0.15
Water	qs to 100

Procedure:
Heat water to 70C. Disperse Carbomer 934. Mix until dissolved,
then neutralize. Dissolve Lexgard M in propylene glycol and add
to the water with the remaining contents of Part B. Separately
melt contents of Part A together. Slowly add Part A to Part B
with agitation. Cool to set point and fill. Full conditioning
develops in 24 hours.
SOURCE: Dr. Madis Laboratories Inc.: Suggested Formulations

Aloe Suntan Lotion

Ingredients:	Percent
A. Water	q.s.
Glycerin	4.0
Glucamate SSG-20	1.5
Carbopol 934	0.1
Preservatives	q.s.
B. Glucate-SS	1.5
Cetyl alcohol	1.0
Cetyl palmitate	1.0
Glyceryl stearate	0.22
P.E.G. 100 stearate	0.28
Stearic acid	4.0
Escalol 507	4.0
Mineral oil	5.0
C. A.M.P.-95	0.32
D. Aloe Veragel 200 Powder	0.1
Water	9.9

Procedure:
   Dissolve the S.D. Aloe Vera in water to prepare gel. Heat
balance of water to 80C and dissolve Carbopol. Add rest of Phase
"A" and keep temperature at 80C. Heat phase "B" to 80C. and add
to "A". Add phase "C". Cool to 35C. and add "D". Mix well.

After Sun Lotion

Ingredients:	Percent
A. Water	500.0 gr.
Carbomer	2.0 gr.
B. Rosswax 2540	6.0 gr.
Rosswax 1824	15.0 gr.
Coconut Oil #6	15.0 gr.
Gms SE	6.0 gr.
Ross Jojoba Oil	4.0 gr.
C. Aloe Vera Liquid 1:1	10.0 gr.
D. Germaben II	6.0 gr.
E. Fragrance	q.s.
F. Triethanolamine	4.5 gr.

Procedure:
   Heat the water in steam jacketed kettle and add the Carbomer
934 with agitation. In a separate jacketed kettle, heat B till
clear. Next add B to the water phase with agitation. Next add
C, then D, then E, fragrance and finally add F. Cool to 130F and
package.
SOURCE: Dr. Madis Laboratories Inc.: Suggested Formulations

Buffered Self Tanning Lotion

Resulting product is a pourable lotion that could be dispensed from an opaque glass or rigid plastic bottle. The formulation utilizes a 5% Dihydroxyacetone (DHA) as the self tanning ingredient. The formulation contains no amines to react with Dihydroxyacetone and the final pH is adjusted to approximately 4, to insure natural colored tan development. The formulation applies easily and evenly. The finished product should be stored away from light and at temperature conditions below 40C to prevent DHA from degrading.

	%w/w
A. Deionized Water	62.18
Methylparaben (Lexgard M)	0.20
2,4-Dichlorobenzyl Alcohol (Myacide SP)	0.20
Propylene Glycol USP	3.00
Glycerin	1.00
Citric Acid USP-FCC (Anhydrous Fine Granular)	0.24
Sodium Citrate USP-FCC	0.28
B. Hydroxyethylcellulose (Cellosize QP-15,000H)	0.20
C. Glyceryl Stearate (Lexemul 515)	2.00
Glyceryl Stearate (and) PEG-100 Stearate (Lexemul 561)	4.00
Sorbitan Stearate (Arlacel 60)	0.60
Propylene Glycol Dicaprylate/Dicaprate (Lexol PG-865)	2.00
Caprylic/Capric Triglyceride (Lexol GT-865)	3.00
Octyl Stearate (Lexol EHS)	1.00
Propylparaben (Lexgard P)	0.10
D. Dihydroxyacetone	5.00
Deionized Water	15.00

Procedure:
1. Combine section "A" heating to 75-80C. Use a propeller mixer for agitation.
2. Sprinkle section "B" into section "A". Increase mixing speed to create a vortex during addition, then slow to highest speed which does not produce a vortex.
3. Combine section "C". Heat to 80-85C with continuous slow mixing.
4. When sections "AB" and "C" are homogeneous and at the designated temperatures add section "C" to sections "AB". Increase mixing speed to a high speed. Begin cooling. Slow mixing speed to highest speed possible that will not cause vortexing.
5. Using gloves, combine section "D" and mix until clear.
6. Add section "D" to sections "ABC" at 35-40C. Continue mixing and cooling.
7. At 30-35C adjust for water loss.
8. Mix batch until it reaches room temperature.
9. Adjust final pH to 4.00-4.10 with Citric Acid or Sodium Citrate solutions, as required.

Physical Properties:
Viscosity: 1,700 cPs @ 24C (Brookfield RVT, TA @ 10 rpm)
pH: 1.0 @ 24C

SOURCE: Inolex Chemical Co.: Formulation 398-101-1

Cooling After Sun Lotion

Ingredients:	%w/w
Phase A:	
Polyglyceryl-3 Methyl Glucose Distearate (Tego Care 450)	2.0
Caprylic/Capric Triglycerides (Tegosoft CT)	6.5
Mineral Oil	5.7
Phase B:	
Glycerin	3.0
Water	71.4
Phase C:	
Mineral Oil	0.8
Carbomer 941	0.2
Ethanol	10.0
Phase D:	
Sodium Hydroxide (10% solution)	0.4
Phase E:	
Fragrance	Q.S.
Preservatives	Q.S.

Procedure:
1. Heat the ingredients of Phase A to 80C.
2. Heat the ingredients of Phase B to 80C.
3. Add A to B or B to A without stirring. 4. Stir.
5. Disperse Carbomer into the oil/ester add to A/B. Homogenize
6. Cool to 35-40C with stirring. 7. Add Ethanol.
8. Add phase D/E. Stir.
9. Mix until viscosity is correct.

Suntan Lotion

Ingredients:	%w/w
Phase A:	
Polyglyceryl-3 Methyl Glucose Distearate (Tego Care 450)	2.0
Decyl Oleate (Tegosoft DO)	6.5
Isopropyl Palmitate (Tegosoft P)	5.7
Octyl Methoxycinnamate	7.0
Oxybenzone	3.0
Phase B:	
Glycerin	3.0
Water	71.4
Phase C:	
Isopropyl Palmitate (Tegosoft P)	0.8
Carbomer 941	0.2
Phase D:	
Sodium Hydroxide (10% solution)	0.4
Phase E:	
Fragrance	Q.S.
Preservatives	Q.S.

Procedure:
1. Heat the ingredients of Phase A to 80C. 2. Heat the ingredients of Phase B to 80C. 3. Add A to B or B to A without stirring. 4. Stir. 5. Disperse Carbomer into the oil/ester add to A/B. Homogenize. 6. Cool to 35-40C with stirring. 7. Add Phase D/E. Stir. 8. Mix until viscosity is correct.

SOURCE: Goldschmidt Chemical Corp.: Suggested Formulations

Daily UV Protection Lotion (Approximate SPF 18)

Dermacryl LT provides increased moisture protection, rub-off resistance and water resistance as well as improved rub-in properties.

Ingredients:	%W/W
Phase A:	
Neo Heliopan AV	7.50
Brij 76	1.00
Cerasynt Q	1.50
Cetyl Alcohol	1.00
Emersol 132	1.50
Tioveil FIN	1.70
Finsolv TN	5.00
DC 334 Fluid	3.00
DC 556 Fluid	1.00
Abil B8852	0.50
Phase B:	
Deionized Water	40.30
Triethanolamine (99%)	4.00
Dermacryl LT	1.00
Neo Heliopan Hydro	4.00
Carbopol 940 (2% Aq. Soln.)	20.00
Phase C:	
Propylene Glycol	3.00
Dry-Flo PC	3.00
Phase D:	
Germaben IIE	1.00

Procedure:
Phase B:
   Combine Triethanolamine 99% and Deionized Water, heat to 60C. Slowly sift in Dermacryl LT, heat to 80C. When complete, sift in Neo Heliopan Hydro and Carbopol 940, mix until complete.

Phase A:
   Combine and heat to 80C. Add Phase A to Phase B at 80C, mix for 15-30 minutes. Cool to 40C. Slurry Dry-Flo PC in Propylene Glycol, add to A and B at 40C and mix thoroughly. Add Phase D. Cool to room temperature and package.

SOURCE: National Starch and Chemical Co.: Formula 7528-149B

European O/W Sun Protection Cream

Ingredients:	%W/W
A Mineral oil	10.00
Isohexadecane	5.00
Octyl Methoxycinnimate	5.00
Cyclomethicone	2.80
Arlamol E	1.20
Behenyl alcohol	2.00
Tocopherol Acetate	5.50
B Arlatone 2121	5.50
Glycerol	4.00
*Preservative	
Water	62.85
C Xanthan gum	0.10
D *Perfume	

*q.s. these ingredients.

Procedure:
Melt Arlatone 2121 and add to remainder of (B) at 80C.
Disperse (C) and (B) at 75C with moderate stirring. Add heated
(A), (E), and (F) to mixture with vigorous stirring. Homogen-
ize mixture between 75C and 65C and cool to 40C with moderate
stirring. Homogenize at 30C to 40C with addition of (D).

O/W Moisturizing Sunscreen Cream

Ingredients:	%W/W
A Octyl dimethyl PABA	7.00
Benzophenone-3	3.00
Mineral oil	5.00
Stearyl alcohol	0.50
Brij 721	2.00
Brij 72	2.00
Dimethicone	0.50
B Water	79.60
Carbomer 940	0.20
C Sodium hydroxide (10% W/W aqueous)	0.20
D *Preservative and Fragrance	

*q.s. these ingredients.

Procedure:
Disperse Carbomer in water and heat (B) to 70C. Heat (A) to
72C. and add (B) to (A) with propeller agitation. Slowly add
(C) and increase speed of the agitation as needed. Add (D)
and replace water lost by evaporation.
SOURCE: ICI Surfactants: Suggested Formulations

Lip Balm with Sunscreen-A

Ingredients:	%w/w
Phase A:	
Petrolatum	40.7
Cetyl Alcohol	4.0
Beeswax	6.0
Carnauba Wax	6.4
Paraffin	16.4
Ozokerite	6.0
Cetyl Dimethicone (Abil Wax 9801)	1.0
Stearyl Dimethicone (Abil Wax 9800)	1.0
Caprylic/Capric Triglycerides (Tegosoft CT)	5.0
Cetearyl Octanoate (Tegosoft Liquid)	3.0
Mineral Oil	5.0
Phase B:	
Octyl Methoxycinnamate	4.0
Benzophenone-3	1.5
Phase C:	
Color	Q.S.
Fragrance	Q.S.

Lip Balm with Sunscreen-B

Ingredients:	%w/w
Phase A:	
Petrolatum	44.5
Cetyl Alcohol	6.0
Beeswax	14.0
Carnauba Wax	2.0
Paraffin	8.0
Ozokerite	3.0
Cetyl Dimethicone (Abil Wax 9801)	0.5
Stearyl Dimethicone (Abil Wax 9800)	0.5
Caprylic/Capric Triglycerides (Tegosoft CT)	8.0
Cetearyl Octanoate (Tegosoft Liquid)	3.0
Mineral Oil	5.0
Phase B:	
Octyl Methoxycinnamate	4.0
Benzophenone-3	1.5
Phase C:	
Color	Q.S.
Fragrance	Q.S.

Procedure:
1. Heat Phase A ingredients together until melted. Begin cooling.
2. Add Phase B, mix until uniform.
3. Add color and fragrance when batch is cooled to a creamy consistency. 4. Mold.
SOURCE: Goldschmidt Chemical Corp.: Suggested Formulations

Non-Chemical Sunscreen Lotion (approx. SPF 15)

This lotion incorporates titanium dioxide as the sunscreen agent. It has a smooth feel upon application and rubs in easily without whitening. The Geahlene 750 adds thickness and may contribute to water resistance and product stability.

Ingredient/Trade Name:	Weight%
A Deionized Water	64.80
Magnesium Aluminum Silicate/Veegum Ultra	0.80
Xanthan Gum/Keltrol	0.25
Methylparaben	0.20
Tetrasodium EDTA/Hamp-Ene 220	0.10
B Mineral Oil (and) Hydrogenated Butylene/Ethylene/ Styrene Copolymer (and) Hydrogenated Ethylene/ Propylene/Styrene Copolymer//Geahlene 750	10.00
C12-15 Alkyl Benzoate/Finsolv TN	10.00
Titanium Dioxide/Micronized TiO2 LA-20	6.00
Phenoxyethanol/Emeressence 1160	0.70
Stearic Acid/Emersol 132	2.50
Glyceryl Stearate SE/Lexemul T	2.00
DEA-Cetyl Phosphate/Amphisol	2.50
Propylparaben	0.10
C Fragrance	0.05

Procedure:
    Disperse Veegum Ultra in rapidly agitated deionized water. Mix well. Add Keltrol. Mix until uniform. Heat to 80C. Add remaining part A ingredients. Homogenize Finsolv TN and Micronized Titanium Dioxide until smooth. Add remaining part B ingredients. Heat to 80-85C. Mix until all the added solids are completely dissolved. Add part B to part A while mixing with good agitation. Mix for 30 minutes. until homogeneous. Cool to 40C. Add part C. Continue mixing and cooling to 30C.

SOURCE: Penreco: Suggested Formulation

O/W After Sun Milk

Ingredients:		%W/W
A	Arlamol S7	10.00
	Arlamol HD	8.00
B	Arlatone 2121	2.00
	Alpantha	0.50
	Glycerol	2.50
	*Preservative	
C	Water	76.78
	Carbopol EDT	0.07
D	*NaOH (10% solution)	
E	Perfume	0.15
	*q.s. these ingredients.	

Procedure:
   Disperse Carbopol in water while stirring. Heat to 80C. Mix the Arlatone 2121 into the heated (C) phase while stirring moderately. Homogenise this mixture until a homogeneous dispersion formed. Heat (A) to 80C and add to (B+C) while stirring intensively. Homogenise the mixture between 75 and 65C. Allow to cool down to 50C while stirring. Add (D) until pH of 6.5 to 7 is reached. Add (E) at 40C while stirring.

W/O After Sun Cream

Ingredients:		%W/W
A	Arlacel 582	2.50
	Arlatone T	0.50
	Magnesium stearate	0.50
	Beeswax	1.50
	Candelilla wax	0.50
	Arlamol HD	15.00
	Sunflower seed oil	3.00
B	a-Tocopherol acetate	1.00
	Tegiloxan 100	0.50
	Oxynex K liquid	0.05
	Aerosil R972	0.30
C	Atlas G-2330	4.00
	MgSO4.7H2O	0.50
	Alpantha	0.50
D	*Preservative	
	Water	69.45
E	Perfume Dinalya 129.051	0.20
	*q.s. these ingredients.	

Procedure:
   Mix (A and B). Mix (C+D). Heat (A+B) and (C+D) separately to 75C. Add (C+D) to (A+B) while stirring intensively. Homogenise for 1 minute. Allow to cool to 35C while stirring. Add (E) while stirring. Homogenise again for 1 minute. Package.

SOURCE: ICI Surfactants: Suggested Formulations

O/W Cream (SPF 5)
[UVA/UVB Ratio 0.9]

Phase A:	%W/W
Glycerine	5.00
Glucamate SSE-20	1.50
Glucate SS	1.50
Glyceryl Monostearate	5.00
Myritol 318	5.00
Spectraveil TG (60% solids dispersion)	10.00
Phase B:	
Veegum Regular	1.50
Kelzan	0.30
Methyl Paraben	0.15
Demineralised water	69.60
Phase C:	
Germall II	0.25
Perfume	0.20

Manufacture:
* Heat both phases to 65-70C mixing with high shear mixer.
* Add phase A to phase B with mixing.
* Stir on the high shear mixer allowing product to cool.
* When below 60C add the Germall II. When below 40C add the perfume * At 35C stop stirring and allow to cool.
    In-Vitro SPF Value: 4-6   In-Vivo SPF Value: 5.9
Formulation Number 22

W/O Sun Protection Cream (SPF 15+)
[UVA/UVB Ratio 0.75]

Phase A:	%W/W
DC200/100cs	2.00
Elfacos C26	5.00
Elfacos E200	2.00
Elfacos ST9	4.00
Cocoa Butter	3.00
Propyl Paraben	0.08
Spectraveil 90/MOTG (40% solids dispersion)	25.00
Phase B:	
Propylene Glycol	5.00
Methyl Paraben	0.10
Demineralised water	53.82

Manufacture:
* Into a suitable vessel weigh out the ingredients of phase A (except the Elfacos C26). Bring the temperature of the oils slowly to 75C with occasional stirring. The Elfacos C26 constitutent is warmed to 85C and added separately.
* Pre-mix the Propylene Glycol and Methyl Paraben. Add to the water, stirring well and heat to 75C.
* When both phases are completely dissolved and at 75-80C, add phase B to phase A while mixing with a high shear mixer. Mix for five minutes and then commence cooling.
* Cool to approximately 25C with slow speed mixing.
* Mix briefly at high speed until smooth.
    In-Vitro SPF Value: 17-21   In-Vivo SPF Value: 20.5
Formulation Number 24
SOURCE: Tioxide Specialties Ltd.: SPECTRAVEIL Custom UV
         Protection

O/W Hydroalcoholic Sunscreen

Ingredients:	%w/w
Phase A:	
Glyceryl Stearate (and) Ceteth-20 (Teginacid H)	4.0
Cetyl Dimethicone (Abil Wax 9801)	2.0
Stearyl Alcohol	2.0
Mineral Oil	3.5
Cyclomethicone (Abil B 8839)	5.0
Stearoxy Dimethicone (Abil Wax 2434)	1.5
Isopropyl Stearate (Tegosoft S)	5.0
Octyl Methoxycinnamate	4.0
Benzophenone-3	2.0
Phase B:	
Propylene Glycol	3.0
Carbomer 934 (1.5% - NaOH Neutralized)	0.3
Water	45.7
Preservatives	Q.S.
Phase C:	
SD Alcohol 40A	20.0
Fragrance	Q.S.

Procedure:
1. Combine the ingredients of Phase A. Heat to 80C.
2. Combine the ingredients of Phase B. Heat to 80C.
3. Mix Phases A & B. Cool to 60C while mixing. Maintain a milky appearance at all times. Homogenize.
4. Cool to 45C with sweep mix.
5. Add Phase C. Rehomogenize. Cool to 25C with sweep mix.

W/O Sunscreen
(SPF 17)

Ingredients:	%w/w
Phase A:	
Cetyl Dimethicone Copolyol (and) Polyglyceryl-4 Isostearate (and) Hexyl Laurate (Abil WE-09)	4.0
Cetyl Dimethicone (Abil Wax 9801)	2.0
Polyethylene Wax	1.0
Benzophenone-3	2.0
Octyl Salicylate	3.0
Octyl Methoxycinnamate	2.0
Sesame Oil	1.0
Isopropyl Palmitate (Tegosoft P)	3.0
Cyclomethicone (Abil B 8839)	4.0
Octyl Stearate (Tegosoft OS)	3.0
Octyl Palmitate (Tegosoft OP)	2.5
Phase B:	
Water	71.4
Sodium Chloride	0.5
Polyquaternium-10	0.3
Germaben II	0.3

Procedure:
1. Add the components of Phase A. Heat while mixing to 80C to incorporate the waxes. Cool to 50C.
2. Heat Phase B to 50C. Add B to A slowly with a low energy mixer. Maintain a smooth milky appearance at all times.
3. Cool to 35C with sweep mixer. Add fragrance. 4. Homogenize
SOURCE: Goldschmidt Chemical Corp.: Formulations BAM-1-97

O/W Moisturizing Sunscreen Cream

Ingredients:	%W/W
A Octyl dimethyl PABA	5.00
Mineral oil	5.00
Stearyl alcohol	0.50
Brij 721	2.00
Brij 72	2.00
Dimethicone	0.50
Water	84.60
B Carbomer 940	0.20
C Sodium hydroxide (10% W/W aqueous)	0.20
D *Preservative and fragrance	

*q.s. these ingredients

Procedure:
Disperse Carbomer in water and heat (B) to 70C. Heat (A) to 72C. and add (B) to (A) with propeller agitation. Slowly add (C) and increase speed of the agitation as needed. Add (D) and replace water lost by evaporation.

W/O Sunscreen Cream

Ingredients:	%W/W
A Arlacel 186	3.00
Sorbo	27.00
B Ceresin wax	1.00
Beeswax	1.00
Mineral oil	11.00
Sunscreen agent	6.00
C Water	51.00
D *Preservative	

*q.s. these ingredients.

Procedure:
Add Sorbo to Arlacel 186 slowly with continuous agitation. Add (B) to (A) and heat to 70C. Add (C) and mix to a uniform dispersion. Heat (D) to 72C. and add to (A,B,C) with agitation. Cool and mill to improve smoothness and shelf life.

SOURCE: ICI Surfactants: Suggested Formulations

O/W Sunscreen Lotion

		Wt.%
1.	A-C Polyethylene 617	1.0
2.	A-C Copolymer 540	1.0
3.	Escalol 507	5.0
4.	Dow 556 Fluid	2.0
5.	Propylene Glycol Dipelargonate	10.5
6.	Hydroxyol	2.0
7.	Ethoxyol 24	1.0
8.	Arlacel 60	1.3
9.	Tween 60	1.8
10.	Propyl-P-Hydroxybenzoate	0.1
11.	Sorbitol	5.0
12.	Carbopol 941	0.5
13.	Germall 115	0.4
14.	Methyl-P-Hydroxybenzoate	0.2
15.	Triethanolamine	0.75
16.	Water	68.45

Procedure:

Disperse Carbopol in water. Weigh 1-10 and heat to 80-90C with slow agitation. Add remaining ingredients to Carbopol/water dispersion, except triethanolamine, and heat to 80-90C. Add the wax phase to the aqueous phase and shear in homomixer. Continue shearing while cooling to 40C, then add triethanolamine, mixing well. Cool to 30C, add perfume, deaerate and package.
Ref: 5189-2-6

O/W Sunscreen Lotion

		Wt.%
1.	A-C Copolymer 580	2.0
2.	Distilled Isopropyl Lanolate	3.0
3.	Escalol 507	5.0
4.	Dow 556 Fluid	2.0
5.	Propylene Glycol Dipelargonate	10.0
6.	Ethoxyol 24	1.0
7.	Arlacel 60	1.0
8.	Tween 60	2.0
9.	Propyl-P-Hydroxybenzoate	0.1
10.	Sorbitol (70%)	5.0
11.	Carbitol 941	0.5
12.	Triethanolamine	0.75
13.	Methyl-P-Hydroxybenzoate	0.2
14.	Water	67.45

Procedure:

Disperse Carbopol in water. Weigh 1-9 and heat to 80-90C with slow agitation. Add remaining ingredients to Carbopol/water dispersion, except triethanolamine, and heat to 80-90C. Add the water phase to the aqueous phase and shear in homomixer. Continue shearing while cooling to 40C, then add triethanolamine, mixing well. Cool to 30C, add perfume, de-aerate and package.
Ref: 5189-2-9

SOURCE: Allied Signal Inc. Suggested Formulations

O/W Sunscreen Lotion

Ingredients:	%W/W
A Mineral oil	24.50
Arlacel 60	1.50
Tween 60	8.50
Benzyl salicylate	2.00
Benzyl cinnamate	2.00
B Water	61.50
*Preservative	
*q.s. these ingredients.	

Procedure:
    Heat (A) to 60C. Heat (B) to 62C. Add (B) to (A) with thorough but gentle stirring. Cool to 25C with gentle stirring.

O/W Sunscreen Lotion

Ingredients:	%W/W
A Mineral oil	18.80
Cetyl alcohol	5.00
Arlacel 60	2.50
Tween 60	7.50
Amyl para-dimethylaminobenzoate	1.20
B Water	65.00
*Preservative	
*q.s. these ingredients.	

Procedure:
    Heat (A) to 55C. and (B) to 60C. Add (B) to (A) with agitation. Stir until cool.

O/W Emollient Sunscreen Lotion

Ingredients:	%W/W
A Octyl dimethyl PABA	5.00
Arlamol E	7.00
Stearyl alcohol	2.50
Dimethicone	1.00
Arlasolve 200	3.10
Brij 72	3.90
B Water	77.10
Carbomer 934	0.20
C Sodium hydroxide (10% W/W aqueous)	0.20
D *Preservative and fragrance	
*q.s. these ingredients.	

Procedure:
    Disperse Carbomer in water and heat (B) to 70C. Heat (A) to 72C. and add (B) to (A) with propeller agitation. Slowly add (C) and increase speed of the agitation as needed. Add (D) and replace water lost by evaporation.

SOURCE: ICI Surfactants: Suggested Formulations

O/W Sunscreen Lotion (SPF 12)

Phase A:	%W/W
Arlacel 165	5.00
Stearyl Alcohol	0.50
Span 60	0.50
Tween 60	0.90
Petrolatum	3.00
DC200/350	1.00
Sweet Almond Oil	5.00
Mineral Oil	8.00
Evening Primrose Oil	0.50

Phase B:	
Deionized water	56.76
Sorbic Acid	0.20
Glycerine	5.00
Tioveil AQ	12.50
Keltrol	0.10

Phase C:	
Deionised water	1.00
Bronopol	0.04

Manufacture:
* Heat the deionised water, sorbic acid and glycerine to 75C,
  stirring occasionally until all the sorbic acid has dissolved.
* Sprinkle in the Keltrol and homogenise until this is fully
  dispersed. (or: paddle stir for 1 hour to hydrate the
  Keltrol). Maintain temperature at 75C.
* Add the Tioveil AQ and homogenise briefly.
* Heat phase A to 75C.
* Add A to B and homogenise for appropriate time. Paddle stir
  to cool.
* Add phase C at 40C. Continue cooling with stirring to 25C.

In-Vitro SPF Value: 12-15

SOURCE: Tioxide Specialties Ltd.: TIOVEIL Physical Sunscreen

O/W Sunscreen Lotion (SPF 15)

Phase A:	%W/W
DC 593 fluid	7.00
Mineral oil	6.00
Crodamol ML	2.00
Cetyl alcohol	2.00
GMS/SE	1.20
Eumulgin B2	0.40
Stearic acid	1.00
Propylparaben	0.05
Phase B:	
Deionised water	58.45
Glycerine	1.00
Triethanolamine	1.00
Aloe vera gel 10:1	0.50
Methylparaben	0.15
Tioveil AQ	18.75
Perfume	0.20
Germall 115	0.30

Manufacture:
* Heat phases A and B separately to 70-75C.
* Add phase A to phase B and homogenize for appropriate time.
* Cool down to 50C with slow stirring, then add perfume
* Continue cooling with stirring. Add Germall at 35C. Stop
  stirring at 30C.
  In-Vitro SPF Value: 17-21   In-Vivo SPF Value: 17.6
  Viscosity: 60000 mPas (Brookfield LVT, Spindle E, 3 rpm)
Formulation Number 9

O/W Sunscreen Lotion (SPF 10)

Phase A:	%W/W
Isopropyl myristate	4.00
Mineral oil	6.50
Grape seed oil	2.50
Stearyl alcohol	1.00
Petrolatum	2.00
Tioveil MOTG	5.00
Phase B:	
Deionised water	62.90
Grilloten LSE87K	0.10
Grilloten PSE141G	6.00
Glycerine	4.00
Allantoin	0.20
d-Panthenol	0.80
Phase C:	
Tioveil AQ	5.00
Phase D:	
Preservative	qs

Manufacture:
* Heat A and B separately to 80C. * Add A to B, stirring inten-
  sively. Add C with intensive stirring. * Homogenize
* Allow to cool for 25C with slow agitation. Add D.
  In-Vitro SPF Value: 12-15   Viscosity: 144 mPas (Brookfield)
Formulation Number 18
SOURCE: Tioxide Specialties Ltd.: TIOVEIL Physical Sunscreens

O/W Sunscreen Lotion (SPF 15)

Phase A:	%W/W
Octyl Palmitate	5.00
Crodamol CAP	5.00
DC200/100cs	3.00
Oxynex 2004	0.05
White Beeswax	3.50
Phase B:	
Arlatone 2121	5.50
Keltrol RD	0.20
Veegum Ultra	0.80
Sodium Lactate (70% soln.)	0.30
Demineralised water	61.65
Tioveil AQ	15.00
Preservatives	qs

Manufacture:
* Mix water and sodium lactate and heat to 80C
* Melt Arlatone 2121 and add this to the hot water phase with vigorous stirring
* Disperse Keltrol and Veegum in the hot water phase
* Add Tioveil AQ with vigorous stirring
* Heat phase A to 75-80C, until fully melted and homogeneous.
* Heat phase A to phase B with high-shear mixing.
* Cool with moderate stirring to 30C.Homogenize briefly when cold
In-Vitro SPF Value: 15-19   Viscosity: 49680 mPas
Formulation Number 19

W/O Sunscreen Lotion (SPF 20)

Phase A:	%W/W
Abil EM90	2.00
Mineral oil	5.00
Arlamol HD	5.00
Caprylic-capric Triglyceride	5.00
Octyl Palmitate	8.00
Arlamol S3	5.00
Naturechem GMHS	0.80
Lunacera M	1.20
Phase B:	
Demineralised water	47.50
Sodium Chloride	0.50
Preservative	qs
Phase C:	
Tioveil MOTG	20.00

Manufacture:
* Heat phases A and B separately to 75C * Heat phase C to >60C
* Slowly add B to A with intensive stirring * Add C with intensive stirring * Cool to 25-30C, maintaining the same stirrer speed * Homogenize
In-Vitro SPF Value: 20-24   Viscosity: 2000 mPas (Brookfield)
Formulation Number 20

SOURCE: Tioxide Specialties Ltd.: TIOVEIL Physical Sunscreens

<u>O/W Vitaminized Broad Spectrum Sunscreen Lotion</u>
<u>(Indicative SPF 8)*</u>

Ingredients:	% w/w
A) Parsol MCX	2.00
Parsol 1789	1.00
Parsol 5000	0.60
Glyceryl Monomyristate	4.00
Ganex V-220	2.00
Cetyl Alcohol Extra	1.00
Dermol 185	10.00
Amphisol K	2.00
Edeta BD	0.10
Butylated Hydroxytoluene	0.05
Phenonip	0.60
B) Deionized Water	34.93
Carbopol 981 1% sol'n	10.00
Propylene Glycol	5.00
C) KOH 10% sol'n	0.64
D) Deionized Water	20.00
Parsol HS	0.60
KOH 10% sol'n	1.08
E) Vitamin E Acetate	2.00
F) Panthenol	2.00
G) Parfex 49915	0.40**

Procedure:
Heat part A) to 85C while stirring. When homogeneous, add
parts B) and C) pre-heated to 75C, while mixing. Add part D)
pre-heated to 75C, while mixing (be sure that the Parsol HS
has been completely dissolved, if traces remain, add a small
quantity of neutralizing base until the solution is clear).
Cool to 40C, add parts E), F) and G). Compensate for water
loss and continue stirring while cooling to ambient temperature.

Remark:
  *SPF determination: IRI Ref. 582627. Test method based on
   FDA/OTC conditions with six subjects.
 **Dermatologically tested perfume for sunscreens.
   The final pH value should be around 7.0 to prevent recry-
   stallization of Parsol HS

SOURCE: Givaudan-Roure: Formulation 38 COS 085

O/W Vitaminized Broad Spectrum Sunscreen Lotion
(Indicative SPF 25+)*

Ingredients:	%w/w
A) Parsol MCX	5.50
Parsol 1789	2.50
Parsol 5000	2.00
Glyceryl Monomyristate	4.00
Ganex V-220	2.00
Cetyl Alcohol Extra	1.00
Dermol 185	10.00
Amphisol K	2.00
Edeta BD	0.10
Butylated Hydroxytoluene	0.05
Phenonip	0.60
B) Deionized Water	24.61
Carbopol 981 1% sol'n	10.00
Propylene Glycol	5.00
C) KOH 10% sol'n	0.64
D) Deionized Water	20.00
Parsol HS	2.00
KOH 10% sol'n	3.60
E) Vitamin E Acetate	2.00
F) Panthenol	2.00
G) Perfex 49915	0.40**

Procedure:
   Heat part A) to 85C while stirring. When homogeneous,
add parts B) and C) pre-heated to 75C, while mixing. Add part
D) pre-heated to 75C, while mixing (be sure that the Parsol
HS has been completely dissolved, if traces remain, add a
small quantity of the neutralizing base until the solution
is clear). Cool to 40C, add Parts E), F), and G). Compensate
for water loss and continue stirring while cooling to ambient
temperature.

Remark:
   *SPF determination: IRI Ref. 582627. Test method based on
   FDA/OTC conditions with five subjects.
   **Dermatologically tested perfume for sunscreens.
   The final pH value should be around 7.0 to prevent recrys-
   tallization of Parsol HS.

SOURCE: Givaudan-Roure: Formula 38 COS 087

O/W Vitaminized Broad Spectrum Sunscreen Lotion
(Indicative SPF 15)*

Ingredients:	%w/w
A) Parsol MCX	3.30
Parsol 1789	1.50
Parsol 5000	1.20
Glyceryl Monomyristate	4.00
Ganex V-220	2.00
Cetyl Alcohol Extra	1.00
Dermol 185	10.00
Amphisol K	2.00
Edeta BD	0.10
Butylated Hydroxytoluene	0.05
Phenonip	0.60
B) Deionized Water	30.86
Carbopol 981 1% sol'n	10.00
Propylene Glycol	5.00
C) KOH 10% sol'n	0.64
D) Deionized Water	20.00
Parsol HS	1.20
KOH 10% sol'n	2.15
E) Vitamin E Acetate	2.00
F) Panthenol	2.00
G) Parfex 49915	0.40**

Procedure:
Heat part A) to 85C while stirring. When homogeneous, add parts B) and C) pre-heated to 75C, while mixing. Add part D) pre-heated to 75C, while mixing (be sure that the Parsol HS has been completely dissolved, if traces remain, add a small quantity of the neutralizing base until the solution is clear). Cool to 40C, add parts E), F) and G). Compensate for water loss and continue stirring while cooling to ambient temperature.

Remark:
  *SPF determination: IRI Ref. 582627. Test method based on FDA/OTC conditions with six subjects.
  **Dermatologically tested perfume for sunscreens. The final pH value should be around 7.0 to prevent recrystallization of Parsol HS.

SOURCE: Givaudan-Roure: Formulation 38 COS 086

Solar Tanning Cream

	Part By Weight
Water Phase:	Soft
Water	427.9
Carbomer 934	3.0
Oil Phase:	
Protox T-25	1.0
Rosswax 573	4.0
GMS SE	4.0
Coconut Oil #76	16.0
Jojoba Oil	4.0
Rosswax 1824	12.0
Escalol 507	25.0
Fragrance	q.s.
Germaben II	5.0
Triethanolamine	4.4

Procedure:
   Disperse the Carbomer 934 in the water. In a second vessel heat the Oil Phase including the Escalol 507 until completely clear. When both phases are ready add the Oil Phase to the Water Phase, add the preservative, fragrance and add the Triethanolamine under high agitation. When fully mixed you may package.

Solar Tanning Cream

	Parts by Weight
Water Phase:	Hard
Water	410.0
Carbomer 934	3.0
Oil Phase:	
Protox T-25	1.0
Rosswax 573	4.0
GMS SE	4.0
Coconut Oil #76	16.0
Jojoba Oil	4.0
Rosswax 1824	12.0
Escalol 507	25.0
Fragrance	q.s.
Germaben II	4.8
Triethanolamine	4.4

Procedure:
   Disperse the Carbomer 934 in the water. In a second vessel heat the Oil Phase including the Escalol 507 until completely clear. When both phases are ready add the Oil Phase to the Water Phase, add the preservative, Fragrance and add the Triethanolamine under high agitation. When fully mixed you may package.
SOURCE: Frank B. Ross Co., Inc.: Suggested Formulations

<u>Solar Tanning Cream</u>
<u>High Protection</u>

	%
Part (A):	
Water	79.9
Carbomer 934	0.6
Part (B):	
Protox T 25	0.2
Rosswax 573	0.8
Rosswax 1824	2.0
Ross Jojoba Oil	0.8
GMS-SE	0.8
Coconut Oil #76	3.0
Escalol 507	7.0
Escalol 567	3.0
Part (C):	
Germaben II	1.0
Part (D):	
Fragrance	q.s.
Part (E):	
Triethanolamine	0.9

Procedure:
    Heat the water in a steam jacketed kettle and add the Carbomer 934 under agitation. Heat Part (B) in a steam jacketed kettle until clear under agitation. When fully mixed add Part (B) to Part (A) under agitation. Then add Part (C) and mix thoroughly. Next add Part (D) and finally add Part (E) with agitation. Cool to 120F and package.

<u>Tanning Jelly</u>

	%
Petrolatum USP	49.0
Mineral Oil #7	20.0
Henkel Cutina-LM	23.9
Ross Jojoba Oil	2.0
Escalol 507	5.0
Propyl Paraben	0.1
Fragrance	q.s.

Procedure:
    Load ingredients in steam jacketed kettle and melt to a liquid state under agitation. When thoroughly mixed, cool to 130F, fragrance and package.

SOURCE: Frank B. Ross Co., Inc.: Suggested Formulations

Solar Tanning Cream
Super Protection

	%
Part (A):	
Water	69.1
Carbomer 934	0.6
Part (B):	
Protox T 25	0.2
Rosswax 573	0.8
Rosswax 1824	2.0
Ross Jojoba Oil	1.0
GMS-SE	0.8
Coconut Oil #76	3.2
Escalol 507	8.0
Escalol 567	4.5
Escalol 557	7.5
Part (C):	
Germaben II	1.0
Part (D):	
Fragrance	q.s.
Part (E):	
Triethanolamine	0.9

Procedure:
   Heat the waxes in a steam jacketed kettle and add the
Carbomer 934 under agitation. Heat Part (B) in a steam jacketed
kettle until clear under agitation. When fully mixed add Part
(B) to Part (A) under agitation. Then add Part (C) and mix
thoroughly. Next add Part (D) and finally add Part (E) with
agitation. Cool to 120F and package.

Jojoba After Sun Lotion

	%
Part A:	
Mineral Oil 60/70	8.2
Modulan	5.0
Rosswax 63-0412	7.6
Propylene Glycol	2.3
Ross Jojoba Oil	1.7
Part B:	
Water	69.7
Aloe Vera Liquid	3.3
Triethanolamine	1.2
Fragrance	q.s.
Germaben II	1.0

Procedure:
   Melt Part (A) and Part (B) in separate vessels to 170F under
agitation. When temperature is reached, mix Part (A) to Part (B)
and cool. Package in containers at below 120F.

SOURCE: Frank B. Ross Co., Inc.: Suggested Formulations

Solar Tanning Lotion-A

Part I:	Parts by Weight
Water	568.0
Carbomer 934	2.0

Part II:	
Rosswax 573	4.0
GMS SE	4.0
Jojoba Oil	4.0
Escalol 507	13.0

Part III:	
Fragrance	q.s.

Part IV:	
Germaben II	6.0

Part V:	
Triethanolamine	4.0

Solar Tanning Lotion-B

Part I:	Parts by Weight
Water	568.0
Carbomer 934	2.0

Part II:	
Rosswax 573	4.0
GMS SE	4.0
Jojoba Oil	4.0
Escalol 507	19.0

Part III:	
Fragrance	q.s.

Part IV:	
Germaben II	6.0

Part V:	
Triethanolamine	4.0

Procedure:
Heat the water with agitation and add the Carbomer 934. In a separate steam jacketed kettle melt Part II until clear. As soon as everything is melted add Part II to Part I with agitation. Then add Part III and Part IV with increased agitation; then add Trithanolamine. Cool to 130F and package.

SOURCE: Frank B. Ross Co., Inc.: Suggested Formulations

Solar Tanning Lotion-C

Part I	Parts by Weight
Water	568.0
Carbomer 934	2.0

Part II:	
Rosswax 573	4.0
GMS SE	4.0
Jojoba Oil	4.0
Escalol 507	25.0

Part III:	
Fragrance	q.s.

Part IV:	
Germaben II	6.0

Part V:	
Triethanolamine	4.0

Solar Tanning Lotion-D

Part I:	
Water	568.0
Carbomer 934	2.0

Part II:	
Rosswax 573	4.0
GMS SE	4.0
Jojoba Oil	4.0
Escalol 507	32.0

Part III:	
Fragrance	q.s.

Part IV:	
Germaben II	6.0

Part V:	
Triethanolamine	4.0

Procedure:
Heat the water with agitation and add the Carbomer 934. In a separate steam jacketed kettle melt Part II until clear. As soon as everything is melted add Part II to Part I with agitation. Then add Part III and Part IV with increased agitation; then add Triethanolamine. Cool to 130F and package.

SOURCE: Frank B. Ross Co., Inc.: Suggested Formulations

<u>Solar Tanning Lotion</u>
<u>High Protection</u>

	Parts by Weight
Water	517.0
Carbomer 934	2.0
Rosswax 573	3.6
Gms SE	3.6
Jojoba Oil	42.0
Escalol 507	18.0
Germaben II	6.0
Fragrance	q.s.
Triethanolamine	3.6

Procedure:
Heat the water with agitation and add the Carbomer 934. In a separate steam jacketed kettle melt the Oil Phase till clear with agitation. Now add the Oil Phase to the Water Phase with agitation, add the Germaben II, fragrance and finally add the Triethanolamine with high agitation. Next cool to 130F and package.

<u>Solar Tanning Lotion</u>
<u>Super Protection</u>

	Parts by Weight
Water	381.5
Carbomer 934	1.5
Rosswax 2540	3.0
GMS-SE	3.0
White Jojoba Oil	3.0
Escalol 507	40.0
Escalol 567	22.5
Escalol 557	37.5
Germaben II	5.0
Fragrance	q.s.
Triethanolamine	3.6

Procedure:
Heat the water in a steam jacketed kettle with agitation and add the Carbomer 934. In a separate steam jacketed kettle melt the Oil Phase until clear with agitation. Now add the Oil Phase to the Water Phase with agitation, then the Germaben II, then the Fragrance and finally add the Triethanolamine with high agitation. Next cool to 130F and package.

SOURCE: Frank B. Ross Co., Inc.: Suggested Formulations

Solar Tanning Oil-A

	%
Cocoanut Oil #76	15.0
Dow Corning #344	16.0
Isopropyl Myristate	13.0
Mineral Oil #7	42.0
Acetulan	8.0
Glucam P-20	2.0
Jojoba Oil	2.0
Escalol 507	2.0
Fragrance	q.s.

Solar Tanning Oil-B

	%
Cocoanut Oil #76	14.0
Dow Corning #344	16.0
Isopropyl Myristate	13.0
Mineral Oil #7	42.0
Acetulan	8.0
Glucam P-20	2.0
Jojoba Oil	2.0
Escalol 507	3.0
Fragrance	q.s.

Solar Tanning Oil-C

	%
Cocoanut Oil #76	13.0
Dow Corning #344	16.0
Isopropyl Myristate	13.0
Mineral Oil #7	42.0
Acetulan	8.0
Glucam P-20	2.0
Jojoba Oil	2.0
Escalol 507	4.0
Fragrance	q.s.

Solar Tanning Oil-D

	%
Cocoanut Oil #76	12.0
Dow Corning #344	16.0
Isopropyl Myristate	13.0
Mineral Oil #7	42.0
Acetulan	8.0
Glucam P-20	2.0
Jojoba Oil	2.0
Escalol 507	5.0
Fragrance	q.s.

Procedure:
   Mix all of the above ingredients in a stainless steel vessel, run thru a filter and package.

SOURCE: Frank B. Ross Co., Inc.: Suggested Formulations

Solar Tanning Stick
White Color

	%
Rosswax 26-1152	15.0
Rosswax 1641	15.0
Rosswax 1824	20.0
Mineral Oil #7	17.5
Dow Silicone 344	8.0
Isopropyl Myristate	6.5
Coconut Oil #76	5.0
Acetulan	4.0
Glucam P-20	2.0
Jojoba Oil	2.0
Escalol 507	5.0
Fragrance	q.s.
Preservative	q.s.

Procedure:
Load the waxes and the oils in a steam jacketed kettle, under agitation until melted. Cool to just before cloudy, add preservatives. Mold in containers. (Note: Capping may be necessary).

Solar Tanning Stick
Tan Color

	%
Rosswax 26-1152	30.0
Rosswax 1824	20.0
Mineral Oil #7	17.5
Dow Silicone 344	8.0
Isopropyl Myristate	6.5
Coconut Oil #76	5.0
Acetulan	4.0
Glucam P-20	2.0
Jojoba Oil	2.0
Escalol 507	5.0
Fragrance	q.s.
Preservative	q.s.

Procedure:
Heat the waxes and the oil in a steam jacketed kettle to 175F under agitation. When mixed fully, cool to just before cloudy, and add Fragrance and Preservative. Mold in containers. (Note: Capping may be necessary).

SOURCE: Frank B. Ross Co., Inc.: Suggested Formulations

Sun Block Soft Cream (O/W)
Tested SPF 26

Ingredients:	%(w/w)
A Arlacel 165	3.00
Eumulgin B 2	1.00
Lanette O	2.00
Myritol 318	4.00
Cetiol OE	3.00
Abil 100	1.00
Bentone Gel MIO	3.00
Neo Heliopan, Type OS	3.00
Neo Heliopan, Type AV	5.00
Neo Heliopan, Type E 1000	5.00
Neo Heliopan, Type MBC	1.50
Cutina CBS	2.00
B Demineralized water	31.70
Veegum ultra	1.00
Glycerin 86%	3.00
Phenonip	0.50
Neo Heliopan, Type Hydro; used as a 15% solution	20.00
neutralized with sodium hydroxide (active: 3.00%)	
Zinc Oxide neutral H&R	10.00
C Perfume oil	0.30

Manufacturing Process:
Part A: Heat up to 75C with thorough agitation.
Part B:
  Heat the water to 90C. Disperse the Veegum in the water during high speed agitation. Then add the other ingredients and disperse the Zinc Oxide neutral H&R in the solution. Put Part B to Part A while stirring.
Part C:
  At 30C add the perfume oil to Part A/B and homogenize the emulsion.

  The pH-value of the finished emulsion should be approx. 7.5 and has to be checked.

SOURCE: Haarman & Reimer: Formulation K 18/2-21302/E

Sun Filter Foam

		Wt%
A.	Miglyol 840	10.0
	Miglyol 812	5.0
	Softisan 649	1.0
	Softigen 767	1.5
	Softigen 701	0.5
	Cetyl alcohol	1.0
	Stearic acid	4.0
	Triethanolamine	2.0
	Neo Heliopan E 1000	3.0
B.	Perfume	qs
	Water	to 100.0

Preparation:
All components under A. are heated to 70-75C. Water is then emulsified in and perfume is incorporated at 30C.
Aerosol filling:
90% Active ingredient
10% Drivosol 27

After Sun Stick

		Wt%
A.	Softigen 767	40.0
	Sodium stearate	8.0
	Glycerol	10.0
	Sugar	8.0
	D-Panthenol, 50%	6.0
	Allantoin	0.2
	Water	24.4
B.	Perfume	0.1
	Ethanol 96%	3.3
	Preservative	qs

Preparation:
The ingredients of A. are melted and dissolved at approx. 60C, the mix is stirred, B. being added at approx. 40C, and when cold it is cast into moulds.

SOURCE: Huls America Inc.: Formulations for Cosmetics: Formulas

Sun Protection Lotion

This medium viscosity lotion utilizes a synergistic Veegum/ Rhodigel blend to help stabilize the emulsion and modify the viscosity. In addition, this formula incorporates A-C 617G polyethylene to provide a luxurious after feel and improve the water resistance of the sun protection film. This product is designed to have an SPF (Sun Protection Factor) of about 15.

Ingredient:	% by Wt.
A Veegum	1.00
Rhodigel	0.20
Glycerin	5.50
Deionized Water	68.80
B A-C 617G Polyethylene	2.00
Glyceryl Monostearate, SE (2)	3.00
Dioctyl Malate (3)	2.00
Cetyl Alcohol	0.50
Mineral Oil	4.00
Steareth-2 (4)	0.30
Steareth-20 (5)	2.70
Benzophenone-3 (6)	3.00
Octyl Dimethyl PABA (7)	7.00
C Preservative, Fragrance	q.s.

(2) Kessco Glycerol Monostearate S.E.
(3) Ceraphyl 45
(4) Brij 72
(5) Brij 78
(6) Escalol 567
(7) Escalol 507

Preparation:
Dry blend Veegum and Rhodigel. Add Veegum/Rhodigel blend to water preheated to 85-90C. Hydrate using maximum available shear until smooth, uniform and free of undispersed particles. Add glycerin and maintain temperature at 85-90C. In a separate container, add all B ingredients and heat to 85-90C until all components are in a liquid state. Stir gently as necessary. Slowly add B to A and homogenize for 5 minutes. Cool emulsion quickly to room temperature with gentle stirring. Add C and mix until uniform.

Consistency:
Flowable Liquid; Viscosity measured after 30 days at room temperature: 1500-1900 cps.

Suggested Packaging: Plastic bottles or tubes.
Comments:
This prototype formula is designed to serve as a guide for the development of new products or improvement of existing ones.

SOURCE: R.T. Vanderbilt Co., Inc.: Formula No. 444

Sun Protection Lotion

This medium viscosity lotion utilizes a synergistic Veegum/ Rhodigel blend to help stabilize the emulsion and modify the viscosity. It is designed to have an SPF (Sun Protection Factor) of about 15.

Ingredient:	% by Wt.
A Veegum	1.00
Rhodigel	0.20
Glycerin	5.50
Deionized Water	70.80
B Glyceryl Monostearate, SE (2)	3.00
Dioctyl Malate (3)	2.00
Cetyl Alcohol	0.50
Mineral Oil	4.00
Steareth-2 (4)	0.30
Steareth-20 (5)	2.70
Benzophenone-3 (6)	3.00
Octyl Dimethyl PABA (7)	7.00
C Preservative, Fragrance	q.s.

(2) Kessco Glycerol Monostearate S.E.
(3) Ceraphyl 45
(4) Brij 72
(5) Brij 78
(6) Escalol 567
(7) Escalol 507

Preparation:
Dry blend Veegum and Rhodigel. Add Veegum/Rhodigel blend to water preheated to 75 to 85C. Hydrate using maximum available shear until smooth, uniform and free of undispersed particles. Add glycerin and maintain temperature at 75 to 85C. In a separate container, add all B ingredients and heat to 75 to 85C until all components are in a liquid state. Stir gently as necessary. Slowly add B to A and homogenize for 5 minutes. Cool emulsion to room temperature with gentle stirring. Add C and mix until uniform.

Consistency: Flowable liquid; Viscosity measured after 30 days at room temperature: 750 to 1000 cps

Comments:
This prototype formula is designed to serve as a guide for the development of new products or improvement of existing ones.

SOURCE: R.T. Vanderbilt Co., Inc.: Formula No. 446

Sunscreen with Amihope

An elegant sunscreen of about SPF-15 using micronized Titanium Dioxide. Amihope LL provides improved spreadability and an excellent afterfeel.

Ingredient/Trade Name:	% by Weight
Part A: Deionized Water	64.30
Synthetic Hectorite	0.80
Disodium EDTA	0.10
Glycerin	6.00
Part B: C12-15 Alkyl Benzoate	15.00
Titanium Dioxide	6.50
Cetearyl Alcohol (and) Ceteareth-20	5.00
PEG-80 Sorbitan Laurate	1.00
Part C: Lauroyl Lysine/Amihope LL	1.00
Part D: Isopropylparaben (and) Isobutylparaben (and)	
Butylparaben	0.10
DMDM Hydantoin (and) Iodopropynyl Butylcarbamate	0.20

Disperse the Synthetic Hectorite in deionized water. Heat to 80C. Add remaining part A ingredients. Mix until uniform. In a separate container, homogenize the C12-15 Alkyl Benzoate and Titanium Dioxide until smooth. Add remaining part B ingredients. Heat to 80C. Add part B to part A with mixing. Mix at 80C for 30 minutes until smooth and uniform. Cool to 70C. Add part D. Mix well. (Homogenize to ensure complete dispersion.) Cool to 40C. Add part E. Continue mixing and cooling to 35C.

Appearance: Off-white, smooth, viscous cream pH: 7.20-7.80
Viscosity: 30,000-36,000 cps (Brookfield #6 @ 10rpm @ 25C)

W/O SPF20 Sunscreen with Elder

Ingredients/Trade Name:	% by Weight
Part A:	
Stearic Acid Triple Pressed	4.00
Cetyl Alcohol	1.00
DEA-Cetyl Phosphate/Amphisol	2.00
Dimethicone(and)Trimethylsiloxysilicate/Dow Corning 593	5.00
Octyl Methoxycinnamate/Parsol MCX	7.50
Benzophenone-3/Uvinul M-40	6.00
Octyl Salicylate/Uvinul O-18	5.00
Octyldodecyl Neopentanoate/Elefac I-205	5.00
Methylparaben (and) Butylparaben (and) Ethylparaben	
(and) Propylparaben/Nipastat	0.20
Di (cholesteryl, behenyl, octyldodecyl) N-Lauroyl-L-	
Glutamic Acid ester/Eldew CL-301	5.00
Part B:	
Deionized Water	57.75
Sodium Carbomer/PNC 400	0.25
Part C:	
Diazolidinyl Urea/Germall II	0.30
Deionized Water	1.00

Combine part A and heat to 75 degrees Centigrade with mixing. Heat deionized water of part B to 75 degrees C. Disperse PNC 400 into the water with high shear mixing until uniform gel is formed. Add part A to part B with mixing. Hold at 75 degrees for 15 minutes. Cool to 40 degrees C and add premixed part C with continued mixing. Cool to 30 degrees C.

SOURCE: Ajinomoto USA, Inc.: Suggested Formulations

Sunscreen Balm, (Cold Processing)
SPF 8 (DIN 67501)*

Ingredients:	%(w/w)
A Demineralized water	72,900
Ethylalcohol (96 vol.%) denatured	8,500
Trilon B liquid	0,100
D-Panthenol	0,500
Germall 115	0,200
Phenonip	0,300
Carbopol 940	0,500
B Sodium hydroxide (10% aq. solution)	2,200
C 1,2-Propylene glycol	3,000
Cremogen Aloe Vera	2,000
D Neo Heliopan, Type AV	6,500
Baysilone Fluid PK 20	3,000
Bisabolol	0,100
Perfume Oil	0,200

*Sun Protection Factor:
Tested mean SPF 8.2 according to the German DIN 67501 method.

Manufacturing Process:
Part A:
   Dissolve Trilon B, Panthenol and Germall 115 in the mixture of alcohol and water. Then slowly add under stirring Carbopol 940 and continue until completely dispersed.

Part B:
   Add slowly the sodium hydroxide solution to part A for neutralization. A transparent high viscid gel will be formed.

Part C:
   Successively add with stirring propylene glycol and Cremogen Aloe Vera into the gel part A/B.

Part D:
   Blend the ingredients and add the mixture with stirring into the gel part A/B/C.

   After complete mixing it is necessary to pass the dispersion through a homogenizer (collloid mill).

   The Sunscreen Balm contains no emulsifier.

SOURCE: Haarman & Reimer: Formulation K 18/5-51132 A/E

Sunscreen Cream

Illustrates the use of Veegum Pro as a suspending agent and viscosity modifier. Veegum Pro effectively thickens and stabilizes the emulsion even at elevated temperatures. This lotion has an estimated SPF of 12 and has a light feel with quick, greaseless rub-in. Benzophenone-3 is a UV-A absorber for protection against tanning radiation. Octyl Methoxycinnamate is a UV-B absorber for protection against burning radiation.

Ingredient:	% by Wt.
A Veegum Pro	1.5
Water	67.7
Propylene glycol	3.0
Triethanolamine	0.6
B Benzophenone-3 (1)	5.0
C12-15 Alcohols Benzoate (2)	7.5
Octyl Methoxycinnamate (3)	7.5
Mineral Oil (and) Lanolin Alcohol (4)	4.0
Stearic acid XXX	2.0
C18-36 Acid (5)	0.2
Glycol Stearate SE (6)	0.5
Cetyl alcohol	0.5
C Preservative, Dye, Fragrance	q.s.

(1) Uvinul M-40
(2) Finsolv TN
(3) Parsol MCX
(4) Ritachol
(5) Synchrowax AW1-C
(6) Cerasynt MN

Procedure:
Heat the water to 75 to 80C, then slowly add the Veegum Pro while agitating at maximum available shear. Mix until smooth. Add remaining ingredients in order shown with careful mixing until smooth, maintain at 75 to 80C. Heat B to 75 to 80C. Add B to A and mix until cool. Add C.

Consistency: Cream

Suggested Packaging: Pump bottle or tube

Comments:
This prototype formula is designed to serve as a guide for the development of new products or improvement of existing ones.

SOURCE: R.T. Vanderbilt Co., Inc.: Formulation No. 421

Sunscreen Cream

		Wt.%
1.	A-C Polyethylene 617	3.0
2.	Beeswax	2.0
3.	Amerchol L-101	5.0
4.	Mineral Oil, 70 s.s.	6.2
5.	Dow Fluid 200, 350 cs.	1.0
6.	2-Ethyl Hexyl Stearate	7.0
7.	Triglycerol Diisostearate	5.5
8.	Escalol 507	5.0
9.	Propyl-P-Hydroxybenzoate	0.1
10.	Sorbitol (70%)	5.0
11.	Sodium Borate, Anhydrous	0.3
12.	Methyl-P-Hydroxybenzoate	0.2
13.	Germall 115	0.3
14.	Water	59.4

Procedure:
Weigh 1-9 and heat to 85C with slow agitation. The blend has a cloud point of approximately 80C. Above the cloud point all waxes will eventually dissolve in the blend. If a higher solvating temperature is used, solvation can be much faster. Hold the wax blend at 85C. Heat 10-14 to 85-90C and stir gently until all has dissolved. Hold at 85C.

Place wax blend in mixing container, add aqueous phase to it and shear with homomixer or colloid mill. At 67C the crude dispersion inverts and a thick creamy emulsion forms. Continue shearing while scraping the sides of the container to make sure the whole content is properly sheared. Add perfume, de-aerate and package.
Ref: 5189-4-1

W/O Sunscreen Cream

		Wt.%
1.	A-C Polyethylene 617	3.0
2.	Beeswax	2.0
3.	Amerchol L-101	5.0
4.	Isopropyl Palmitate	6.2
5.	Dow 200 Fluid, 350 cs.	1.0
6.	2-Ethyl Hexyl Stearate	7.0
7.	Triglycerol Diisostearate	5.5
8.	Escalol 507	5.0
9.	Propyl-P-Hydroxybenzoate	0.1
10.	Sorbitol (70%)	5.0
11.	Sodium Borate, Anhydrous	0.3
12.	Methyl-P-Hydroxybenzoate	0.2
13.	Germall 115	0.3
14.	Water	59.4

SOURCE: Allied Signal Inc.: Suggested Formulations

Sunscreen Cream
Water-in-Oil (Expected SPF 15+)

A)

Laurylmethicone copolyol	2.00%
Carnation Mineral Oil	5.00
Octyl p-methoxy cinnamate	7.50
Benzophenone-3	2.00
Titanium dioxide (small particle size)	4.00
Glyceryl monooleate	1.00
C12-15 alcohols benzoate	5.00
Hydrogenated castor oil	0.50

B)

Water	q.s.
Sodium chloride	2.00
Sorbitol (70%)	3.00

C)
Preservative

Procedure:
Combine A using moderate agitation. Combine B. Add B to A very slowly using very vigorous agitation, heating to melt hydrogenated castor oil. Add C. Pass through homogenizer to insure small uniform particles are produced. Package.

Skin Bronzing Gream

L-Tyrosine	0.20%
Phenylalanine	0.10
Aspartic acid	0.10
Escin	0.05
Copper gluconate	0.01
Hydrolyzed collagen	1.50
Carrot oleoresin	0.15
St. John's Wort extract	1.00
Dihydroxyacetone	0.50
Cetyl alcohol	1.00
Preservative	q.s.
Emulsifiers, thickening agents	4.00
Carnation Mineral Oil	15.00
Lanolin	3.00
Stearin	3.00
Triethanolamine (85%)	1.00
Water	q.s. to 100.00

SOURCE: Witco Corp.: Petroleum Specialties Group: Suggested Formulations

Sunscreen Cream

	Wt.%
1. A-C Polyethylene 617	3.0
2. Beeswax	2.0
3. Amerchol L-101	5.0
4. Isopropyl Palmitate	6.2
5. Dow 200 Fluid, 350 cs.	1.0
6. 2-Ethyl Hexyl Stearate	7.0
7. Triglycerol Diisostearate	5.5
8. Escalol 507	5.0
9. Propyl-P-Hydroxybenzoate	0.1
10.Sorbitol (70%)	5.0
11.Sodium Borate, Anhydrous	0.3
12.Methyl-P-Hydroxybenzoate	0.2
13.Germall 115	0.3
14.Water	59.4

Procedure:
   Weigh 1-9 and heat to 85C with slow agitation. The blend has
a cloud point of approximately 80C. Above the cloud point all
waxes will eventually dissolve in the blend. If a higher solvating temperature is used, solvation can be much faster. Hold
the wax blend at 85C. Heat 10-14 to 85-90C and stir gently
until all has dissolved. Hold at 85C.

   Place wax blend in mixing container, add aqueous phase to it
and shear with homomixer or colloid mill. At 67C the crude dispersion inverts and a thick creamy emulsion forms. Continue
shearing while scraping the sides of the container to make sure
the whole content is properly sheared. Add perfume, de-aerate
and package.
Ref: 5189-4-9

O/W Sunscreen Lotion

	Wt.%
1. A-C Copolymer 580	2.0
2. Distilled Isopropyl Lanolate	3.0
3. Escalol 507	5.0
4. Dow 556 Fluid	2.0
5. Propylene Glycol Dipelargonate	10.0
6. Ethoxyol 24	1.0
7. Arlacel 60	1.0
8. Tween 60	2.0
9. Propyl-p-Hydroxybenzoate	0.1
10.Sorbitol (70%)	5.0
11.Carbopol 941	0.5
12.Triethanolamine	0.75
13.Methyl-p-Hydroxybenzoate	0.2
14.Water	67.45

SOURCE: Allied Signal Inc.: Suggested Formulations

Sun Screen Cream W/O

		Wt%
A.	Miglyol Gel B	20.0
	Imwitor 780K	10.0
	Alugel DF 30	3.0
B.	Hard paraffin	3.0
	Paraffin oil	5.0
	Eusolex 6300	4.0
	Antioxidants	qs
C.	Perfume	qs
	Preservative	qs
D.	Eusolex 232	6.0
	Triethanolamine	5.0
	Mowiol 10-98	3.0
	Water	100.0

Preparation:
The ingredients of A. are mixed and heated to 80C. Those of
B. are brought to the same temperature and heated to A. D. is
heated to approx. 75C and emulsified in A. + B. Perfume is
incorporated at approx. 30C.

Sun Screen Cream O/W

		Wt%
A.	Imwitor 960 flakes	10.0
	Miglyol 840	8.0
	Lanette N	6.0
	Neo Heliopan E 1000	3.0
B.	1,2-Propylene glycol	3.0
	Hygroplex HHG	5.0
	Water	to 100.0
C.	Perfume	qs
	Preservative	qs

Preparation:
The ingredients of A. are heated to 75-80C. Those of B. are
heated to the same temperature and emulsified in A. Perfume is
added at approx. 30C.

SOURCE: Huls America Inc.: Formulations for Cosmetics: Formulas

Sun Screen Cream W/O, Oily

	Wt%
A. Miglyol 840 Gel B	20.0
Softisan 649	5.0
Imwitor 780 K	5.0
Paraffin oil	8.0
Neo Heliopan E 1000	3.0
Hard paraffin	3.0
B. Magnesium sulphate	2.0
Water	to 100.0
C. Perfume	qs
Preservative	qs

Preparation:
The ingredients of A. are mixed and heated to 75-80C. Those of B. are brought to the same temperature and emulsified in A. Perfume is incorporated at 30C.

Self Tanning Lotion

	Wt%
A. Miglyol 812	5.0
Marlowet TA 25	1.5
Marlowet TA 6	1.5
Cremophor EL	1.0
1,2-Propylene glycol	5.0
Cetyl alcohol	2.5
B. Water	to 100.0
C. Perfume	qs
D. Dihydroxyacetone	5.0
Water	5.0
Preservative	qs

Preparation:
The ingredients of A. are heated to 75-80C. Those of B. are brought to the same temperature and emulsified in A. D. is dissolved and added, together with C, at approx. 30C.

SOURCE: Huls America Inc.: Formulations for Cosmetics: Formulas

Sunscreen Cream

Ingredients:	%W/W
A Octyl methoxycinnimate	7.00
Benzophenone-3	3.00
Arlamol E	5.00
Stearyl alcohol	0.50
Dimethicone, 350 cps.	0.50
Brij 721S	2.00
Brij 72	2.00
B Water, deionized	69.60
Carbomer 940	0.20
C Water, deionized	10.00
Sodium hydroxide (10% W/W aqueous)	0.20

Procedure:
   Heat (A) to 60 deg. C and (B) to 65 deg. C. Add (B) to (A) and stir with propeller agitation. Add (C) below 60 deg. C. Stir to 35 deg. C. Add water to replace loss by evaporation and package.

Sunscreen Cream

Ingredients:	%W/W
A Octyl methoxycinnimate	5.00
Benzophenone-3	0.50
Arlamol E	5.00
Stearyl alcohol	0.50
Dimethicone, 350 cps.	0.50
Brij 721S	2.00
Brij 72	2.00
B Water, deionized	74.10
Carbomer 940	0.20
C Water, deionized	10.00
Sodium hydroxide (10% W/W aqueous)	0.20

Procedure:
   Heat (A) to 60 deg. C and (B) to 65 deg. C. Add (B) to (A) and stir with propeller agitation. Add (C) below 60 deg. C. Stir to 35 deg. C. Add water to replace loss by evaporation and package.

SOURCE: ICI Surfactants: Suggested Formulations

Sunscreen Cream

Ingredient:	%W/W
Part A:	
Deionized Water	78.16
Rheolate 5000	0.3
Propylene Carbonate	2.5
Part B:	
Panalene	8.0
Silicone 7207	1.0
Promulgen D	0.5
Ceraphyl 494	2.0
Part C:	
AMP-95	0.24
Part D:	
Euxyl K-400	0.3
Finsolv TN	2.0
Escalol 507	5.0

Manufacturing Directions:
1. Combine the ingredients in Part A by slowly sifting in the polymer to the water, mixing for 20 minutes, then adding the propylene glycol. Heat to 80C.
2. Combine all ingredients in Part B, then heat to 78C.
3. Add Part B to Part A while stirring. Mix for 10 minutes, then add Part C.
4. Cool to 40C before adding Part D. Package at Room Temperature.
Formula TS-330

Suntan Cream

Materials:	% by Weight
Part A:	
1. Deionized Water	73.50
2. Glycerine	2.50
3. Triethanolamine	0.70
4. Methyl Paraben	0.10
5. Bentone LT rheological additive	0.50
Part B:	
6. Ceraphyl 424	3.00
7. Isocetyl Stearate	2.00
8. Acetol	2.50
9. Cocoa Butter	1.00
10.Hystrene 5016	4.00
11.Lexemul 503	6.80
12.Escalol 507	3.00
13.Propyl Paraben	0.10
Part C: Fragrance	0.30

Manufacturing Directions:
1. Part A-In a stainless steel steam jacketed kettle, add items 1 to 4 and heat to 60C. Using a homomixer, add item 5 slowly to avoid lumps and mix for 20 minutes or until homogeneous. Heat to 80C.
2. Part B-In a separate vessel, add items 6 to 13 and heat to 80C. Mix until completely melted and homogeneous.
3. Add Part B slowly in Part A at 80C using sweep blades.
4. At 50C, add Part C, homogenize and fill units.
Formula TS-256
SOURCE: Rheox, Inc.: Suggested Formulations

Sunscreen Gel (approx. SPF 10)

A clear sunscreen gel designed to provide UV protection while leaving the skin soft and conditioned. Geahlene provides viscosity, emolliency, and water repellency to this formula.

Ingredient/Trade Name:	Weight%
A Mineral Oil (and) Hydrogenated Butylene/Ethylene/ Styrene Copolymer (and) Hydrogenated Ethylene/ Propylene/Styrene Copolymer//Geahlene 750	73.80
Octyldodecyl Neopentanoate/Elefac I-205	14.00
Tocopherol Acetate/Vitamin E Acetate	0.50
Fragrance	0.10
B Octyl Methoxycinnamate/Parsol MCX	7.50
Benzophenone-3/Uvinul M-40	2.00
Octyl Salicylate/Uvinul O-18	2.00
Propylparaben	0.10
C D&C Green No. 6	q.s.

Procedure:
Mix A until homogeneous. Heat B to 60C until clear. Add B to A with moderate mixing. Add C as necessary.

Sunscreen Gel (approx. SPF 12)

A clear sunscreen gel with excellent spreadability and a pleasant, emollient after feel. Geahlene provides viscosity, lubricity, and water repellency to this formula.

Ingredient/Trade Name:	Weight%
A Mineral Oil (and) Hydrogenated Butylene/Ethylene/ Styrene Copolymer (and) Hydrogenated Ethylene/ Propylene/Styrene Copolymer//Geahlene 750	65.00
Isopropyl Isostearate/Prisorine 2021	20.80
Tocopherol Acetate/Vitamin E Acetate	0.50
Fragrance	0.10
B Octyl Methoxycinnamate/Parsol MCX	7.50
Benzophenone-3/Uvinul M-40	4.00
Octyl Salicylate/Uvinul O-18	2.00
Propylparaben	0.10
C D&C Red No. 4 Aluminum Lake	q.s.

Procedure:
Mix A until homogeneous. Heat B to 60C until clear. Add B to A with moderate mixing. Add C as necessary.

SOURCE: Penreco: Suggested Formulations

Sunscreen Gel
SPF 6 (FDA-OTC)*

Ingredients:	%w/w
A Ethylalcohol (96 Vol. %) denatured	5,000
Demineralized water	60,350
1,2-Propylene glycol	5,000
Germaben II	1,000
Allantoin	0,100
D-Panthenol	0,500
Carbopol 940	1,100
B Demineralized water	5,000
Triethanolamine	2,200
C Demineralized water	15,000
Neo Heliopan, Type Hydro	2,000
Triethanolamine	1,200
D Cremophor NP 14	1,200
Perfume Oil	0,300
E Sunset Yellow 307009	0,050
1% aq. solution	

*Sun Protection Factor:
  Tested SPF 5.9 on a five subject panel according to the
  FDA-OTC proposed method.

Manufacturing Process:
Part A:
  Dissolve propylene glycol, Germaben II, Panthenol and Allantoin in the alcohol-water solution. Then slowly add while
stirring Carbopol 940 and continue until completely dispersed.
Part B:
  Dilute the neutralizing agent with water and add slowly into
part A for neutralization. A high viscid gel will be formed.
Part C:
  Add Neo Heliopan, Type Hydro while stirring into the water
to get a suspension. Continue stirring and add slowly triethan-
olamine until the neutralization of Neo Heliopan, Type Hydro
is complete and a clear solution is obtained. Before adding the
solution into the gel part A/B, it is recommendable to filter the
solution.
Part D:
  Blend the perfume oil with the solubilizer and add the mixture while stirring into the gel A/B/C.
Part E:
  While stirring, add the colorant to the transparent gel.

  The pH value of the gel should be approx. 7.2-7.5. At a pH
value below 7 Neo Heliopan, Type Hydro could precipitate.

SOURCE: Haarman & Reimer: Formula K 18/5-51282 D/E

Sunscreen Lotion

This formula demonstrates the rheological and stabilizing properties of Cera Bellina and of Siliconyl Beeswax. Incorporated into this formula is the sunscreen Neo Heliopan AV (Octyl methoxycinnamate) giving this formula an appropriate SPF value. The product has high gloss and excellent skin feel. The rheological properties of Cera Bellina allow for this product to be packaged in convenient tubes or squeeze bottles.

Oil Phase:

Neo Heliopan AV (H&R)	5.00%
Squalane (Centerchem)	2.00%
Cera Bellina (Pg-3 Beeswax; Koster Keunen)	4.00%
Glycerol Monostearate (Henkel)	0.50%
Light Mineral Oil (Witco)	1.50%
Isostearic Acid (Unichema)	4.00%
Liquapar (Sutton)	0.40%
Octyl Palmitate (Inolex)	5.00%
Shea Butter (Koster Keunen)	4.00%
Emulsifying Wax (Koster Keunen)	1.50%
Vitamin E (BASF)	0.10%
Vitamin A Palmitate (BASF)	0.10%
Stearic Acid (Unichema)	0.50%
Ceresine 130/135 (Koster Keunen)	0.70%
Hexanediol Behenyl Beeswax (Koster Keunen)	1.00%
Siliconyl Beeswax (Koster Keunen)	1.40%
Propyl Paraben (Sutton)	0.20%
Elefac I 205 (Bernel)	1.40%
Sunflower Oil (Lipo)	3.00%
Propylene Glycol Dioctanate (Inolex)	5.00%
Isopropyl Palmitate (Unichema)	3.00%
Diisostearyl Malate (Bernel)	1.20%
Gamma Orzanol (Koster Keunen)	1.00%

Water Phase:

Water (Distilled)	41.70%
1,3-Butylene Glycol (Hoechst)	3.00%
Carboxymethyl Cellulose (Hercules)	0.30%
Keltrol TF (Kelco)	0.30%
Methyl Paraben (Sutton)	0.30%
Sodium Borate (Borax)	0.60%

Water Phase II:

Water (Distilled)	5.00%
Tea 99% (Dow)	0.80%

Active Phase:

Silicone 245 (Dow)	1.00%
Phenonip (Nipa)	0.50%

Heat the water phase to 80C under agitation ensuring that the entire phase is solubilized. Heat and mix the oil phase to 82C. Slowly add the oil phase to the water phase, cool to 70C and add water phase II. Cool to 40 and add the active phase. Cool to room temperature.

It is easy to alter the sunscreen to suit your preference, without changing the consistency. The emulsion viscosity can easily be altered by changing the oil and/or wax concentration. Orange Wax (Koster Keunen) can be added to naturally enhance SPF.

SOURCE: Koster Keunen, Inc.: Suggested Formulations

Sunscreen Lotion (approx. SPF 15

This nongreasy sunscreen is easily applied and has a dry after feel. The Geahlene 750 adds viscosity and richness to the lotion, and should contribute to water resistance.

Ingredient/Trade Name:	Weight%
A Deionized Water	69.80
Acrylates C10-30 Alkyl Acrylate Cross-	
polymer/Carbopol 1382	0.20
Methylparaben	0.15
Tetrasodium EDTA/Hamp-Ene 220	0.10
B Triethanolamine, 99%	0.20
C Mineral Oil (and) Hydrogenated Butylene/Ethylene/	
Styrene Copolymer (and) Hydrogenated Ethylene/	
Propylene/Styrene Copolymer//Geahlene 750	10.00
Octyl Methoxycinnamate/Parsol MCX	7.50
Benzophenone-3/Uvinul M-40	4.00
Octyl Salicylate/Uvinul O-18	2.00
Cetyl Alcohol/Lanette 16	1.00
Propylparaben	0.05
Phenoxyethanol/Emeressence 1160	0.60
Stearic Acid/Emersol 132	2.00
Potassium Cetyl Phosphate/Amphisol K	2.00
D Diazolidinyl Urea/Germall II	0.30
Fragrance	0.10

Procedure:
Disperse Carbopol 1382 in rapidly agitated deionized water. Mix until completely lump-free. Heat to 80C. Add remaining part A ingredients. Add part B to part A. Mix until uniform. In a separate container, heat part C to 80-85C. Mix until all the solids are dissolved. Add part C to part A. Mix for 30 minutes until homogeneous. Cool to 40C. Add part D. Continue mixing and cooling to 30C.

SOURCE: Penreco: Suggested Formulation

Sunscreen Lotion

Ingredients:	%W/W
A Octyl methoxycinnimate	7.00
Benzophenone-3	3.00
Petrolatum	25.00
Dimethicone	3.00
Brij 721S	1.20
Brij 72	3.80
B Water, deionized	46.60
Carbomer 934	0.20
C Water, deionized	10.00
Sodium hydroxide (10% W/W aqueous)	0.20

Procedure:
   Heat (A) to 60 deg. C and (B) to 65 deg. C. Add (B) to (A)
and stir with propeller agitation. Add (C) below 60 deg. C.
Stir to 35 deg. C. Add water to replace loss by evaporation
and package.

O/W Emollient Sunscreen Lotion

Ingredients:	%W/W
A Octyl dimethyl PABA	7.00
Benzophenone-3	3.00
Arlamol E	7.00
Stearyl alcohol	2.50
Dimethicone	1.00
Arlasolve 200	3.10
Brij 72	3.90
B Water	72.10
Carbomer 934	0.20
C Sodium hydroxide (10% W/W aqueous)	0.20
D *Preservative and fragrance	

   *q.s. these ingredients.

Procedure:
   Disperse Carbomer in water and heat (B) to 70C. Heat (A) to
72C. and add (B) to (A) with propeller agitation. Slowly add
(C) and increase speed of the agitation as needed. Add (D) and
replace water lost by evaporation.

SOURCE: ICI Surfactants: Suggested Formulations

Sunscreen Lotion (o/w)
SPF 6(DIN 67501)*

Ingredients:	%(w/w)
A Demineralized water	25,000
Carbopol 934	0,400
Sodium hydroxide, (10% aq. solution)	2,050
B Tegin M	2,500
Tagat S	1,950
Lanette O	1,800
Paraffin oil (70 mPas)	2,500
Isopropyl myristate	2,000
Myritol 318	3,000
Neo Heliopan, Type AV	2,500
Cetiol MM	2,000
Abil 100	0,500
Phenonip	0,200
C Demineralized water	36,850
Phenonip	0,300
Neo Heliopan, Type Hydro	3,350
used as a 15% aq. solution neutralized with sodium hydroxide corresponding 0,5% active Neo Heliopan, Type Hydro	
1,2-Propylene glycol	2,000
D Demineralized water	10,000
Phosphoric acid, Disodium salt	0,580
Phosphoric acid, Monopotassium salt	0,120
E Perfume Oil	0,400

*Sun Protection Factor
Tested mean SPF 6 according to the German DIN 67501 method.

Manufacturing Process:
Part A:
   Disperse the Carbopol in water using high speed agitation.
Mix to form a uniform dispersion free from lumps. With stirring
add to the dispersion the sodium hydroxide solution to form a
high viscid gel.
Part B: Heat to 75C.
Part C:
   Heat to 85C. Add part C to part B while stirring rapidly.
Mix while cooling to 60C and add the Carbopol gel part A.
Part D:
   Dissolve Disodium Phosphate and Potassium Phosphate in
water at 60C. Add the buffer solution to the emulsion at 55-60C.
Part E:
   Cool while stirring to 35-40C and add the perfume oil.
Continue stirring to room temperature.

The pH-value of the finished emulsion should be 7.5 and
must be controlled.

SOURCE: Haarman & Reimer: Formulation K 18/1-51093 D/E

Sunscreen Lotion (o/w)
SPF 11 (DIN 67501)*

Ingredients:	%(w/w)
A Tegin M	3,500
Tagat S	1,500
Lanette O	2,700
Paraffin oil (70mPas)	2,000
Isopropyl myristate	2,000
Myritol 318	3,000
Neo Heliopan, Type E 1000	3,000
Cetiol MM	2,000
Abil 100	0,500
Phenonip	0,200
B Demineralized water	52,500
Carbopol 934	0,400
Sodium hydroxide (10% aq. solution)	2,050
Neo Heliopan, Type Hydro used as a 15% aq. solution	13,350
neutralized with sodium hydroxide corresponding 2% active Neo Heliopan, Type Hydro	
Phenonip	0,200
C Demineralized water	10,000
Phosphoric acid, Disodium salt	0,580
Phosphoric acid, Monopotassium salt	0,120
D Perfume Oil	0,400

*Sun Protection Factor:
   Tested mean SPF 11.8 according to the German DIN 67501 method.

Manufacturing Process:
Part A: Heat to 75C.
Part B:
   Disperse the Carbopol well in water using high speed agitation. Mix to form a uniform dispersion free from lumps. Add with stirring to the dispersion Phenonip and the sodium hydroxide solution. Then add the neutralized solution of the 15% Neo Heliopan, Type Hydro and heat to 85C. Add part B slowly with thorough agitation to part A. Then cool with stirring to 60C.
Part C:
   Dissolve Dicalcium Phosphate and Potassium Phosphate in water at 60C (buffer system). Add part C to part A/B and cool with stirring to 35-40C.
Part D:
   Add the perfume oil and continue stirring to room temperature.

   The pH value of the finished emulsion should be 7.5 and must be controlled.

SOURCE: Haarman & Reimer: Formulation K 18/1-51097/E

Sunscreen Lotion (O/W)
SPF 4 (DIN 67501)

Ingredients:	%(w/w)
A Arlatone 983 S	1,500
Brij 76	1,000
Lanette O	1,200
Paraffin oil (70 mPas)	2,000
Cegesoft C24	3,000
Eutanol G	5,000
Baysilone Fluid M10	1,000
Neo Heliopan, Type E 1000	3,500
Solbrol P	0,080
B Demineralized water	52,370
1,2 Propylene glycol	2,000
Solbrol M	0,200
Germall 115	0,150
C Demineralized water	25,000
Carbopol 934	0,300
Sodium hydroxide (10% solution in water)	1,300
Perfume Oil	0,400

Manufacturing Process:
Part A: Heat up to 75C.
Part B: Heat up to 85C. Add part B to part A while stirring.
    Cool with stirring to 55C.
Part C: Disperse the Carbopol in the water using high speed
 agitation. Mix to form a uniform dispersion free from lumps.
 Add sodium hydroxide solution while stirring to form a high
 viscid gel. Add part C to part A/B while stirring. At 35C add
 the fragrance and cool down while stirring to room temperature.
 The pH value of the finished emulsion should be 7-7.5.

SOURCE: Haarman & Reimer: Formula K 18/1-51 090 /E

Sunscreen Lotion (o/w)
SPF 18 (FDA-OTC)*

Ingredients:	%(w/w)
A Neo Heliopan, Type AV	7,500
Neo Heliopan, Type MA	5,000
Isopropyl myristate	3,000
Myrj 52	2,000
Cutina MD	3,000
Promulgen D	2,000
Antaron V-220	4,000
Baysilone Fluid M 10	2,000
B Demineralized water	45,830
Carbopol 940; used as a 2% aq. solution	15,000
1,2-Propylene glycol	2,000
Trilon BD	0,100
C Neo Heliopan, Type Hydro used as a 30% aq. solution neutralized with triethanolamine corresponding 2% active Neo Heliopan Type Hydro	6,670
D Triethanolamine	0,500
E Germaben II	1,000
Perfume Oil	0,400

*Sun Protection Factor:
Tested mean SPF value 18.25 on a five subject panel
according to the FDA-OTC proposed method.

Manufacturing Process:
In a suitable vessel weigh phase A and heat to 75C with
agitation. In another vessel able to contain the entire batch,
weigh phase B and heat to 75C with agitation. Slowly add phase
A to phase B, mix for 10 minutes and add phase C. Mix until
uniform and start cooling with continuous agitation. Cool to 40C
and add phases D and E. Continue cooling with agitation to
28-25C and package.

SOURCE: Haarman & Reimer: Formulation K 18/1-51677/E

Sunscreen Lotion (w/o)
SPF 10 (DIN 67501)*

Ingredients:	%(w/w)
A Arlacel 481	3,500
Arlacel 989	1,500
Eutanol G	2,250
Isopropyl isostearate	2,250
Paraffin oil (70 mPas)	3,340
Neo Heliopan, Type E 1000	4,000
Neo Heliopan, Type AV	4,000
Neo Heliopan, Type BB	1,000
Baysilone Fluid M 10	1,000
Vaseline Type Merkur 1546	7,500
Amerchol L 101	1,000
Solbrol P	0,080
B Demineralized water	65,380
Solbrol M	0,200
Glycerin 86%	2,000
Magnesium sulfate	0,500
C Perfume Oil	0,500

*Sun Protection Factor:
 Tested mean SPF 10.6 according to the German DIN 67501
 method.

Manufacturing Process:
Part A: Heat to 75C.
Part B:
   Heat to 85-90C. Add part B to part A slowly under thorough
stirring. At 70C homogenize the emulsion for a few minutes.
Part C:
   Continue cooling under stirring and add the perfume oil at
35C into the emulsion. Then continue stirring to room temp-
erature.

SOURCE: Haarman & Reimer: Formulation K 18/1-51098/E

Sunscreen Lotion (o/w) (cold processing)
SPF 6 (FDA-OTC)*, waterproof**

Ingredients:		%(w/w)
A	Demineralized water	77,500
	1,2-propylene glycol	2,000
	Arosol	0,500
	Solbrol M	0,250
	Solbrol P	0,100
	Trilon B liquid	0,100
B	Neo Heliopan, Type AV	3,000
	Neo Heliopan, Type E 1000	3,000
	Paraffin oil (70 mPas)	6,500
	Cetiol S	5,000
	Lameform TGI	1,000
	Perfume oil	0,300
	Pemulen TR-1	0,250
	Carbopol 954	0,050
C	Triethanolamine	0,450

*Sun Protection Factor(SPF):
The SPF was determined on a five subject panel test according
to the FDA-OTC proposed method.
**Waterproof:
The procedure for determining "waterproof" effectiveness was
20 minutes moderate activity in water. The SPF before water
immersion was 5.7 and after water immersion 6.1.

Manufacturing Process:
Part A:
  Dissolve Arosol, Soltrol P and M in propylene glycol. Add the
mixture and Trilon B liquid to the water.
Part B:
  Mix all ingredients (without Permulen and Carbopol). Disperse
Carbopol and Pemulen very carefully with high speed agitation.
Then add part B to part A while stirring. Stir for 45 minutes.
Part C:
  Add triethanolamine while stirring and stir until homogeneous.
The pH-value of the finished emulsion should be approx. 7 and
has to be controlled.

SOURCE: Haarman & Reimer: Formulation K 18/1-51 669 C/E

Sunscreen Lotion
SPF 30 (FDA-OTC)*

Ingredients:	%(w/w)
A Neo Heliopan, Type AV	7,500
Neo Heliopan, Type BB	6,000
Neo Heliopan, Type 303	9,000
Ganex WP 660	3,000
Lanette O	0,500
Abil 100	2,000
Protachem SQI	0,200
B Deionized water	52,150
Carbopol 940 used as a 2% aq. solution	5,000
Pemulen TR-1 used as a 2% aq. solution	10,000
1,2-Propylene glycol	3,000
C Triethanolamine	0,250
D Germaben II	1,000
Perfume Oil	0.400

*Sun Protection Factor:
  Tested SPF 31.5 on a five subject panel according to the
  FDA-OTC proposed method.

Manufacturing Process:
  In a suitable vessel weigh phase A, heat up to 80C-85C and
mix until uniform. In another vessel able to contain the entire
batch, weigh phase B, heat to 80C and mix until uniform. Slowly
add phase A to phase B with agitation and mix for ten minutes.
Slowly add phase C and start cooling. Cool with continuous agitation to 40C and add phase D. Continue cooling with agitation
to 28C-25C. Mill and package.

SOURCE: Haarman & Reimer: Formulation H 101-47-5

Sunscreen Lotion (O/W)
Tested SPF 10

Ingredients:	%(w/w)
A Arlacel 165	3.00
Eumulgin B 2	1.00
Lanette O	1.15
Myritol 318	4.00
Cetiol OE	5.00
Abil 100	1.00
Bentone Gel MIO	3.00
Cutina CBS	1.00
Neo Heliopan, Type AV	3.00
Neo Heliopan, Type E 1000	3.00
Neo Heliopan, Type MBC	1.00
Neo Heliopan, Type BB	0.70
B Demineralized water	63.05
Phenonip	0.50
Glycerin 86%	3.00
Veegum ultra	1.00
Natrosol 250 HHR	0.30
Zinc Oxide neutral H&R	5.00
C Perfume oil	0.30

Manufacturing Process:
Part A: Heat up to 75C with thorough agitation.
Part B:
 Add Phenonip and glycerin to the water. Heat up to 95C. Then add the Veegum ultra and the Natrosol during high speed agitation. Then disperse Zinc Oxide Neutral H&R in the water phase. Add Part B to Part A while stirring. Cool down while stirring to room temperature.
Part C:
 At 30C add the perfume oil to Part A/B and homogenize the emulsion.
 The pH-value of the finished emulsion should be approx. 7 and has to be checked.

Instruction:
 In EEC countries the use of more than 0.5% Benzophenone-3 in sunscreen products is liable to declare: contains Oxybenzone.

SOURCE: Haarman & Reimer: Formula K 18/1-21287/E

Sunscreen Lotion (O/W)
Tested SPF 16

Ingredients:	%(w/w)
A Arlacel 165	3.00
Eumulgin B2	1.00
Lanette O	1.00
Myritol 318	4.00
Cetiol OE	5.00
Abil 100	1.00
Bentone Gel MIO	3.00
Neo Heliopan, Type AV	4.00
Neo Heliopan, Type E 1000	4.00
Cutina CBS	1.00
B Demineralized water	49.60
Glycerin 86%	3.00
Phenonip	0.50
Neo Heliopan, Type Hydro; used as a 15% solution	13.30
neutralized with sodium hydroxide (active: 2.00%)	
Veegum ultra	1.00
Natrosol 250 HHR	0.30
Zinc Oxide neutral H&R	5.00
C Perfume oil	0.30

Manufacturing Process:
Part A: Heat up to 75C with thorough agitation.
Part B:
   Add Phenonip, glycerin and the Neo Heliopan, Type Hydro
solution to the water. Heat up to 95C. Add Veegum ultra and
Natrosol during high speed agitation. Then add Zinc Oxide
neutral H&R and disperse. Add Part B while stirring to Part A.
Continue stirring to toom temperature.
Part C:
   At 30C add the perfume oil to Part A/B and homogenize the
emulsion.
   The pH-value of the finished emulsion should be approx.
7.5 and has to be checked.

SOURCE: Haarman & Reimer: Formulation K 18/1-21284/E

Sunscreen Lotion (w/o)
SPF 6 (FDA-OTC), waterproof*

Ingredients:	%(w/w)
A Abil EM 90	2,000
Tegin T 4753	1,000
Cetiol S	17,000
Paraffin oil (34 mPas)	8,500
B Demineralized water	58,450
Neo Heliopan, Type Hydro used as a 30% aq solution	10,000
neutralized with triethanolamine corresponding 3%	
active Neo Heliopan, Type Hydro	
1,2-Propylene glycol	2,000
Triethanolamine	0,350
Phenonip	0,400
C Perfume Oil	0,300

*Sun Protection Factor:
Tested mean SPF value before swim 6.32 and after swim 5.76 on a five subject panel according to the FDA-OTC proposed method.

Manufacturing Process:
Part A: Heat to 70C.
Part B:
Dissolve the ingredients in the water. The pH value of the water phase part B should be 7.5 and must be controlled. Now add part B while slowly stirring to part A within a time of 3-5 minutes (hot/cold procedure).
Part C:
Continue stirring, at 30C add the fragrance. Then pass the emulsion through a homogeniser (colloid mill).

SOURCE: Haarman & Reimer: Formulation K 18/1-51566 A/E

Sunscreen Lotion (W/O)
Tested SPF 11

Ingredients:	%(w/w)
A Dehymuls HRE 7	4.00
Lameform TGI	4.00
Pemulgin 3220	1.50
Isopropyl isostearate	8.00
Cetiol OE	7.50
Baysilone Fluid M 10	1.00
Vitamin E Acetate	0.50
Neo Heliopan, Type AV	4.00
Neo Heliopan, Type E 1000	4.00
Neo Heliopan, Type MBC	1.00
Neo Heliopan, Type BB	1.50
B Demineralized water	54.10
Glycerin 86%	3.00
Phenonip	0.50
Zinc Oxide neutral H&R	5.00
C Perfume oil	0.40

Manufacturing Process:
Part A: Heat up to 75C with thorough agitation.
Part B:
   Heat up to 90C (without the Zinc Oxide neutral H&R). Then add and disperse the Zinc Oxide neutral H&R. Add Part B to part A slowly while stirring. Continue stirring to room temperature.
Part C:
   At 40C add the perfume oil to part A/B and homogenize the emulsion.

Instruction:
   In EEC countries the use of more than 0.5% Benzophenone-3 in sunscreen products is liable to declare: contains Oxybenzone.

SOURCE: Haarman & Reimer: Formulation K 18/1-51835/E

Sunscreen Moisturizing Cream

This formula demonstrates the stabilizing properties of
Siliconyl Beeswax. Incorporated into this formula is the sunscreen Neo Heliopan AV (Octyl methoxycinnamate) giving this
formula an appropriate SPF value. The cream has high gloss
and excellent skin feel.

Oil Phase:
Cera Bellina (Pg-3 Beeswax, Koster Keunen)	6.0%
Elefac I 205 (Bernel)	1.4%
Glycerol Monostearate (Henkel)	1.4%
Isopropyl Palmitate (Unichema)	2.0%
Ceresine 130/135 (Koster Keunen)	1.3%
Propyl Paraben (Sutton)	0.2%
Squalane (Centerchem)	1.8%
Jojoba oil (Jojoba Growers)	1.0%
Liquapar (Sutton)	0.4%
Neo Heliopan AV (H&R)	5.0%
Vitamin E (BASF)	1.0%
Octyl Palmitate (Inolex)	2.5%
Shea Butter (Koster Keunen)	2.5%
Isostearic acid (Unichema)	3.0%
Lipopeg 100-S (Lipo)	1.5%
Siliconyl Beeswax (Koster Keunen)	3.0%
Emulsifying Beeswax (Koster Keunen)	1.7%
Hexanediol Behenyl Beeswax (Koster Keunen)	1.0%
Gamma Orzanol (Koster Keunen)	1.0%

Water Phase:
Water (Distilled)	44.3%
Butylene Glycol (Hoechst)	3.0%
Sodium Borate (Borax)	0.8%
Methyl Paraben (Sutton)	0.3%
Aloe Vera Gel 1:1 (Active Organics)	3.0%
Glycerin 99%	4.0%
Keltrol TF (Kelco)	0.3%

Water Phase II:
TEA 99% (Dow)	1.0%
D.I. Water	5.0%

Active Phase:
Silicone 245 (Dow)	1.0%
Phenonip (Nipa)	0.5%

Procedure:
Heat the water phase to 80C under agitation ensuring that the
entire phase is solubilized. Melt and mix the oil phase to 82C
is maintained. At the last minute added the gamma orzanol to
the oil phase, allow to disperse then slowly add the oil phase
to the water phase while mixing. Allow to cool to 45C and add
the silicone and the phenonip. Cool to room temperature.

Adaptation of formula and its influence on the product:
It is easy to alter the sunscreen to suit your preference,
without changing the consistency. The emulsion viscosity can
easily be altered by changing the oil and/or wax concentration.
Orange Wax (Koster Keunen) can be added to naturally enhance the
SPF.

SOURCE: Koster Keunen, Inc.: Suggested Formulations

Tanning Accelerator

Part A:	Parts by Weight
Water	500.0
Carbomer 934	2.0

Part B:	
Rosswax 1824	15.0
Rosswax 2540	6.0
GMS-SE	6.0
Ross Jojoba Oil	4.0
Escalol 507	12.0
Coconut Oil #76	25.0
Unipertan P-24	3.0

Part C:	
Germaben II	6.0

Part D:	
Fragrance	q.s.

Part E:	
Triethanolamine	4.5

Procedure:
In a steam jacketed kettle heat the water and add the Carbomer 934 until fully dispersed under agitation. In a separate steam jacketed kettle melt the Oil Phase. When fully melted, add the Oil Phase to the Water Phase under agitation. Then add the Germaben II, then the fragrance and finally add the TEA with high agitation until smooth. Cool to 130F and package.

Sun Screen Stick

Ingredients:	%
Ross White Bleached Beeswax	20.0
Ross Pure Refined Candelilla Wax	16.1
Ross Pure #1 Yellow Carnauba Wax	4.0
Petrolatum	16.1
IPM	10.0
Mineral Oil 60/70	32.6
Ross Jojoba Oil	1.2
Amerscreen P	q.s.

Procedure:
Heat all ingredients in a steam jacketed kettle to 170F under agitation. When fully mixed cool to 145F and package. (Note: Capping may be necessary).

SOURCE: Frank B. Ross Co., Inc.: Suggested Formulations

Titanium Dioxide Based Waterproof Sunscreen
SPF 12(Tested)

Dermacryl 79 provides water resistance, increased moisture protection and rub-off resistance.

Ingredients:	% W/W
Phase A:	
Ceraphyl ICA	7.00
Finsolv TN	2.00
Emersol 132	2.00
Myrj 52S	2.00
Abil B8852	1.00
Cetyl Alcohol	1.00
Cerasynt SD	0.50
Armeen DM 18D	2.00
Dermacryl-79	2.00
Tioveil FIN	10.00
Phase B:	
Deionized Water	59.30
Carbomer 941 (2% Aq. Soln)	10.00
Methylparaben	0.15
Propylparaben	0.10
Triethanolamine (99%)	0.80
Phase C:	
Germall II	0.15
Phase D:	
Fragrance	Q.S.

Procedure:
   Combine Phase B and heat to 80C. In a separate vessel combine Phase A except for Dermacryl-79 and Titanium Dioxide and heat to 80C. Slowly sift in Dermacryl-79 with constant stirring until completely dissolved. Add Titanium Dioxide with constant stirring until completely dispersed. Add Phase A to Phase B at 80C and mix for 30 minutes. Cool to 40C and add Phase C and Phase D, homogenize, cool to room temperature and package.

SOURCE: National Starch and Chemical Co.: Formula 7023-24A

Ultra Violet Absorbing Sunscreen

Veegum Ultra, magnesium aluminum silicate, is used to thicken and stabilize this sunscreen emulsion. Two ultra violet absorbers are used to achieve an estimated SPF (Sun Protection Factor) of approximately 15. This smooth, flowable lotion spreads easily and dries quickly, leaving non-tacky after-feel.

Ingredient:	% by Wt.
A Veegum Ultra (Magnesium Aluminum Silicate)	1.50
Deionized Water	70.50
Glycerin	5.50
B PEG-150 Distearate	3.00
Dioctyl Malate*	2.00
Mineral Oil	4.00
Cetyl Alcohol	0.50
Benzophenone-3	3.00
Octyl Dimethyl PABA	7.00
Steareth-2	0.90
Steareth-20	2.10
C Preservative, Fragrance	q.s.

*Ceraphyl 45

Procedure:
Heat the water to 55C. Slowly add Veegum Ultra to the water while stirring with a propeller mixer at 700 rpm. Increase the mixer speed to 1500-1700 rpm and mix for 30 minutes, maintaining temperature at 55C. Add glycerin and mix for 5 minutes. Mix B ingredients and heat to 60C. Add B to A while mixing at 1500-1700 rpm. Continue mixing for 30 minutes. Avoid air entrapment. Slow mixing speed to 1000 rpm and continue mixing while cooling to 35C. Add C and mix until uniform. Package.

Product Charactertistics:
Viscosity: 5000-6000 cps
pH: 5.0+-0.2

Comments:
This prototype formula is designed to serve as a guide for the development of new products or the improvement of existing ones.

SOURCE: R.T. Vanderbilt Co., Inc.: Formula No. 448

Un-Buffered Self Tanning Cream

Resulting product is a soft cream that could be dispensed from a flexible tube or a jar. The formulation utilizes 5% Dihydroxyacetone (DHA) as the self tanning ingredient. The formulation contains no amines to react with Dihydroxyacetone. The formulation applies easily and evenly and forms a moderately thick film allowing maximum tan color development. The finished product should be stored away from light and at temperature conditions below 40C to prevent DHA from degrading.

	%w/w
A. Deionized Water	63.10
Methylparaben (Lexgard M)	0.20
2,4-Dichlorobenzyl Alcohol (Myacide SP)	0.20
Propylene Glycol USP	1.00
B. Hydroxyethylcellulose (Cellosize QP-15,000H)	0.40
C. Glyceryl Stearate (Lexemul 515)	6.00
Glyceryl Stearate (and) PEG-100 Stearate (Lexemul 561)	3.00
Caprylic/Capric Triglyceride (Lexol GT-865)	4.00
Isobutyl Stearate (Lexol BS)	2.00
Propylparaben (Lexgard P)	0.10
D. Dihydroxyacetone	5.00
Deionized Water	15.00

Procedure:
1. Combine section "A" heating to 75-80C. Use a propeller mixer for agitation.
2. Sprinkle section "B" into section "A". Increase mixing speed to create a vortex during addition, then slow to highest speed which does not produce a vortex.
3. Combine section "C". Heat to 80-85C with continuous slow mixing.
4. When sections "AB" and "C" are homogeneous and at the designated temperatures add section "C" to sections "AB". Increase mixing speed to a high speed. Begin cooling. Slow mixing speed to highest speed possible that will not cause vortexing.
5. Using gloves, combine section "D" and mix until clear.
6. Add section "D" to sections "ABC" at 35-40C. Continue mixing and cooling.
7. At 33-35C adjust for water loss.
8. Mix batch until it reaches 30.

Physical Properties:
   Viscosity: 20,000 cPs @ 25C (Brookfield RVT, TA @ 10 rpm)
           24 hour sample
   pH: 5.0 @ 24C

SOURCE: Inolex Chemical Co.: Formulation 398-105-2

Waterproof Sun Protection Lotion
(with 8% Titanium Dioxide)
SPF 15

Ingredients:	%w/w
Phase A:	
Cetyl Dimethicone Copolyol (and) Polyglyceryl-4	
Isostearate (and) Hexyl Laurate (Abil WE-09)	5.0
Octyl Stearate (Tegosoft OS)	12.0
Cyclomethicone (Abil B 8839)	8.0
Cetyl Dimethicone (Abil Wax 9801)	3.0
Hydrogenated Castor Oil	0.5
Microcrystalline Wax	1.0
Mineral Oil	2.0
Phase B:	
Titanium Dioxide	8.0
Phase C:	
Water	60.0
Sodium Chloride	0.5
Phase D:	
Fragrance, Preservatives	Q.S.

1. Combine the ingredients of Phase A. Heat to 80C to melt and disperse the waxes. Cool to 60C.
2. Add the Titanium Dioxide. Disperse and mill the pigment. Cool to 50C.
3. Mix Phase C. Heat to 50C. Add to Phase A/B slowly with low energy stirrer. Maintain a milky appearance at all times.
4. Cool to 35C. Add Phase D. Homogenize.

W/O Sunscreen Lotion
SPF 22

Ingredients:	%w/w
Phase A:	
Cetyl Dimethicone Copolyol (and) Polyglyceryl-4	
Isostearate (and) Hexyl Laurate (Abil WE-09)	5.0
Cetyl Dimethicone (Abil Wax 9801)	1.0
Octyl Palmitate (Tegosoft OP)	1.0
Octyl Stearate (Tegosoft OS)	5.1
Mineral Oil	1.5
Beeswax	1.2
Hydrogenated Castor Oil	0.8
Phase B:	
Octyl Methoxycinnamate	5.0
Titanium Dioxide	5.0
Cyclomethicone (Abil B 8839)	4.4
Phase C:	
Water	69.2
Sodium Chloride	0.8
Preservatives	Q.S.

1. Heat Phase A to 85C to melt and disperse waxes.
2. Cool Phase A to 50C. Add B to A slowly with a low energy mixer. Maintain a smooth milky appearance at all times.
3. Roller mill to reduce particle size of Titanium Dioxide.
4. Cool to 50C. 5. Heat Phase C to 50C. Add Phase C while mixing. 6. Cool to 35C and homogenize.

SOURCE: Goldschmidt Chemical Corp.: Formulas BJS-1-45/GRH-2-37

Waterproof Sunscreen Gel (Tested SPF 30)

Dermacryl LT provides water resistance, increased moisture protection and rub-off resistance as well as improved rub-in properties.

Ingredients:	%W/W
Anhydrous Ethanol	65.50
Klucel MF	1.00
Dermacryl-79	1.00
Finsolv TN	10.00
Dow Corning 344 Fluid	3.00
Abil B8852	1.00
Neo Heliopan Type AV	7.50
Neo Heliopan Type OS	5.00
Neo Heliopan Type BB	6.00

Procedure:
    Dissolve Hydroxypropyl Cellulose in SD Alcohol 40. When complete, slowly sift in Dermacryl-79. Combine remaining ingredients, then add to alcohol mixture. Stir until uniform.
Formula 7761-133

Waterproof Sunscreen Spray Mist
SPF 16 (Tested)

Dermacryl LT provides water resistance, increased moisture protection and rub-off resistance.

Ingredients:	%W/W
Phase A:	
Anhydrous Ethanol	64.20
Deionized Water	5.00
Triethanolamine (99%)	0.30
Dermacryl-79	1.00
Phase B:	
Neo Heliopan AV	7.50
Neo Heliopan BB	3.00
Neo Heliopan MA	3.50
DC 344 Fluid	5.00
Finsolv TN	10.00
Vitamin E Acetate	0.50
Phase C:	
Fragrance	Q.S.

Procedure:
    Combine Phase A except Dermacryl-79. Slowly sift in Dermacryl-79. Mix until complete. Add Phase B. Mix until complete. Add Phase C. Mix until complete and package.
Formula 7172-44

SOURCE: National Starch and Chemical Co.: Suggested Formulas

Waterproof Sunscreen W/O Emulsion
SPF 22 (Static)
SPF 21 (Waterproof)

Ingredients:	%w/w
Phase A:	
Cetyl Dimethicone Copolyol (and) Polyglyceryl-4	
Isostearate (and) Hexyl Laurate (Abil WE-09)	4.0
Cetyl Dimethicone (Abil Wax 9801)	2.0
Hydrogenated Castor Oil	0.5
Microcrystalline Wax	0.5
Benzophenone-3	2.0
Octyl Methoxycinnamate	3.0
Sesame Oil	1.0
Isopropyl Palmitate (Tegosoft P)	3.0
Cyclomethicone (Abil B 8839)	4.0
Octyl Stearate (Tegosoft OS)	3.0
Octyl Palmitate (Tegosoft OP)	2.5
Phase B:	
Water	73.4
Sodium Chloride	0.5
Polyquaternium-10	0.3
Germaben II	0.3

1. Add the components of Phase A. Heat while mixing to 80C to incorporate the waxes. Cool to 50C.
2. Heat Phase B to 50C. Add B to A slowly with a low energy mixer. Maintain a smooth milky appearance at all times.
3. Cool to 35C with sweep mixer. Add fragrance. 4. Homogenize.

Waterproof Sunscreen W/O Emulsion
SPF 15

Ingredients:	%w/w
Phase A:	
Cetyl Dimethicone Copolyol (and) Polyglyceryl-4	
Isostearate (and) Hexyl Laurate (Abil WE-09)	4.0
Cetyl Dimethicone (Abil Wax 9801)	1.0
Polyethylene Wax	0.5
Hydrogenated Castor Oil	0.5
Octyl Methoxycinnamate	3.0
Sesame Oil	1.0
Isopropyl Palmitate (Tegosoft P)	2.0
Cyclomethicone (Abil B 8839)	4.0
Octyl Stearate (Tegosoft OS)	3.0
Octyl Palmitate (Tegosoft OP)	2.5
Isopropyl Myristate (Tegosoft M)	1.0
C12-15 Alkyl Benzoate	2.5
Phase B:	
Water	73.9
Sodium Chloride	0.5
Polyquaternium-10	0.3
Germaben II	0.3

1. Add the components of Phase A. Heat while mixing to 80C to incorporate the waxes. Cool to 50C.
2. Heat Phase B to 50C. Add B to A slowly with a low energy mixer. Maintain a smooth milky appearance at all times.
3. Cool to 35C with sweep mixer. Add fragrance. 4. Homogenize.

SOURCE: Goldschmidt Chemical Corp.: Formulas BAM-2-1/BAM-2-15

Waterproof W/O Sunscreen Lotion Base
SPF 3.5

Ingredients:	%w/w
Phase A:	
Cetyl Dimethicone Copolyol (and) Polyglyceryl-4	
Isostearate (and) Hexyl Laurate (Abil WE-09)	5.0
Mineral Oil	5.0
Octyl Stearate (Tegosoft OS)	6.0
Cyclomethicone (Abil B 8839)	4.0
Cetyl Dimethicone (Abil Wax 9801)	1.0
Isopropyl Myristate (Tegosoft M)	4.0
Hydrogenated Castor Oil	0.8
Microcrystalline Wax	1.2
Phase B:	
Water	72.2
Sodium Chloride	0.8
Preservatives	Q.S.
Phase C:	
Fragrance	Q.S.

Procedure:
1. Add the components of Phase A. Heat while mixing to 80C to incorporate the waxes. Cool to 50C.
2. Heat Phase B to 50C. Add B to A slowly with a low energy mixer. Maintain a smooth milky appearance at all times.
3. Cool to 35C with sweep mixer. Add fragrance.
4. Homogenize.
Formulation BJS-1-47

Waterproof W/O Sunscreen Lotion Base
SPF 3.7

Ingredients:	%w/w
Phase A:	
Cetyl Dimethicone Copolyol (and) Polyglyceryl-4 Iso-	
stearate (and) Hexyl Laurate (Abil WE-09)	5.0
Mineral Oil	5.4
Octyl Stearate (Tegosoft OS)	6.4
Cyclomethicone (Abil B 8839)	4.3
Cetyl Dimethicone (Abil Wax 9801)	1.0
Isopropyl Myristate (Tegosoft M)	4.3
Hydrogenated Castor Oil	0.8
Microcrystalline Wax	1.2
Phase B:	
Water	70.8
Sodium Chloride	0.8
Preservatives	Q.S.
Phase C:	
Fragrance	Q.S.

Procedure:
1. Add the components of Phase A. Heat while mixing to 80C to incorporate the waxes. Cool to 50C.
2. Heat Phase B to 50C. Add B to A slowly with a low energy mixer. Maintain a smooth milky appearance at all times.
3. Cool to 35C with sweep mixer. Add fragrance. 4. Homogenize.
Formulation BJS-1-15
SOURCE: Goldschmidt Chemical Corp.: Suggested Formulations

Waterproof W/O Sunscreen Lotion
SPF 21

Ingredients:	%w/w
Phase A:	
Cetyl Dimethicone Copolyol (and) Polyglyceryl-4	
Isostearate (and) Hexyl Laurate (Abil WE-09)	5.0
Mineral Oil	5.3
Octyl Stearate (Tegosoft OS)	4.2
Cyclomethicone (Abil B 8839)	4.2
Isopropyl Myristate (Tegosoft M)	4.2
Almond Oil	2.0
Hydrogenated Castor Oil	0.8
Microcrystalline Wax	1.2
Octyl Methoxycinnamate	3.0
Phase B:	
Water	69.3
Sodium Chloride	0.8
Preservatives	Q.S.
Phase C:	
Fragrance	Q.S.

Procedure:
1. Add the components of Phase A. Heat while mixing to 80C to incorporate the waxes. Cool to 50C.
2. Heat Phase B to 50C. Add B to A slowly with a low energy mixer. Maintain a smooth milky appearance at all times.
3. Cool to 35C with sweep mixer. Add fragrance.
4. Homogenize.

Waterproof W/O Sunscreen Lotion
SPF 15

Ingredients:	%w/w
Phase A:	
Cetyl Dimethicone Copolyol (Abil EM-90)	2.0
Mineral Oil	5.9
Octyl Stearate (Tegosoft OS)	4.7
Cyclomethicone (Abil B 8839)	4.7
Cetyl Dimethicone (Abil Wax 9801)	1.0
Isopropyl Myristate (Tegosoft M)	4.7
Almond Oil	2.0
Hydrogenated Castor Oil	0.8
Microcrystalline Wax	1.2
Octyl Methoxycinnamate	3.0
Phase B:	
Water	69.2
Sodium Chloride	0.8
Preservatives	Q.S.
Phase C:	
Fragrance	Q.S.

Procedure:
1. Add the components of Phase A. Heat while mixing to 80C to incorporate the waxes. Cool to 50C.
2. Heat Phase B to 50C. Add B to A slowly with a low energy mixer. Maintain a smooth milky appearance at all times.
3. Cool to 35C with sweep mixer. Add fragrance. 4. Homogenize.
SOURCE: Goldschmidt Chemical Corp.: Formulas BJS-1-35/BJS-1-53

Waterproof W/O Sun Protection Lotion
SPF 17

Ingredients:	%w/w
Phase A:	
Cetyl Dimethicone Copolyol (Abil EM-90)	2.0
Caprylic/Capric Triglyceride (Tegosoft CT)	4.0
Octyl Palmitate (Tegosoft OP)	5.5
Hydrogenated Castor Oil	0.4
Synthetic Wax or Beeswax	0.6
Cyclomethicone (Abil B 8839)	7.0
Cetyl Dimethicone (Abil Wax 9801)	1.0
Octyl Methoxycinnamate	5.0
Benzophenone-3	1.5
Menthyl Anthranilate	3.0
Phase B:	
Water	73.9
Sodium Chloride	0.6
Preservatives	Q.S.
Fragrance	Q.S.

Procedure:
1. Add the components of Phase A. Heat while mixing to 80C to incorporate the waxes. Cool to 50C.
2. Heat Phase B to 50C. Add B to A slowly with a low energy mixer. Maintain a smooth milky appearance at all times.
3. Cool to 35C with sweep mixer. Add fragrance.
4. Homogenize.

Waterproof Sun Protection Lotion
(with 8% Titanium Dioxide)
SPF 12

Ingredients:	%w/w
Phase A:	
Cetyl Dimethicone Copolyol (Abil EM-90)	2.0
Octyl Stearate (Tegosoft OS)	13.6
Cyclomethicone (Abil B 8839)	9.1
Cetyl Dimethicone (Abil Wax 9801)	3.0
Hydrogenated Castor Oil	0.5
Microcrystalline Wax	1.0
Mineral Oil	2.3
Phase B:	
Titanium Dioxide	8.0
Phase C:	
Water	60.0
Sodium Chloride	0.5
Phase D:	
Fragrance, Preservatives	Q.S.

Procedure:
1. Combine the ingredients of Phase A. Heat to 80C to melt and disperse the waxes. Cool to 60C.
2. Add the Titanium Dioxide. Disperse and mill the pigment. Cool to 50C.
3. Mix Phase C. Heat to 50C. Add to Phase A/B slowly with low energy stirrer. Maintain a milky appearance at all times.
4. Cool to 35C. Add Phase D. Homogenize.

SOURCE: Goldschmidt Chemical Corp.: Formula BJS-1-43/BJS-1-33

Water Resistant Sunscreen

		% By Weight
A	Veegum Pro	1.0
	Water	83.2
B	Glycerin	1.5
	Triethanolamine	0.6
C	Amerscreen P	4.0
	Cetyl alcohol	0.8
	Stearic acid XXX	2.0
	SS4267 Fluid	4.0
	Isopropyl myristate	2.0
	Synchrowax AW1 C	0.4
	Cerasynt MN	0.5
	Preservative	q.s.

Procedure:
Heat the water to 70 to 75C, then slowly add the Veegum PRO while agitating at maximum available shear. Mix until smooth. Add B to A with slow agitation until smooth. Maintain A/B at 70 to 75C; heat C to 70 to 75C. Add C to A/B and mix until cool.

Consistency: Medium viscosity lotion.

Suggested Packaging: Squeeze or pump bottles.

Comments: Veegum PRO effectively stabilizes the emulsion
even at elevated temperatures. This lotion, using silicone
fluid to provide an emollient water resistant protective
barrier, imparts a non-greasy feel. The Amerscreen P at 4%
provides an estimated Sun Protection factor of 8.

SOURCE: R.T. Vanderbilt Co., Inc.: Formulation No. 418

Water Resistant Sunscreen

Resulting product is a viscous sunscreen lotion which can be dispensed from a flexible tube or soft bottle. The formulation exhibits exceptional water resistance which is contributed by Lexorez 100. The formula applies easily and leaves a non-greasy film on the skin. The sunscreen active ingredient is Octyl Methoxycinnamate. The SPF value of the formulation is unknown.

	%w/w
A. Deionized Water	71.45
Propylene Glycol USP	7.50
Methylparaben (Lexgard M)	0.20
Tetrasodium EDTA (Hamp-ene Na4)	0.05
B. Carbomer (Carbopol ETD 2055)	0.15
C. Sodium Hydroxide (10% w/w)	QS
D. Glyceryl Stearate (Lexemul 515)	4.00
Glyceryl Stearate (and) PEG-100 Stearate (Lexemul 561)	1.00
Cetyl Alcohol NF	0.50
Aluminum Starch Octenylsuccinate (Dry Flo-PC)	1.00
Caprylic/Capric Triglyceride (Lexol GT-865)	3.00
Isopropyl Palmitate (and) Isopropyl Myristate (and) Isopropyl Stearate (Lexol 3975)	1.00
Tocopherol (Copherol F-1300)	0.05
Propylparaben (Lexgard P)	0.10
Octyl Methoxycinnamate (Escalol 557)	8.00
Glycerin/Diethylene Glycol/Adipate Crosspolymer (Lexorez 100)	2.00

Procedure:
1. Combine all ingredients in phase "A" and heat to 75-80C. Agitate at low speed.
2. Slowly add phase "B" to phase "A". Agitate at a higher speed until all Carbomer is fully dispersed.
3. Combine phase "D" in a separate vessel and heat to 80-85C. Add phase "D" to phase "AB". Mix for five minutes and start to cool.
4. At 45C adjust for water loss. Mix until homogeneous. Then adjust for viscosity using 10% NaOH (9,500+-500 cps using spindle #6 @ 50rpm, 25C). The final product has a pH of approximately 6.0.

SOURCE: Inolex Chemical Co.: Formulation 383-208

W/O SPF15 Sunscreen with Ajidew and Eldew

Ingredients/Trade Name:	% by Weight
Part A:	
Di-(Cholesteryl, behenyl, octyldodecyl)	
N-Lauroyl-L-glutamic acid ester/Eldew CL-301	2.00
Octyldodecyl Neopentanoate/Elefac I-205	10.00
Dimethicone(and)Trimethylsiloxysilicate/Dow Corning 593	0.50
Propylparaben	0.05
Methylparaben	0.15
Phenoxyethanol/Emeressence 1160	0.60
Benzophenone-3/Uvinul M-40	4.00
Octyl Methoxycinnamate/Parsol MCX	7.50
Part B:	
Polyglyceryl-4 Isostearate (and) Cetyl Dimethicone	
Copolyol (and) Hexyl Laurate/Abil WE09	5.00
Part C:	
Deionized Water	61.40
Sodium Chloride	0.80
Sodium PCA/Ajidew N-50	1.00
Partially Deacetylated Chitin (1.0%)/Marine Dew	2.00
Propylene Glycol	5.00

Procedure:
Pre-melt part A at 50 degrees Centigrade. Add part B to part A.
Premix part C until homogeneous. Slowly add part C into part A
and B mixture with high shear mixing.
   Appearance: White, smooth, shiny lotion.   pH: 6.0-6.5
   Viscosity: 10,000-20,000 (RVT#6 @ 10rpm @ 25 degrees C)

Sunscreen Lip Balm with Eldew

Ingredient/Trade Name:	% by Weight
Part A:	
Carnauba/Carnauba Wax #1 Yellow	4.00
Ozokerite/Ozokerite Wax JH1680	4.00
Synthetic Beeswax/Synthetic Beeswax JH-1508	10.00
Propylparaben	0.15
Cholesteryl/Behenyl/Octyldodecyl Lauroyl Glutamate/	
Eldew CL-301	10.00
Part B:	
Castor Oil	26.55
PPG-3 Hydrogenated Castor Oil/Hetester HCP	15.00
Glyceryl Triacetyl Hydroxystearate/Hetester HCA	15.00
Octyl Methoxycinnamate/Parsol MCX	7.50
Benzophenone-3/Uvinul M-40	5.00
Octyl Salicylate/Uvinul O-18	2.00
Tocopheryl Acetate/Vitamin E Acetate	0.60
Part C:	
Bisabolol (Synthetic)	0.20

Procedure:
   Heat Part A ingredients to 80 degrees Centigrade. Mix until
all solids are completely dissolved. Add Part B ingredients.
Mix until uniform. Cool to 65 degrees Centigrade. Add Part C.
Mix well. Fill.
SOURCE: Ajinomoto USA, Inc.: Suggested Formulations

W/O SPF 15 Sunscreen with Ajidew and Eldew

Ingredients/Trade Name: % by Weight
Part A:
Di(Cholesteryl, behenyl, octyldedecyl) N-Lauroyl-L-
 glutamic acid ester/Eldew CL-301   2.00
Octyldodecyl Neopentanoate/Elefac I-205   10.00
Dimethicone(and)Trimethylsiloxysilicate/Dow Corning 593   0.50
Propylparaben   0.05
Methylparaben   0.15
Phenoxyethanol/Emeressence 1 160   0.60
Benzophenone-3/Uvinul M-40   4.00
Octyl Methoxycinnamate/Parsol MCX   7.50

Part B:
Polyglyceryl-4 Isostearate (and) Cetyl Dimethicone
 Copolyol (and) Hexyl Laurate/Abil WE09   5.00

Part C:
Deionized Water   61.40
Sodium Chloride   0.80
Sodium PCA/Ajidew N-50   1.00
Partially Deacetylated Chitin (1.0%)/Marine Dew   2.00
Propylene Glycol   5.00

Procedure:
    Pre-melt part A at 50 degrees Centigrade. Add part B to part
A. Premix part C until homogeneous. Slowly add part C into part
A and B mixture with high shear mixing.
    Appearance: White, smooth, shiny lotion   pH: 6.0-6.5
    Viscosity: 10,000-20,000 (RVT #6 @ 10 rpm @ 25 degrees C)

Procedure:
    Heat Part A ingredients to 80 degrees Centigrade. Mix until
all solids are completely dissolved. Add Part B ingredients.
Mix until uniform. Cool to 65 degrees Centigrade. Add Part C.
Mix well. Fill.

SOURCE: Ajinomoto USA, Inc.: Suggested Formulation

W/O Sunscreen Lotion
SPF-18

Ingredients:	%w/w
Phase A:	
Cetyl Dimethicone Copolyol (and) Polyglyceryl-4	
Isostearate (and) Hexyl Laurate (Abil WE-09)	5.0
Cetyl Dimethicone (Abil Wax 9801)	1.0
Octyl Palmitate (Tegosoft OP)	4.5
Caprylic/Capric Triglycerides (Tegosoft CT)	3.0
Cyclomethicone (Abil B 8839)	5.5
Hydrogenated Castor Oil	0.4
Synthetic Wax	0.6
Octyl Methoxycinnamate	5.0
Benzophenone-3	1.5
Menthyl Anthranilate	3.0
Phase B:	
Water	69.7
Sodium Chloride	0.8
Preservatives	Q.S.

Procedure:
1. Add the components of Phase A. Heat while mixing to 85C to incorporate the waxes. Cool to 50C.
2. Heat Phase B to 50C. Add B to A slowly with a low energy mixer. Maintain a smooth milky appearance at all times.
3. Cool to 35C with sweep mixer. Add fragrance. 4. Homogenize.
Formula GRH-2-31

W/O Sunscreen Lotion
SPF-17

Ingredients:	%w/w
Phase A:	
Cetyl Dimethicone Copolyol (and) Polyglyceryl-4-	
Isostearate (and) Hexyl Laurate (Abil WE-09)	5.0
Cetyl Dimethicone (Abil Wax 9814)	3.0
Octyl Stearate (Tegosoft OS)	12.0
Cyclomethicone (Abil B 8839)	8.0
Mineral Oil	2.0
Hydrogenated Castor Oil	0.5
Microcrystalline Wax	1.0
Phase B:	
Titanium Dioxide	8.0
Phase C:	
Water	59.7
Sodium Chloride	0.8
Preservatives	Q.S.

Procedure:
1. Heat Phase A to 85C to melt and disperse waxes.
2. Cool Phase A to 50C. Add B to A slowly with a low energy mixer. Maintain a smooth milky appearance at all times.
3. Roller mill to reduce particle size of Titanium Dioxide.
4. Cool to 50C.
5. Heat Phase C to 50C. Add Phase C while mixing.
6. Cool to 35C and homogenize.
Formula GRH-2-33
SOURCE: Goldschmidt Chemical Corp.: Suggested Formulations

Section XII
Miscellaneous

Aloe Medicated Gel
With Alcohol (SDA 39C)

Formula:	%	Per 2000g
l-Menthol	6.00	120.00
SDA 39C	8.00	160.00
Tween 60	3.00	60.00
Aloe Moist	83.00	1660.00

Procedure (With alcohol, SDA 39C):
1. Add Aloe Moist to large beaker. Fit with stirrer.
2. In a separate container, mix the menthol, alcohol and Tween and stir until clear.
3. With slow-medium agitation add the mixture to the Aloe Moist.
4. Stir until a milky white and homogeneous.
5. Fill into HDPE jars.

Aloe Medicated Gel
Without Alcohol

Formula:	%	Per 75g
l-Menthol	3.00	2.25g
Tween 60	12.00	9.00g
Aloe Moist	79.70	59.78g
3% Carbopol 940	5.00	3.75g
TEA	0.30	0.225g
FDC Yellow #5, 1%	qs	2 drops
FDC Blue #1, 1%	qs	2 drops

Procedure (Without alcohol):
1. Mix l-menthol and Tween 60 together and heat to about 45C.
2. Heat Aloe Moist to 55-58C - add colors.
3. Add the menthol/Tween mix to the Aloe Moist with stirring.
4. Add the 3% Carbopol 940 and mix well.
5. Add the TEA and mix well.
6. Cool to room temp with slow agitation.

SOURCE: Terry Laboratories, Inc.: Suggested Formulations

Cetylpyridinium Chloride Mouthwash

Ingredients:	%W/W
A Cetylpyridinium chloride, NF	0.10
Citric Acid, USP	0.10
Peppermint oil, USP	0.07
Eucalyptus oil, NF	0.02
Clove oil, USP	0.05
Tween 60	0.30
Alcohol, USP	10.00
B *Color	
Sorbo, Sorbitol solution, USP	20.00
Water, deionized	69.36

*q.s. these ingredients.

Procedure:
Dissolve the cetylpyridinium chloride and citric acid in water. Add the peppermint oil, eucalyptus oil and clove oil to the Tween 20. Add the alcohol slowly with agitation and add the water mixture. Add the Sorbo, water and color to make the desired volume.

Spearmint Mouthwash

Ingredients:	%W/W
A Tween 60, polysorbate 60	0.32
Spearmint flavor oil	0.24
B Arlasolve DMI	10.00
Ethanol (190 proof)	10.00
C Water	69.14
Sodium saccharin USP	0.10
Sodium benzoate	0.20
D Sorbo, 70% sorbitol solution	10.00

Procedure:
Mix (A) well. Mix (B) well. Add (B) to (A), mix well. Mix (C) well. Add (D) to (C), mix well. Add (CD) to (AB) and mix well. Formula PC 8207

SOURCE: ICI Surfactants: Suggested Formulations

Clear Gel without DEA

This formula contains no diethanolamine, yet is able to produce a microemulsion gel which exhibits clarity and possesses a manageable set point (80C), using potassium hydroxide in place of the DEA. Lower ethoxylates like the Volpos and Crodafoses used here are the emulsifiers of choice for clear gel formulations, since less ethoxylated materials have been shown to cause less irritation. The powerful emulsifying properties of these esters yield an extremely efficient system in which a 1:1 ratio of emulsifier:oil concentration is achieved - a balance that minimizes irritation potential and optimizes gel formation.

Ingredients:	%
Part A:	
Deionized Water	52.6
Propylene Glycol	12.0
Part B:	
Volpo 3 (Oleth-3)	7.0
Volpo 5 (Oleth-5)	4.0
Crodafos N3A (Oleth-3 Phosphate)	2.0
Crodafos N10A (Oleth-10 Phosphate)	4.0
Mineral Oil (70ssu)	17.0
Part C:	
KOH (25% Aq. Soln.)	1.4

Procedure:
Combine ingredients of Part A with mixing and heat to 90-95C. Combine ingredients of Part B with mixing and heat to 90-95C. Add Part C to Part A with mixing. Add Part A/C to Part B with mixing using a propeller blade mixer. Avoid aeration as much as possible. Cool to desired fill temperature, cooling at a 2C/min rate. Set point approximately 80C.
    Emulsifier:Oil Ratio: 1:1
    Set Point: 80C
Formula CG-25

Silicone Emulsion

This product uese Incroquat Behenyl TMS to form a stable emulsion (1 month @ 50C), specifically with Fluid 200 (200 cs) but also suitable for other silicone fluids, for Volatile Silicone and dimethicone.

Ingredients:	%
Part A:	
Deionized Water	72.0
Incroquat Behenyl TMS (Behentrimonium Methosulfate (and) Cetearyl Alcohol)	7.0
Silicone Fluid (200cps)	20.0
Part B:	
Germaben II	1.0

Procedure:
Combine ingredients of Part A with mixing and heat to 80-85C. Cool to 40C with mixing and add Part B. Cool with mixing to desired fill temperature.
Formula SC-231
SOURCE: Croda Inc.: Suggested Formulations

Clear Ringing Gel

Ingredients:	%W/W
A Mineral oil, Carnation brand	11.00
Brij 93	6.00
Arlasolve 200L	27.80
B Propylene glycol	5.00
Sorbo	7.00
Water, deionized	43.20

Procedure:
Heat (A) to 90C and (B) to 95C. Add (B) to (A) with moderate stirring. Remove from heat and continue to stir as the mixture cools. When the mixture turns clear (about 65C) pour into tubes or jars.

Clear Ringing Gel

Ingredients:	%W/W
A Arlamol E	5.00
Mineral oil, Carnation brand	6.00
Brij 93	6.00
Arlasolve 200L	27.80
B Propylene glycol	5.00
Sorbo	7.00
Water, deionized	43.20

Procedure:
Heat (A) to 90C and (B) to 95C. Add (B) to (A) with moderate stirring. Remove from heat and continue to stir as the mixture cools. When the mixture turns clear (about 65C) pour into tubes or jars.
Formula PC 8180

Clear Gel Base

Ingredients:	%W/W
Arlasolve DMI	96.00
Hyroxypropyl cellulose	4.00

Procedure:
Slowly sprinkle the hydroxypropyl cellulose in the Arlasolve DMI as you rapidly agitate. Stir until clear. A gel will form upon standing. Gel strength varies with thickener concentration. Active ingredients should be dissolved in Arlasolve DMI prior to addition of hydroxypropyl cellulose.

SOURCE: ICI Surfactants: Suggested Formulations

Cream Sachet

Ingredients:	%W/W
A Arlacel 165	16.00
Spermaceti	5.00
Cetyl alcohol	5.00
Isopropyl myristate	3.00
B Atlas G-2162	2.00
Water, deionized	59.00
*Preservative	
C Fragrance	10.00

q.s. these ingredients

Procedure:
Heat (A) to 75C. Heat (B) to 77C. Add (B) to (A) slowly with moderate but thorough agitation. Add (C) at 45C. Stir until room temperature and package.

Liquid Cologne

Ingredients:	%W/W
Perfume oil	5.00
Tween 20	21.50
Water	73.50
*Preservative	

*q.s. these ingredients.

Procedure:
Mix the perfume oil and Tween 20. Slowly add this mixture to the water and add preservative. Stir until a clear product results.

Solid Cologne

Ingredients:	%W/W
A Stearic acid, triple pressed	5.00
Sorbo	2.00
Alcohol, SD-40	83.00
B Sodium hydroxide to saponify the 5% stearic acid	
Water, deionized	5.00
C Perfume oil	5.00

Procedure:
Heat (A) to 65C and (B) to 70C. Add (B) to (A) with agitation. Add (C) at 55-60C. Pour into heated molds. Cool slowly to prevent formation of air pockets and to achieve the greatest translucency. Unmold, wrap in foil and package in air tight containers.

SOURCE: ICI Surfactants: Suggested Formulations

High-Luster Toothpaste

Ingredients:	% by Weight
Part A:	
(1) Sorbitol (70% solution)	30.00
(2) Deionized Water	17.52
Part B:	
(3) Kaopolite SF	24.25
(4) Dicalcium Phosphate Dihydrate	24.25
(5) Sodium Lauryl Sulfate	2.00
(6) Carboxymethyl Cellulose (medium viscosity)	0.80
(7) Trimagnesium Phosphate	0.50
(8) Korthix H	0.25
(9) Sodium Saccharin	0.20
(10) Methyl Paraben	0.18
(11) Propyl Paraben	0.05
Part C:	
(12) Flavoring Oil	q.s.

Procedure:

Mix (1) and (2). In a separate container, thoroughly dry-blend ingredients (3) through (11). Add Part B slowly to Part A. Mix thoroughly. Add Part C, mill twice and deaerate.

Follow recommended handling practices of the supplier of each product used.

Denture Cleaner

Ingredients:	% by Weight
(1) Deionized Water	30.8
(2) Korthix H	1.5
(3) Carboxymethyl Cellulose (medium viscosity)	0.5
(4) Sorbitol (70% solution)	10.0
(5) Glycerin	10.0
(6) Sodium Saccharin	0.2
(7) Kaopolite 1147	45.0
(8) Sodium Lauryl Sulfate	2.0
(9) Flavor, Preservative & Colorant	q.s.

Procedure:

Mix (2) into (1) with high shear. Add (3) and continue mixing at high shear until well dispersed. Reduce speed and add (4), (5), and (6). Slowly add (7) and continue to mix until well dispersed. Then add (8). Add (9) and mill twice and deaerate.

Follow recommended handling practices of the supplier of each product used.

SOURCE: Kaopolite, Inc.: Suggested Formulations

Hydrophilic O/W Ointment Base

Ingredients:	%W/W
A Stearyl alcohol	25.00
Petrolatum	25.00
B Propylene glycol	12.00
Myrj 52	5.00
Methyl paraben & propyl paraben	0.40
Water, deionized	32.60

Procedure:
Heat (A) to 70C. Heat (B) to 72C. Add (B) to (A) with agitation. Stir until set.

Emollient Anhydrous Ointment

Ingredients:	%W/W
G-1726	10.00
Arlamol E	40.00
Arlacel 186	50.00

Procedure:
Heat all ingredients until liquid and stir until room temperature.

W/O Ointment Base

Ingredients:	%W/W
Petrolatum	54.00
Arlacel 83, sorbitan sesquioleate	6.00

Procedure:
Heat (A) to 70C and (B) to 72C. Add (B) to (A) slowly with moderate, but thorough agitation. Stir until room temperature.

Washable Anhydrous Base

Ingredients:	%W/W
Petrolatum	95.00
Tween 61 or Tween 81 or Myrj 52	5.00

Procedure:
Heat all ingredients until liquid and stir until room temperature.

Washable Anhydrous Base

Ingredients:	%W/W
Hydrophilic Petrolatum, U.S.P.	50.00
Tween 60	50.00

Procedure:
Heat all ingredients until liquid and stir until room temperature.

SOURCE: ICI Surfactants: Suggested Formulations

Jojoba Massage Oil

	%
Mineral Oil	61.5
Isopropyl Palmitate	24.0
Coconut Oil #76	5.0
Jojoba Oil	2.0
Almond Oil Sweet	2.0
Acetulan	2.0
Glucam P-20	1.0
Dow Corning Silicone 344	2.0
Vitamin E	0.5
Fragrance	q.s.

Procedure:
   Load all ingredients into a stainless steel kettle. Warm slightly until clear with agitation, add Fragrance and package.

Massage Oil

	%
Solulan P B 5	3.0
Dow Corning Silicone #344	16.0
Emerest 2314	13.0
Drakeol #9	31.0
Coconut Oil	31.0
Escalol 507	3.0
Ross Jojoba Oil	3.0
Perfume Nova Rome DE 51	q.s.

Procedure:
   Load all ingredients into a vessel. Warm slightly until clear under agitation and package.

SOURCE: Frank B. Ross Co., Inc.: Suggested Formulations

Micro-Emulsions with Aloe Vera and d-Limonene

The following formulas are clear, microemulsions of d-Limonene and Aloe Vera. They represent the range of d-Limonene and Aloe Vera concentrations commonly used as degreasing solvent cleaners. These formulas are generally used by diluting in water (1->16 oz./gallon depending on the use requirements). The formulas are clear and exhibit a bloom effect upon dilution.

	%w/w
d-Limonene	10.0
Monamulse DL-1273ᵛ	18.0
Isopropanol	10.0
Water	42.0
Aloe Vera Gel (Terry AG002)	20.0

	%w/w
d-Limonene	35.0
Monamulse DL-1273	28.0
Isopropanol	5.0
Water	12.0
Aloe Vera Gel (Terry AG002)	20.0

	%w/w
d-Limonene	65.0
Monamulse DL-1273	23.0
Aloe Vera Gel (Terry AG002)	12.0

Procedure:
    In all cases, mix Monamulse DL-1273 with d-Limonene until completely dissolved and of uniform consistency. Add the Isopropanol, mixing until homogeneous. Pre-mix Aloe Vera Gel with water. Slowly add to batch with stirring until uniform.

SOURCE: Terry Laboratories, Inc.: Suggested Formulations

Modified Petrolatum

Ingredients:	%W/W
White petrolatum USP	75.00
Arlamol ISML, Isosorbide monolaurate	20.00
G-695, glycerol monooleate	5.00

Procedure:
  Add phase (A) together and stir with slight heat. Add phase (B) and heat to 80C and stir in phase (C). Cool while stirring and add phase (D).
Formula CP 1051

Modified Petrolatum

Ingredients:	%W/W
Petrolatum, White Protopet	60.00
Span 40, sorbitan monopalmitate	24.00
Arlatone G, PEG 25 hydrogenated castor oil	16.00

Procedure:
  Melt and mix.
Formula CP 1095

Modified Petrolatum

Ingredients:	%W/W
Petrolatum, White Protopet	50.00
Span 40, sorbitan monopalmitate	30.00
Arlatone G, PEG 25 hydrogenated castor oil	20.00

Procedure:
  Melt and mix
Formula CP 1096

Washable Oil

Ingredients:	%W/W
A Arlamol E	71.40
Atlas G73500	14.30
B Brij 30	9.50
C Dow Corning 344 cyclomethicone	4.80

Procedure:
  Mix (B) well. Add (B) to (A) and mix well. Add (C) to (AB) and mix well.
Formula CP 1117

SOURCE: ICI Surfactants: Suggested Formulations

Ointment Base

Ingredients:	%W/W
A Cetyl alcohol	7.00
Arlacel 165	5.00
B Sorbo	5.00
Water, deionized	73.00
C Arlasolve DMI	10.00
D *Preservative	

*q.s. these ingredients.

Procedure:
Heat (A) to 72C and (B) to 75C. Add (B) to (A) slowly with good agitation and add (C) at 35C and mix thoroughly. Add (D). Replace water lost due to evaporation. Package.

Anhydrous Ointment

Ingredients:	%W/W
Arlamol ISML, isosorbide monolaurate	25.00
Arlacel 186, glycerol monooleate	7.00
Petrolatum, Perfecta USP	68.00
*Perfume or actives	

*q.s. these ingredients.

Procedure:
Add phase (A) together and stir with slight heat. Add phase (B) and heat to 80C and stir in phase (C). Cool while stirring and add phase (D).

Zinc Oxide Ointment-Anhydrous

Ingredients:	%W/W
Arlamol ISML, isosorbide monolaurate	20.00
Arlacel 186, glycerol monooleate	5.60
Petrolatum, Perfecta USP	54.40
Zinc oxide, USP	20.00
*Perfume or actives	

Procedure:
Add phase (A) together and stir with slight heat. Add phase (B) and heat to 80C and stir in phase (C). Cool while stirring and add phase (D).
Formula CP 1035

SOURCE: ICI Surfactants: Suggested Formulations

O/W Ointment Base

Ingredients: %W/W
A Petrolatum   15.00
   Mineral oil   15.00
   Spermaceti   5.00
   Arlacel 40, sorbitan monopalmitate   5.00
   Tween 40, polyoxyethylene sorbitan monopalmitate   5.00
B Water, deionized   55.00
   *Preservative
   *q.s. these ingredients.

Procedure:
Heat (A) to 70C. Heat (B) to 72C. Add (B) to (A) slowly with moderate stirring. Stir to 35C and add water lost due to evaporation.

O/W Ointment Base

Ingredients: %W/W
A Cetyl alcohol   20.00
   Mineral oil   20.00
   Arlacel 80, sorbitan monooleate   0.50
   Tween 80, polyoxyethylene sorbitan monooleate   4.50
B Water, deionized   55.00
   *Preservative
   *q.s. these ingredients.

Procedure:
Heat (A) to 70C. Heat (B) to 72C. Add (B) to (A) slowly with moderate stirring. Stir to 35C and add water lost due to evaporation.

O/W Ointment Base

Ingredients: %W/W
A Stearyl alcohol   15.00
   Beeswax   5.00
   Mineral oil   15.00
   Arlacel 80, sorbitan monooleate   1.25
   Tween 80, polyoxyethylene sorbitan monooleate   3.75
B Water, deionized   60.00
   *Preservative
   *q.s. these ingredients.

Procedure:
Heat (A) to 70C. Heat (B) to 72C. Add (B) to (A) slowly with moderate stirring. Stir to 35C and add water lost due to evaporation.

SOURCE: ICI Surfactants: Suggested Formulations

Oil-In-Water-In-Silicone Emulsion

Oil-in-water Emulsion Phase:

Water	61.54%
Pantethine (80% aq. sol.)	0.10
Methylparaben	0.20
Carbomer 940	0.10
Glycerine	2.50
Sodium alkyl polyether sulfonate	1.25
Blandol Mineral Oil	1.75
Cholesterol	1.00
Cetyl palmitate	0.20
PEG-22 dodecylglycol copolymer	0.20
Ethylparaben	0.10
Propylparaben	0.15

Silicone Phase:

Cyclomethicone/dimethicone (90:10)	9.50
Cyclomethicone/dimethiconol (13:87)	5.00
Cyclomethicone	3.00
Phenyldimethicone	1.00
Pareth 15-3	2.00
Octylmethoxycinnamate	7.00
Benzophenone	0.50
C12-15 alcohols benzoate	2.85
Color	0.03
Fragrance	0.03

Protein-Lecithin Complex Emulsion

Carnation Mineral Oil	45.00%
Protopet 1S Petrolatum	10.00
Cetyl alcohol	1.50
Glyceryl monostearate	3.00
Casein-lecithin complex	3.00
Propylene glycol	5.00
Fragrance	0.50
Preservative	qs
Water	qs to 100.0

Skin Growth Accelerator

5-Methoxy-6,7-methylenedioxy-Isoflavone-4-O-B-D-glucoside	5.00%
Stearyl alcohol	18.00
Lanolin	20.00
Polyoxyethylene monooleate	0.25
Glyceryl monostearate	0.25
Protopet 1S Petrolatum	40.00
Water	16.50

SOURCE: Witco Corp.: Petroleum Specialties Group: Suggested
        Formulations

Pharmaceutical Cream Base

Ingredients: %W/W
A Stearyl Alcohol   6.00
   Arlacel 165   6.00
B Sorbo, Sorbitol Solution USP   5.00
   Water   83.00
Procedure:
   Heat (A) to 70C. Heat (B) to 72C. Add (B) to (A) with agitation. Stir until set.

Pharmaceutical Cream Base

Ingredients: %W/W
A Stearic Acid   10.00
   Stearyl Alcohol   5.00
   Tween 60   3.00
   Span 60   1.00
B Water   81.00
Procedure:
   Heat (A) to 70C. Heat (B) to 72C. Add (B) to (A) with agitation. Stir until set.

Pharmaceutical Cream Base

Ingredients: %W/W
A Stearic Acid   10.00
   Stearyl Alcohol   5.00
   Corn Oil   3.00
   Isopropyl Myristate   2.00
   Myrj 52   3.00
   Span 60   2.00
B Sorbo, Sorbitol Solution USP   5.00
   Water   70.00
Procedure:
   Hear (A) to 70C. Heat (B) to 72C. Add (B) to (A) with agitation. Stir until set.

Pharmaceutical Ointment Base

Ingredients: %W/W
A Cetyl Alcohol   20.00
   Corn Oil   20.00
   Tween 80   4.50
   Span 80   0.50
B Water   55.00
Procedure:
   Heat (A) to 70C. Heat (B) to 72C. Add (B) to (A) with agitation. Stir until set.

SOURCE: ICI Surfactants: Suggested Formulations

Washable Oil

Ingredients:	%W/W
Arlamol S3	80.00
Brij 30	10.00
Arlatone T	10.00

Procedure:
    Mix well at room temperature.
Formula CP 1111

Washable Oil

Ingredients:	%W/W
Arlamol E	50.00
Atlas G73500	50.00

Procedure:
    Mix well at room temperature.
Fromula CP 1112

Washable Oil

Ingredients:	%W/W
Arlamol S3	60.00
Isopropyl palmitate	30.00
Brij 93	5.00
Brij 30	4.60
Water	0.40

Procedure:
    Mix well at room temperature.
Formula CP 1115

Washable Oil

Ingredients:		%W/W
A	Arlamol E	50.00
	Atlas G73500	25.00
B	Brij 30	20.00
C	Dow Corning 344 cyclomethicone	5.00

Procedure:
    Mix (B) well. Add (B) to (A) and mix well. Add (C) to (AB)
and mix well.
Formula CP 1116

SOURCE: ICI Surfactants: Suggested Formulations

Washable Oil

Ingredients:	%W/W
A Arlamol E	79.50
B Brij 30	20.00
Water	0.50

Procedure:
Mix (B) well. Add (B) to (A) and mix well.
Formula CP 1109

Washable Oil

Ingredients:	%W/W
A Arlamol E	70.00
B Arlacel 186	10.00
Arlatone T	20.00

Procedure:
Mix (B) well, using heat to clarify the Arlacel 186. Mix (A&B)
together at room temperature and stir well.
Formula CP 1110

Washable Oil

Ingredients:	%W/W
Arlamol E	65.00
Arlacel 186	15.00
Brij 93	20.00

Procedure:
Mix well at room temperature.
Formula CP 1113

Washable Oil

Ingredients:	%W/W
Arlamol E	60.00
Isopropyl palmitate	30.00
Brij 93	10.00

Procedure:
Mix well at room temperature.
Formula CP 1114

SOURCE: ICI Surfactants: Suggested Formulations

Wound Healing Accelerator

Polyoxyethyleneglycol monostearate	2.00%
Glyceryl monostearate	5.00
Stearic acid	5.00
Behenyl alcohol	1.00
Blandol Mineral Oil	1.00
Glyceryl trioctanoate	5.00
Kojic acid dipalmitate	2.00
Preservative and fragrance	qs
1,3 Butylene glycol	5.00
Allantoin	0.10
Water	qs to 100.00

Massage Gel

2-Hexadecyl phosphate arginine salt	1.20%
Water	3.60
Glycerin	14.00
Dipropylene glycol	8.70
Isostearyl diglyceride	36.00
Carnation Mineral Oil	35.50
Cetyl alcohol	0.40
Stearyl alcohol	0.60
Fragrance	q.s.
Preservative	q.s.

Petrolatum Stick

This high petrolatum stick offers the well known benefits of petrolatum in an easy to use form.

Protopet 1S Petrolatum	85.0%
Syncrowax HGLC	12.0
Syncrowax ERCL	3.0

Procedure:
Combine the petrolatum and the syncrowaxes. Mix while heating to 80C. Cool to 65-70C and pour. Chill the molds or the cases to 40C prior to pouring.

SOURCE: Witco Corp.: Petrolatum Specialties Group: Suggested Formulations

W/O Ointment Base

Ingredients: %W/W
A Mineral oil   50.00
   Beeswax   10.00
   Lanolin   3.10
   Arlacel 83, sorbitan sesquioleate   1.00

B Borax   0.70
   Water, deionized   35.20
   *Preservative

*q.s. these ingredients.

Procedure:
Heat (A) to 70C and (B) to 72C. Add (B) to (A) slowly with moderate, but thorough agitation. Stir until room temperature.

W/O Ointment Base

Ingredients: %W/W
A Petrolatum   40.00
   Mineral oil   15.00
   Beeswax   4.00
   Arlacel 83, sorbitan sesquioleate   6.00

B Water, deionized   35.00
   *Preservative

*q.s. these ingredients.

Procedure:
Heat (A) to 70C and (B) to 72C. Add (B) to (A) slowly with moderate, but thorough agitation. Stir until room temperature.

W/O Ointment Base

Ingredients: %W/W
A Petrolatum   54.00
   Arlacel 83, sorbitan sesquioleate   6.00

B Water, distilled   40.00
   *Preservative

*q.s. these ingredients.

Procedure:
Heat (A) to 70C and (B) to 72C. Add (B) to (A) slowly with moderate, but thorough agitation. Stir until room temperature.

SOURCE: ICI Surfactants: Suggested Formulations

Section XIII
Trade–Named Raw Materials

RAW MATERIALS	CHEMICAL DESCRIPTION	SOURCE
Abil AV20	Phenyl trimethicone	Goldschmidt
Abil B8839	Cyclomethicone	Goldschmidt
Abil B8851	Dimethicone copolyol	Goldschmidt
Abil B8852	Dimethicone copolyol	Goldschmidt
Abil B9950	Dimethicone propyl PG betaine	Goldschmidt
Abil B88183	Dimethicone copolyol	Goldschmidt
Abil EM-90	Cetyl dimethicone copolyol	Goldschmidt
Abil WE09	Polyglyceryl-4 Isostearate (and) Cetyl Dimethicone Copolyol (and) Hexyl Laurate	Goldschmidt
Abil OSW12	Cyclomethicone (and) Dimeth-iconol (and) Dimethicone	Goldschmidt
Abil Quat 3270	Quaternium-80	Goldschmidt
Abil Quat 3272	Quaternium-80	Goldschmidt
Abil Quat 3474	Quaternium-80	Goldschmidt
Abil S201	Dimethicone/Sodium PG Propyl Dimethicone Thiosulfate Copolymer	Goldschmidt
Abil Wax 2434	Stearoxy dimethicone	Goldschmidt
Abil Wax 9800	Stearyl dimethicone	Goldschmidt
Abil Wax 9801	Cetyl dimethicone	Goldschmidt
Abil Wax 9809	Stearyl methicone	Goldschmidt
Abil Wax 9814	Cetyl dimethicone	Goldschmidt
Abil WE-09	Cetyl Dimethicone Copolyol (and) Polyglyceryl-4 Isostear-ate (and) Hexyl Laurate	Goldschmidt
Abil 100	Dimethicone	Goldschmidt
Abil 350	Dimethicone	Goldschmidt

RAW MATERIALS	CHEMICAL DESCRIPTION	SOURCE
Abil 500	Dimethicone	Goldschmidt
Abil 1000	Dimethicone	Goldschmidt
A-C Copolymer 400	Ethylene-vinyl acetate copolymer	Allied Sig-
A-C Copolymer 540	Copolymer	Allied Sig-
A-C Copolymer 580	Copolymer	Allied Sig-
A-C Polyethylene 9A	Polyethylene	Allied Sig-
A-C Polyethylene 617	Polyethylene	Allied Sig-
A-C Polyethylene 617A	Polyethylene	Allied Sig-
A-C Polyethylene 617G	Polyethylene	Allied Sig-
Acetol	Acetylated lanolin alcohol	Henkel
Acetol 1706	Acetylated lanolin alcohol	Henkel
Acetulan	Acetylated lanolin alcohol	Amerchol
Acritamer 940	Carbomer 940	RITA
Acritamer 941	Carbomer 941	RITA
Acrysol ICS-I	Acrylate/Steareth-20/Meth-acrylate Copolymer	Rohm&Haas
Active #4		Blew
Acuscrub 44	Scrub agent	Allied Sig-
Acuscrub 50	Scrub agent	Allied Sig-
Acylglutamate HS-11	Surfactant	Ajinomoto
Adogen MA-108	Dimethyl stearamine	Witco
Adol 52	Cetyl alcohol	Witco
Adol 52 NF	Cetyl alcohol	Witco
Adriano O/235970	Fragrance	
Aerosil R972	Fumed silica	Degussa

RAW MATERIALS	CHEMICAL DESCRIPTION	SOURCE
AGI Talc	Talc	Whittaker
Ajidew N-50	Sodium PCA	Ajinomoto
Ajidew T-50	Pyrrolidonecarboxylic acid	Ajinomoto
Alcolec BS	Single bleached lecithin	Amer Lech
Alconate SBR-3	Sodium petroleum sulfonate	Witco
Aldo HMS	Glyceryl monostearate	Lonza
Aldo MCT	Caprylic capric triglyceride	Lonza
Alfol 18	Stearyl alcohol	Vista
Alkamide DL 203/S	Alkanolamide	Rhone-
Alkamide DC 212/S	Alkanolamide	Rhone-
Alkamide DIN 295/S	Alkanolamide	Rhone-
Alkamide LE	Alkanolamide	Rhone-
Alkamul EGMS		
Allantoin		ICI
Aloe Extract	Aloe	Terry
Aloe Moist	Aloe	Terry
Aloe Vera 200X	Aloe	Dr. Madis
Aloe Veragel 200	Aloe	Dr. Madis
Aloe Vera Gel Decolorized 1X		Terry
Aloe Vera Liquid	Aloe	Dr. Madis
Aloe Veragel Lipoid	Aloe	Dr. Madis
Aloe Veragel Liquid 1:1		Dr. Madis
Aloe Veragel Liquid Concentrate 1:40		Dr. Madis
Alpine Floralistic	Fragrance	Alpine

RAW MATERIALS	CHEMICAL DESCRIPTION	SOURCE
Alpine 165-049	Fragrance	Alpine
Alugel DF30	Aluminum hydroxide	Guilini
Amerchol CAB	Sterol emulsifier/stabilizer	Amerchol
Amerchol H-9	Sterol emulsifier/stabilizer	Amerchol
Amerchol L-101	Mineral oil and lanolin alcohol	Amerchol
Amerchol 400	Sterol emulsifier/stabilizer	Amerchol
Amerlate	Isopropyl lanolate	Amerchol
Amerlate P	Lanolin fatty acid/ester	Amerchol
Amerscreen P	Ethyl dihydroxypropyl PABA	Amerchol
Amihope LL	Lauroyl lysine	Ajinomoto
Amisoft CA	Cocoyl glutamate	Ajinomoto
Amisoft CS-11	Sodium cocoyl glutamate	Ajinomoto
Amisoft CT-12	TEA-Cocoyl glutamate	Ajinomoto
Amiter LG-OD	Di-octyldodecyl lauroyl glutamate	Ajinomoto
Amiter LGOD-2		Ajinomoto
Amiter LGS-2	Disteareth-2 lauroyl glutamate	Ajinomoto
Ammonyx KP	Olealkonium chloride	Stepan
AMP	2-amino-2-methyl-1-propanol	Angus
AMP-95	2-amino-2-methyl-1-propanol	Angus
Amphisol	DEA-Cetyl phosphate	Givaudan
Amphisol A		Givaudan
Amphisol K	Potassium cetyl phosphate	Givaudan
Ampholyt JA140	Sodium lauroamphoacetate	Huls
Ampholyt JB130	Cocamidopropyl betaine	Huls

RAW MATERIALS	CHEMICAL DESCRIPTION	SOURCE
Amphomer LV-71		
Amphotensid B4		Zschimmer
Amphotensid GB 2009		Zschimmer
Amphotensid 9M		Zschimmer
Antaron V-220		ISP
Antil 141 Liquid	Propylene glycol (and) PEG-55 propylene glycol oleate	Goldschmidt
Antil 141B	PEG-55 propylene glycol oleate	Goldschmidt
Antil 141S	PEG-55 propylene glycol oleate	Goldschmidt
Antil HS60	Cocamidopropyl Betaine (and) Glyceryl Laurate	Goldschmidt
Antil 171	PEG-18 Glyceryl glycol cocoate/ oleate	Goldschmidt
APT		Centerchem
Aquasol 104	Aloe vera gel	CLE
Arlacel-C	Emulsifying agent	ICI
Arlacel 20	Emulsifying agent	ICI
Arlacel 40	Sorbitan monopalmitate	ICI
Arlacel 60	Sorbitan monostearate	ICI
Arlacel 80	Sorbitan monooleate	ICI
Arlacel 83	Sorbitan sesquioleate	ICI
Arlacel 85	Emulsifying agent	ICI
Arlacel 165	Glyceryl Stearate (and) PEG 100 Stearate	ICI
Arlacel 186	Glycerol monooleate	ICI
Arlacel 481	Emulsifying agent	ICI

RAW MATERIALS	CHEMICAL DESCRIPTION	SOURCE
Arlacel 582	Emulsifying agent	ICI
Arlacel 780	Emulsifying agent	ICI
Arlacel 989	Emulsifying agent	ICI
Arlacel 1689	Emulsifying agent	ICI
Arlacel DOA	Emulsifying agent	ICI
Arlamol E	PPG-15 stearyl ether	ICI
Arlatone G	PEG 25 hydrogenated castor oil	ICI
Arlamol HD	Emulsifying agent	ICI
Arlamol ISML	Isosorbide monolaurate	ICI
Arlamol M812	Emulsifying agent	ICI
Arlamol S3	Emulsifying agent	ICI
Arlamol S7	Emulsifying agent	ICI
Arlasolve DMI	Emulsifying agent	ICI
Arlasolve 200 Liquid	Isoceteth-20	ICI
Arlatone G	PEG 25 hydrogenated castor oil	ICI
Arlatone PQ220	Polyquaternium-19	ICI
Arlatone T	Sunscreen agent	ICI
Arlatone 985	Sunscreen agent	ICI
Arlatone 2121	Sunscreen agent	ICI
Armeen DM18D	Aliphatic amines	Akzo
A-SM	Stearamine oxide	Nippon Oil
AT Series	Iron oxides	
Atlas G-1726	Beeswax derivative	ICI
Atlas G-1795		ICI

RAW MATERIALS	CHEMICAL DESCRIPTION	SOURCE
Atlas G-2162		ICI
Atlas G-2330		ICI
Atlas G-3570		ICI
Atlas G 73500		ICI
Atmul 84S		
Aubygel x-125		
Avocado Oil		Lipo
Avocado Special		Dragoco
Baby Powder #165-848	Fragrance	
Baysilone Fluid M10	Silicone	Mobay
Baysilone Fluid PK20	Silicone	Mobay
Beauty O/239870	Perfume oil	
Bee's Milk		Koster
Bentone EW	Hectorite clay	Rheox
Bentone Gel CAO	Rheological additive	Rheox
Bentone Gel IPM	Rheological additive	Rheox
Bentone Gel LOI	Rheological additive	Rheox
Bentone Gel MIO	Rheological additive	Rheox
Bentone Gel SS71	Rheological additive	Rheox
Bentone Gel VS-5	Rheological additive	Rheox
Bentone Gel VS-5 PC	Rheological additive	Rheox
Bentone LT	Rheological additive	Rheox
Bentone MA	Rheological additive	Rheox

RAW MATERIALS	CHEMICAL DESCRIPTION	SOURCE
Bermocoll E230		Whittaker
Bermocoll E481		Whittaker
Beta Glucam 15% Coarse		Koster
Biodynes TRF	Tissue respiratory factors	Brooks
Biolac	Lactic acid, 88%	
Bio-sulphur CLR		CLR
Bioterge AS-40	Sodium C14-C16 olefin sulfonate	Stepan
Birch (water) Special		Dragoco
Blandol	Mineral oil	Witco
Bouquet Eau de Mer PC 916.315	Perfume	
Brij	Oleth-10	ICI
Brij 20	Emulsifying agent	ICI
Brij 30	Emulsifying agent	ICI
Brij 35	Emulsifying agent	ICI
Brij 35SP	Emulsifying agent	ICI
Brij 52	Ceteth-2	ICI
Brij 56	Emulsifying agent	ICI
Brij 58	Emulsifying agent	ICI
Brij 72	Steareth-2	ICI
Brij 76	Steareth-10	ICI
Brij 78	Steareth-20	ICI
Brij 93	Emulsifying agent	ICI
Brij 97	Emulsifying agent	ICI
Brij 700	Emulsifying agent	ICI

RAW MATERIALS	CHEMICAL DESCRIPTION	SOURCE
Brij 721	Steareth-21	ICI
Brij 721S	Emulsifying agent	ICI
Bronopol	Bactericide	Inolex
Busan 1504	Microbiocide	Buckman
C19-011	D&C Red No. 7	Sun
C19-012	D&C Red No. 6	Sun
C65-4429	F&D Yellow No. 5	Sun
Cab-O-Sil M-5	Silica	Cabot
Cab-O-Sil TS-530	Silica	Cabot
CAE	DCA ethyl cocoyl arginate	Ajinomoto
Camomile Special		Dragoco
Caprylic/Capric Triglyceride		Bernel
Carbopol EDT	Carbomer	Goodrich
Carbopol ETD 2055	Carbomer	Goodrich
Carbopol 934	Carbomer 934	Goodrich
Carbopol 940	Carbomer 940	Goodrich
Carbopol 941	Carbomer 941	Goodrich
Carbopol 954	Carbomer 954	Goodrich
Carbopol 980	Carbomer 980	Goodrich
Carbopol 981	Carbomer 981	Goodrich
Carbopol 1382	Carbomer 1382	Goodrich
Carnation Mineral Oil		Witco
Castorwax	Hydrogenated castor oil	CasChem

RAW MATERIALS	CHEMICALS DESCRIPTION	SOURCE
Cegosoft C24		
Cellosize QP 100MH	Hydroxyethylcellulose	Union Car
Cellosize QP-4400-H	Hydroxyethylcellulose	Union Car
Cellosize QP-15,000-H	Hydroxyethylcellulose	Union Car
Celquat H-100	Polyquaternium-4	NatStarch
Celquat SC-240	Polyquaternium-10	NatStarch
Cera Albalate 103	Behenyl beeswaxates	Koster
Cera Bellina	Pg-3 Beeswax	Koster
Ceralan		Amerchol
Ceraphyl 41	C12-15 alcohols lactate	VanDyk
Ceraphyl 45	Dioctyl malate	VanDyk
Ceraphyl 50	Emulsifier	VanDyk
Ceraphyl 50S	Emulsifier	VanDyk
Ceraphyl 60	Quaternium-22	VanDyk
Ceraphyl 65	Quaternium-26	VanDyk
Ceraphyl 368	Octyl palmitate	VanDyk
Ceraphyl 375	Isostearyl neopentanoate	VanDyk
Ceraphyl 424	Emulsifier	VanDyk
Ceraphyl 494	Emulsifier	VanDyk
Ceraphyl 847	Octyldodecyl stearoyl stearate	VanDyk
Ceraphyl ICA	Emulsifier	VanDyk
Cerasynt MN	Glycol stearate SE	VanDyk
Cerasynt PA	Propylene glycol stearate	VanDyk
Cerasynt Q	Glyceryl stearate SE	VanDyk

RAW MATERIALS	CHEMICAL DESCRIPTION	SOURCE
Cerasynt 840	PEG-20 stearate	Van Dyk
Ceresine Wax 130/135		Koster
Ceresine Wax 140/150		Koster
Cetal	Cetyl alcohol	Amerchol
Cetiol MM	Emollient esters	Henkel
Cetiol OE	Emollient esters	Henkel
Cetiol S	Emollient esters	Henkel
Cetiol SN	Emollient esters	Henkel
Chamomile LS	Safflower oil (and) chamomile extract	
Chiara O/238927	Fragrance	
Chlorhydrol, 50%	Aluminum chlorhydrate, 50%	Reheis
Cholesterol NF		Rita
Chroma-lite Black	Pearlescent pigment	VanDyk
Chroma-lite Dark Blue	Pearlescent pigment	VanDyk
Citroflex-2	Plasticizer	Morflex
Clearlan	Lanolin	Henkel
CMC 7MF	Cellulose gum	Aqualon
Coconut Oil #76		Welch
Collagen CLR		CLR
Collasol	Collagen	Croda

Color No. 3170 Pur Oxy Yellow B.C.
Color No. 3315 Pur Oxy Umber B.C.
Color No. 2511 Lo-Micron Pink B.C.
Color No. 3121-D & C Red No. 21
Color No. 7055 Pur Oxy Yellow B.C.
Color No. 7153 Lo-Micron Pink B.C.

RAW MATERIALS	CHEMICAL DESCRIPTION	SOURCE
Copherol F-1300	Tocopherol	Henkel
Corona Pure New Lanolin		
Cosmetic Tan C33-130	Iron oxide	Sun
Cosmowax J	Cetearyl alcohol (and) cetear-eth-20	Croda
Courage O/243101	Fragrance	
CP 1052	Shampoo base	ICI
Cremogen Aloe Vera	Propylene glycol (and) ethoxy-diglycol (and) aloe extract	Haarman
Cremophor A6	Surfactant	BASF
Cremophor A25	Surfactant	BASF
Cremophor EL	Surfactant	BASF
Cremophor HP14	Surfactant	BASF
Cremophor RH410	Surfactant	BASF
Crodacel QM	Cocodimonium hydroxypropoxy cellulose	
Crodacid B	Behenic acid	Croda
Crodacol C-70	Cetyl alcohol	Croda
Crodacol C-95	Cetyl alcohol	Croda
Crodacol CS-50	Cetearyl alcohol	Croda
Crodacol S-95	Stearyl alcohol	Croda
Crodacol S-95NF	Stearyl alcohol	Croda
Crodafos CES	Cetearyl alcohol (and) cetearyl phosphate	Croda

RAW MATERIALS	CHEMICAL DESCRIPTION	SOURCE
Crodafos N3 Neutral	DEA Oleth-3 phosphate	Croda
Crodafos N3A	Oleth-3 phosphate	Croda
Crodafos N10A	Oleth-10 phosphate	Croda
Crodalan LA	Cetyl acetate (and) acetylated lanolin alcohol	Croda
Crodamol CAP	Fatty acid ester	Croda
Crodamol DA	Fatty acid ester	Croda
Crodamol ML	Fatty acid ester	Croda
Crodamol MM	Myristyl myristate	Croda
Crodamol PMP	PPG-2 myristyl ether propionate	Croda
Crodamol PTC	Pentaerthyrityl tetracaprylate/ caprate	Croda
Crodamol PTIS	Pentaerythrityl tetraisostearate	Croda
Crodamol SS	Cetyl esters	Croda
Crodamol W	Stearyl heptanoate	Croda
Crodapearl Liquid	Sodium laureth sulfate (and) hydroxyethyl stearamide-MIPA	Croda
Crodasinic LS-30	Sodium lauroyl sarcosinate	Croda
Cromoist CS	Chondroitin sulfate (and) hydrolyzed protein	Croda
Crodesta F10	Sugar ester	Croda
Crodesta F110	Sugar ester	Croda
Cropeptide W	Hydrolyzed wheat protein (and) wheat oligosaccharides	Croda
Croquat HH	Cocodimonium hydroxypropyl hydrolyzed hair keratin	Croda
Croquat L	Lauryldimonium hydroxypropyl collagen	Croda

RAW MATERIALS	CHEMICAL DESCRIPTION	SOURCE
Crosultaine C-50	Cocamidopropyl hydroxysultaine	Croda
Crotein AD	Hydrolyzed animal protein	Croda
Crotein SPA	Hydrolyzed animal protein	Croda
Crothix	PEG-150 pentaerythrityl tetrastearate	Croda
Crovol A-70	PEG-60 almond glycerides	Croda
Crovol PK-70	PEG-45 palm kernel glycerides	Croda
Crystal O	Castor oil	CasChem
Cutina CBS	Cream base	Henkel
Cutina CP	Cream base	Henkel
Cutina GMS	Cream base	Henkel
Cutina LM	Cream base	Henkel
Cutina MD	Cream base	Henkel
Cyclomide DC 212/S	Cocamide DEA	Cyclo
Cycloryl WAT		Cyclo
Cycloteric BET C-30	Cocamidopropyl betaine	Cyclo
DC-556 Silicone		Dow Corning
DC-593 Silicone		Dow Corning
D&C Red #7 Ca Lake (3107)		Thomasset
D&C Red #9 (31-3009)		Thomasset
Deosafe 75128 N/I	Perfume	Haarman
Dermacryl LT	Skin care polymer	Nat. Starch
Dermacryl-79	Skin care polymer	Nat. Starch

RAW MATERIALS	CHEMICAL DESCRIPTION	SOURCE
Dermol 185	Isostearyl neopentanoate	
Diaformer Z-A	Methacrylol ethyl betaine/ methacrylates copolymer	Sandoz
Diaformer Z-301	Methacrylol ethyl betaine/ methacrylates copolymer	Sandoz
Diaformer Z-400	Methacrylol ethyl betaine/ methacrylates copolymer	Sandoz
Diahold A-503	AMP-Acrylates Copolymer	Sandoz
Dionil OC/K	PEG-4 Oleamide	Huls
D-Limonene	Dipentene	Florida
Dow Corning 2-7224	Silicone	Dow Corning
Dow Corning Q2-5000	Dimethicone copolyol	Dow Corning
Dow Corning 190	Dimethicone copolyol	Dow Corning
Dow Corning 193	Dimethicone copolyol	Dow Corning
Dow Corning 200	Dimethicone	Dow Corning
DC 200 Fluid; 50 vis	Dimethicone	Dow Corning
Dow Corning 200; 100	Dimethicone	Dow Corning
Dow Corning 200; 200	Dimethicone	Dow Corning
Dow Corning 344 Fluid	Cyclomethicone	Dow Corning
Dow Corning 345 Fluid	Cyclomethicone	Dow Corning
Dow Corning 556	Phenyl trimethicone	Dow Corning
Dow Corning 593	Dimethicone (and) trimethyl- siloxysilicate	Dow Corning
Dow Corning 929	Amodimethicone (and) nonoxy- nol-10 (and) tallow-trimonium chloride	Dow Corning
Dow Corning 1401 Fluid		Dow Corning

RAW MATERIALS	CHEMICAL DESCRIPTION	SOURCE
Dowicil 200	Preservative	Dow
Dragocid		Dragoco
Drakeol 9	Mineral oil	Penreco
Drakeol 10B	Mineral oil, naphthenic	Penreco
Drakeol 35	Mineral oil	Penreco
Dry Flo PC	Aluminum starch octenylsuccinate	Nat Starch
Duveen Toilet Soap Base		Duveen
Dynacerin 660	Oleyl erucate	Huls
Dynasan 110	Tricaprin	Huls
Dynasan 114	Trimyristin	Huls
Eastman AQ 38S	Polymer	Eastman
Eastman AQ 55S	Polymer	Eastman
Eastman Vitamin E TPGS (20%)		Eastman
Eastman Vitamin E Acetate 6-81		Eastman
Elastosol Animal Collagen & Elastin		Croda
Eldew CL-301	Di-(cholesteryl, behenyl, octyldodecyl) N-Lauroyl-L-glutamic acid ester	Ajinomoto
Elefac I-205	Octyldodecyl neopentanoate	Bernel
Elfacos C26	Polymer for cosmetics	Akzo
Elfacos E200	Polymer for cosmetics	Akzo
Elfacos ST9	Polymer for cosmetics	Akzo
Emalex CC-168	Cetyl octanoate	Nihon
Emalex GM-5	POE (5) glyceryl monostearate	Nihon

RAW MATERIALS	CHEMICAL DESCRIPTION	SOURCE
Emalex GMS-7CAE	Glyceryl monostearate, self emulsifying	Nihon Emul
Emalex GMS-45RT	Glyceryl monostearate, self emulsifying (HLB5)	Nihon Emul
Emalex HC-5	POE (5) hydrogenated castor oil	Nihon Emul
Emalex S.T.G.-R	Hydrogenated oil	Nihon Emul
Emeressence 1160	Phenoxyethanol	Henkel
Emerest 2000	Glyceryl stearate	Henkel
Emerest 2314	Ester	Henkel
Emerest 2316	Ester	Henkel
Emerest 2350	Glycol stearate	Henkel
Emerest 2388	Propylene glycol dipelargonate	Henkel
Emerest 2400	Glyceryl stearate	Henkel
Emerest 2407	Glyceryl stearate SE	Henkel
Emerest 2452	Polyglyceryl di-isostearate	Henkel
Emersol 132	Stearic acid	Henkel
Emersol 213		Henkel
Emerson-1323		Henkel
Emery IPP		Henkel
Emery 622	Coconut acid	Henkel
Emery 916	Glycerin pure	Henkel
Emery 1723		Henkel
Escalol 507	Octyl dimethyl PABA	Van Dyk
Escalol 557	Octyl methoxycinnamate	Van Dyk
Escalol 567	Benzophenone-3	Van Dyk

RAW MATERIALS	CHEMICAL DESCRIPTION	SOURCE
Estalan 430	Cosmetic emollient ester	Lanaetex
Estol EHP 1543	Cosmetic ester	Unichema
Estol 1550	Cosmetic ester	Unichema
Ethomeen C-25	Ethoxylated aliphatic amines	Akzo
Ethoxylated Carnauba		Koster
Ethoxyol 24	Ethoxylated oleyl alcohol	Lanaetex
Eumulgin B1	Cosmetic emulsifier O/W	Henkel
Eumulgin B2	Cosmetic emulsifier O/W	Henkel
Euperlan PK-771	Pearlshine concentrate	Henkel
Eusolex 232	Sunscreen	EM Indust
Eusolex 6300	Sunscreen	EM Indust
Eutanol G	Octyldodecanol	Henkel
Ewalan ODE 50		HE Wagner
Extract 52		Zschimmer
Extrapon Hamamelis Special		Dragoco
Extrapon 3-Special		Dragoco
Fancol ALA	Acetylated lanolin alcohol	Fanning
Finquat CT	Quaternium-75	Finetex
Finsolv TN	C12-15 alcohols benzoate	Finetex
Flamenco Super Blue	Cosmetics pigment	Mearl
Flamenco Superpearl	Cosmetics pigment	Mearl
Flamenco Velvet-100	Cosmetics pigment	Mearl
Flexricin 9	Fatty acid ester	Caschem

RAW MATERIALS	CHEMICAL DESCRIPTION	SOURCE
Floral Spice NP49	Fragrance	
Florasynth AB5697	Powder type	
Fluilan	Liquid lanolin	Croda
Fomblin HC/R		Brooks
Forestall		ICI
Forlan L	R.I.T.A. blend	Rita
Fougere Lavender	Fragrance	Fritzsche
Fragrance DE-47		
Fragrance DO-60		Novarome
Fragrance-Finesse Type		
Fragrance-Floral #169-120		
Fragrance GG44		Novarome
Fragrance GK-17		Novarome
Fragrance GK-19		Novarome
Fragrance GK-21		Novarome
Fragrance GP-58		Novarome
Fragrance H-45756		Robertet
Fragrance 2991H		IFF
Fragrance #X3110		Roure
Fragrance #1-67		
Fragrance #189-7124		
Fragrance #163-478		

RAW MATERIALS	CHEMICAL DESCRIPTION	SOURCE
G-9600 1053	Shampoo base	ICI
Gaffix VC-213	Vinylcaprolactam/PVP/Dimethyl- aminoethyl methacrylate copolymer	ISP
Gafquat 755	Polyquaternium-11	ISP
Gamma Orzanol		Koster
Ganex V-220	PVP/Eicosene copolymer	ISP
Ganex WP660	Alkylated vinyl pyrrolidone copolymer	ISP
Geahlene 750	Mineral oil (and) hydrogenated butylene/ethylene/styrene copolymer (and) hydrogenated ethylene/propylene/ styrene copolymer	
Gemtone Mauve Quartz	Cosmetic pearl powder	Mearl
Germaben II	Cosmetic preservative	Sutton
Germaben II-E	Cosmetic preservative	Sutton
Germall II	Diazolidinyl urea	Sutton
Germall 115	Imidazolidinyl urea	Sutton
Geropon SBFA-30		Rhone
Ginkgo Biloba		Vernin
Givaudan #PSC 10,435/6 Fragrance		Givaudan
Glucam E-10	Ethoxylated methyl glucosides	Amerchol
Glucam P-20	Propoxylated methyl glucosides	Amerchol
Glucamate DOE-120	PEG-120 methyl glucose dioleate	Amerchol
Glucamate SSE-20	Ethoxylated methyl glucosides	Amerchol
Glucamate SSG-20	Alkylated methyl glucoside ester	Amerchol
Glucate DO	Methyl glucoside ester	Amerchol
Glucate SS	Methyl glucoside ester	Amerchol

RAW MATERIALS	CHEMICAL DESCRIPTION	SOURCE
Glucopon 425 CS		Amerchol
Glucopon 600 CSUP		Amerchol
Glucopon 625 CSUP		Amerchol
Glucquat 100		Amerchol
Glydant	DMDM hydantoin	Lonza
Glydant Plus		Lonza
Glypure 70%	Glycolic acid	
GMS SE	Glycerol Monostearate SE	Stepan
Grillocam E-20	Methyl gluceth-20	Rita
Grilloten LSE-65K	Sucrose cocoate	Rita
Grilloten LSE-87K	Sucrose cocoate	Rita
Grilloten PSE 141G	Sucrose stearate	Rita
Hair Complex Aquosum		CLR
Hamamelis Special		Dragoco
Hamp-ene Na4	Tetrasodium EDTA	Hampshire
Hamp-ene 220	Tetrasodium EDTA	Hampshire
Herbal Fragrance SL 79-1224		PFW
Hetester HCA	Glyceryl triacetyl hydroxy-stearate	Bernel
Hetester HCP	PPG-3 hydrogenated castor oil	Bernel
Hi-Sil T-600		PPG
Horse-chestnut Special		Dragoco
Hostacerin PN73		Hoechst

RAW MATERIALS	CHEMICAL DESCRIPTION	SOURCE
Hostaphat KL340N		Hoechst
Hydrogenated Polyisobutene		Amoco
Hydrolyzed Silk Protein		Ikeda
Hydrotriticum WAA	Wheat amino acids	Croda
Hydrotriticum 2000	Wheat protein hydrolyzates	Croda
Hydroxyol		Malmstrom
Hygroplex HHG		CLR
Hystar CG	Hydrogenated starch hydrolysate	Lonza
Hystrene 5016	Fatty acid	Humko
Igepon AC78	Anionic surfactant	ISP
Imwitor 780K	Isostearyl diglyceryl succinate	Huls
Imwitor 900	Glyceryl stearate	Huls
Imwitor 960 Flakes	Glyceryl stearate SE	Huls
Incrocas 40	PEG-40 castor oil	Croda
Incrodet TD-7C	Trideceth-7 carboxylic acid	Croda
Incromate OLL	Olivamidopropyl dimethylamine lactate	Croda
Incromate SEL	Sesamidopropyl dimethylamine lactate	Croda
Incromectant AMEA-70	Acetamide MEA	Croda
Incromectant AMEA-100	Acetamide MEA	Croda
Incromectant LAMEA	Acetamide MEA (and) lactamide MEA	Croda
Incromide ALD	Almondamide DEA	Croda
Incromide BAD	Babassuamide DEA	Croda

RAW MATERIALS	CHEMICAL DESCRIPTION	SOURCE
Incromide CA	Cocamide DEA	Croda
Incromide CAC	Cocamide DEA cocoyl sarcosine	Croda
Incromide LR	Lauramide DEA	Croda
Incromide WGD	Wheat germamide DEA	Croda
Incromine BB	Behenamidopropyl dimethylamine	Croda
Incromine Oxide WG	Wheat germamidopropyl oxide	Croda
Incronam AL-30	Almondamidopropyl betaine	Croda
Incronam WG-30	Wheat germamidopropyl betaine	Croda
Incronam 30	Cocamidopropyl betaine	Croda
Incropol CS-20	Ceteareth-20	Croda
Incropol SC-20	Ethoxylated alcohol	Croda
Incroquat BA-85	Babassuamidopropalkonium chloride	Croda
Incroquat Behenyl TMS	Behentrimonium methosulfate (and) cetearyl alcohol	Croda
Incroquat CR Conc.	Cetearyl alcohol (and) PEG-40 castor oil (and) stearalkonium chloride	Croda
Incroquat O-50	Olealkonium chloride	Croda
Incroquat S-85	Stearalkonium chloride	Croda
Indopol H-100	Viscous polybutene	Amoco
Irgasan DP300	Bacteriostat	Ciba-Geigy
Iron Oxide Brown 7058		Warner-Jenk
Iron Oxide Yellow 7055		Warner-Jenk
Isopar H	Isoparaffin solvent	Exxon

RAW MATERIALS	CHEMICAL DESCRIPTION	SOURCE
Isopropylan 50	Isopropyl ester of lanolin	Amerchol
Ivarlan 3230	Lanolin oil	Brooks
Ivy Special		Dragoco
J-13-AT	Talc	
Jergens Type R-30318	Fragrance	
Kaopolite SF	Polishing powder	Kaopolite
Kaopolite TLC	Polishing powder	Kaopolite
Kaopolite 1147	Polishing powder	Kaopolite
Karion F		E. Merck
Kathon CG	Methylchlorisothiazolinone (and) Methylisothiazolinone	Rohm&
Keltrol	Xanthan gum	Kelco
Keltrol F	Xanthan gum	Kelco
Keltrol RD	Xanthan gum	Kelco
Keltrol T	Xanthan gum	Kelco
Keltrol TF	Xanthan gum	Kelco
Kelzan	Xanthan gum	Kelco
Kelzan AR	Xanthan gum	Kelco
Kessco Ethylene Glycol Distearate		Stepan
Kessco Glyceryl Monostearate		Stepan
Kessco-653	Fatty ester	Stepan
Kester Wax K-48	Spermaceti	Koster
Kester Wax 62		Koster

RAW MATERIALS	CHEMICAL DESCRIPTION	SOURCE
Kester Wax 100		Koster
Kester Wax 85		Koster
Klucel GF	Hydroxypropylcellulose	Aqualon
Klucel MF	Hydroxypropylcellulose	Aqualon
Korthix H	Thickening agent	Kaopolite
Kytamer KC	Polyquaternium-29	Amerchol
Lamepon S	Protein surfactant	Henkel
Laneto 50	PEG-75 lanolin	Rita
Laneto 100	PEG-75 lanolin	Rita
Lanette O	Cetearyl alcohol	Henkel
Lanette N	Cream base o/w	Henkel
Lanette 16	Cetyl alcohol	Henkel
Lanogene	Lanolin alcohol & mineral oil	Amerchol
Lanolin USP		Rita
Lanolin-X-tra Deodorized		Rita
Lanthanol LAL	Foaming agent	Stepan
Lantrol	Liquid lanolin	Henkel
Lexaine C	Cocoamidopropyl betaine	Inolex
Lexamine S-13	Stearamidopropyl dimethylamine	Inolex
Lexate IL	Concentrate/blend	Inolex
Lexate PX	Petrolatum (and) lanolin alcohol	Inolex
Lexein QX-3000	Quaternium-76 hydrolyzed collagen	Inolex
Lexemul AR	Glyceryl stearate (and) stearamidoethyl diethylamine	Inolex

RAW MATERIALS	CHEMICAL DESCRIPTION	SOURCE
Lexemul CS-20	Cetearyl alcohol (and) cet-eareth-20	Inolex
Lexemul EGMS	Glycol stearate	Inolex
Lexemul GDL	Glyceryl dilaurate	Inolex
Lexemul P	Emulsifier	Inolex
Lexemul T	Glyceryl stearate SE	Inolex
Lexemul 55G	Glyceryl stearate	Inolex
Lexemul 515	Glyceryl stearate	Inolex
Lexemul 561	Glyceryl stearate (and) PEG-100 stearate	Inolex
Lexgard B	Butyl paraben	Inolex
Lexgard Bronopol	2-bromo-2-nitropropane-1,3-diol	Inolex
Lexgard M	Methyl paraben	Inolex
Lexgard P	Propyl paraben	Inolex
Lexol BS	Isobutyl stearate	Inolex
Lexol EHS	Octyl stearate	Inolex
Lexol GT-865	Caprylic/capric triglyceride	Inolex
Lexol PG 8/10	Emollient	Inolex
Lexol PG-865	Propylene glycol dicaprylate/dicaprate	Inolex
Lexol 3975	Isopropyl palmitate (and) isopropyl myristate (and) isopropyl stearate	Inolex
Lexorez 100	Glycerine/diethylene glycol/adipate crosspolymer	Inolex
Lexquat AMG-BEO	Behenamidopropyl PG dimonium chloride	Inolex

RAW MATERIALS	CHEMICAL DESCRIPTION	SOURCE
Lexquat AMG-IS	Isostearamidopropyl-PG-dimonium chloride	Inolex
Lime Blossom Special		Dragoco
Lipacide PCO	Palmitoyl hydrolyzed animal protein	Lipo
Lipofruit Cucumber		Lipo
Liponate SS	Stearyl stearate	Lipo
Lipovol A	Natural vegetable oil	Lipo
Lipovol ALM	Almond oil	Lipo
Lipovol P	Natural vegetable oil	Lipo
Lipopeg 100-S	Polyoxyethylene fatty acid ester	Lipo
Lipoxol 600 MED	PEG-12	Huls
Liquapar	Preservative	Sutton
Liquid Coconut Soap, 40%		Laurel
Locron L		Henkel
Lo Micron Black	Iron oxides (and) talc	Whittaker
Lo Micron Pink	Iron oxides (and) talc	Whittaker
Lo Micron Yellow	Iron oxides (and) talc	Whittaker
Lonzest 143/S	Myristyl propionate	Lonza
Luviskol VA64	Polyvinylpyrrolidone	BASF
Luviskoll VA73W	PVP/VA copolymer	BASF
Luvitol EHO	Synthetic oil	BASF
Lytron 295	Opacifier	Morton

RAW MATERIALS	CHEMICAL DESCRIPTION	SOURCE
Mackadet BSC	Baby shampoo concentrate	McIntyre
Mackadet CBC	Conditioner concentrate	McIntyre
Mackadet LCB	Conditioner concentrate	McIntyre
Mackadet SBC-8	Mild blend	McIntyre
Mackadet WGS		McIntyre
Mackadet 40K		McIntyre
Mackalene NLC	Lactates	McIntyre
Mackalene 116	Cocamidopropyl dimethylamine lactate	McIntyre
Mackalene 316	Stearamidopropyl dimethylamine lactate	McIntyre
Mackalene 326	Stearamidopropyl morpholine lactate	McIntyre
Mackalene 426	Isostearamidopropyl morpholine lactate	McIntyre
Mackalene 716	Wheat germamidopropyl dimethyl- amine lactate	McIntyre
Mackam CET	Cetyl betaine	McIntyre
Mackam WGB	Wheat germamidopropyl betaine	McIntyre
Mackam 2C	Cocoamphodiacetate	McIntyre
Mackam 35	Cocoamidopropyl betaine	McIntyre
Mackam 35HP	Cocoamidopropyl betaine	McIntyre
Mackamide AME-100	Acetamide MEA	McIntyre
Mackamide C	Cocamide DEA	McIntyre
Mackamide CMA	Cocamide MEA	McIntyre
Mackamide CS	Cocamide DEA	McIntyre
Mackamide LLM	Lauramide DEA	McIntyre

RAW MATERIALS	CHEMICAL DESCRIPTION	SOURCE
Mackanate OM	Disodium oleamido MEA sulfo-succinate	McIntyre
Mackanate OP	Disodium oleamido MIPA sulfosuccinate	McIntyre
Mackamide PK	Palmkernalamide DEA	McIntyre
Mackamide PKM	Palmkernalamide MEA	McIntyre
Mackamide S	Soyamide DEA	McIntyre
Mackamine CAO	Cocamidopropylamine oxide	McIntyre
Mackamine WGO	Wheat germamidopropylamine oxide	McIntyre
Mackanate CP	Disodium cocamido MIPA sulfo-succinate	McIntyre
Mackanate DC-30	Disodium dimethicone copolyol sulfosuccinate	McIntyre
Mackanate EL	Disodium laureth sulfosuccinate	McIntyre
Mackanate LA	Diammonium lauryl sulfosuccinate	McIntyre
Mackanate LO-Special	Disodium lauryl sulfosuccinate	McIntyre
Mackanate WGD	Disodium wheatgermamido PEG-2 sulfosuccinate	McIntyre
Mackernium SDC-25	Stearalkonium chloride	McIntyre
Mackernium SDC-85	Stearalkonium chloride	McIntyre
Mackernium 007	Polyquaternium 7	McIntyre
Mackester EGDS	Glycol distearate	McIntyre
Mackester SP	Glycol stearate modified	McIntyre
Mackine 301	Stearamidopropyl Dimethylamine	McIntyre
Mackol 70NS	Sodium laureth sulfate	McIntyre
Mackol 16	Cetyl alcohol	McIntyre
Mackol 1618	Cetearyl alcohol	McIntyre

RAW MATERIALS	CHEMICAL DESCRIPTION	SOURCE
Mackpearl LV	Pearl agent	McIntyre
Mackpro NLP	Natural lipid protein	McIntyre
Mackpro NSP	Oleyl/palmityl/palmitolamido-propyl/silk hydroxypropyl dimonium chloride	McIntyre
Mackstat DM	DMDM hydantoin	McIntyre
Macol CPS	Polyoxyethylene fatty ether	PPG
Macrospherical 95	Aluminum chlorohydrate	Reheis
Magnabrite	Stabilizing, suspending agent	Am Colloid
Magnabrite S	Magnesium aluminum silicate	Whittaker
Maprofix TLS-500	Triethanolamine lauryl sulfate	Onyx
Marcol 70	Mineral oil	Exxon
Marcol 130	Mineral oil	Exxon
Marine-Dew	Partially deacetylated chitin	Ajinomoto
Marlamide DF1218	Cocamide DEA	Huls
Marlamid M1218	Cocamide DEA	Huls
Marlamid PG20	Cocamide MEA glycol ditallowate	Huls
Marlazin KC 30/50	Cocotrimonium chloride	Huls
Marlinat CM 105	Sodium laureth-11 carboxylate	Huls
Marlinat DFK 30	Sodium lauryl sulfate	Huls
Marlinat DFN 30	Ammonium lauryl sulfate	Huls
Marlinat SL 3/40	Disodium laureth sulfosuccinate	Huls
Marlinat 242/28	Sodium laureth sulfate	Huls
Marlinat 242/70	Sodium laureth sulfate	Huls
Marlon PS65	Sodium C13-C17 alkane sulfonate	Huls

RAW MATERIALS	CHEMICAL DESCRIPTION	SOURCE
Marlophor MO3-acid	Laureth-3 phosphate	Huls
Marlophor T10Na-salt	Sodium disteareth-10 phosphate	Huls
Marlowet LMA2	Laureth-2	Huls
Marlowet R11/K	PEG-11 castor oil	Huls
Marlowet R40/K	PEG-40 castor oil	Huls
Marlowet TA6	Ceteareth-6	Huls
Marlowet TA25	Ceteareth-25	Huls
Masil SF-V Fluid	Cyclomethicone	PPG
Masil 280	Dimethicone copolyol	PPG
Masil 656	Silicone	PPG
Maypon 4C	Potassium cocoyl-hydrolyzed collagen	Inolex
Mazox CAPA	Cocamidopropyl amine oxide	PPG
Mearlin-AC	Powdered pearl pigment	Mearl
Merkur 1546	Vaseline type	
Merquat 100	Polyquaternium-6	Calgon
Merquat 550	Polyquaternium-7	Calgon
Methocel E4M	Hydroxypropyl methylcellulose	Dow
Methocel F4M	Hydroxypropyl methylcellulose	Dow
Methocel 40-100	Hydroxypropyl methylcellulose	Dow
Micro-dry	Aluminum chlorohydrate powder	Reheis
Microencapsulated Vitamin E HC487		
Micronized TiO2 LA-20	Titanium dioxide	
Miglyol Gel B	Caprylic/capric triglyceride (and) stearalkonium (and) propylene carbonate hectorite	Huls

RAW MATERIALS	CHEMICAL DESCRIPTION	SOURCE
Miglyol 812	Caprylic/capric triglyceride	Huls
Miglyol 829	Caprylic/capric/diglyceryl succinate	Huls
Miglyol 840	Propylene glycol dicaprylate	Huls
Mineral Colloid BP	Montmorillonite	ECC
Mineral Oil #7		Penreco
Mineral Oil #9		Penreco
Miranol C2M Conc. NP	Cocoamphocarboxyglycinate	Rhone
Miranol Ester PO-LM4	Polypentaerythrityl tetra-laurate	Rhone
Miranol MHT	Lauroamphoglycinate (and) trideceth sulfate	Rhone
Miranol 2MHT Modified	Imidazoline amphoteric	Rhone
Mirasheen 202		Rhone
Mirataine BET C-30	Cocamidopropyl betaine	Rhone
Mirataine BET O-30	Oleamidopropyl betaine	Rhone
Mirataine CB	Surface active agent	Rhone
Modulan	Acetylated lanolin	Amerchol
Monafax 160	Phosphate ester	Mona
Monafax 785	Phosphate ester	Mona
Monamate LNT-40	Diammonium lauryl sulfosuccinate	Mona
Monamid CMA	Cocamide MEA	Mona
Monamid 150ADD	Cocamide DEA	Mona
Monamid 705	Cocamide DEA	Mona
Monamid 716	Lauramide DEA	Mona
Monamid 1089	Alkanolamide	Mona

RAW MATERIALS	CHEMICAL DESCRIPTION	SOURCE
Monamine 1255	Alkanolamide	Mona
Monamulse DL-1273		Mona
Monaquat TG	Dihydroxyethyl dihydroxypropyl stearammonium chloride	Mona
Monaterge 1164	Sodium lauryl sulfate (and) disodium lauryl sulfosuccinate	Mona
Monateric CAB	Cocamidopropyl betaine	Mona
Monateric CLV		Mona
Monateric COAB		Mona
Mowiol 10-98	Polyvinyl alcohol resin	Hoechst
Mulsifan CPA		Zschimmer
Mulsifan RT 203/80		Zschimmer
Mulsifan RT275		Zschimmer
Multiwax W-835	Microcrystalline wax	Witco
Myacide SP	2,4-dichlorobenzyl alcohol	Inolex
Myritol PG	Propylene glycol dicaprylate	
Myrj 52	Polyoxethylene stearate	ICI
Myrj 52S	PEG-50 stearate	ICI
Myvatex 60	Emulsifier	
Myvatex Texture Lite	Emulsifier	
Myverol 18-06	Hydrogenated soy glyceride	Eastman
Natrosol 250 HHR	Hydroxyethyl cellulose	Aqualon
Natrosol 250 HR	Hydroxyethyl cellulose	Aqualon
Naturechem GMHS	Castor oil derivative	Caschem

RAW MATERIALS	CHEMICAL DESCRIPTION	SOURCE
Nature Harvest 165-050 Fragrance		
Neobee M-5	Caprylic/capric triglyceride	Stepan
Neo Heliopan, AV	Octyl methoxy cinnamate	Haarman
Neo Heliopan, BB	Benzophenone-3	Haarman
Neo Heliopan, Type E1000		Haarman
Neo Heliopan, Hydro	Phenylbenzimidazole sulfonic acid	Haarman
Neo Heliopan, MA	Menthyl anthranilate	Haarman
Neo Heliopan, MBC		Haarman
Neo Heliopan, OS		Haarman
Neo Heliopan, 303	Octocryline	Haarman
Nikkol Lecinol S-10-M		Nikkol
Nikkol MYS-55	PEG-55 stearate	Nikko
Nikkol WCB		Nikko
Nimlesterol D		Malmstrom
Ninol AA-62 Extra	Foam stabilizer	Stepan
Ninol 49-CE	Cocamide DEA	Stepan
Nipastat	Parabens	Nipa
Novarome DE-47 Fragrance		Novarome
Novarome DE51	Perfume	Novarome
Novarome Fragrance CD-69		Novarome
Noville #84504	Fragrance	Noville

RAW MATERIALS	CHEMICAL DESCRIPTION	SOURCE
Octipirox	Piroctone alamine	Hoechst
Orgasol 2002D Natural	Extra Cos Nylon-12	Atochem
Oxaban A	Preservative	Angus
Oxetal VD20		Zschimmer
Oxynex K Liquid	Antioxidant	Zschimmer
Oxynex 2004	Antioxidant	Zschimmer
Oxypon 288		Zschimmer
Oxypon 328		Zschimmer
Oxypon 2145		Zschimmer
Oyster Nut Oil		Koster
Ozokerite Waxes		Int Wax
Ozokerite Waxes		Koster
Ozokerite Waxes		Ross
Padimate O	Octyl dimethyl PABA	
Panalane	Hydrogenated polyisobutene	Amoco
Panthenol	Dl-Panthenol Cosmetic Grade	Roche
Paragon	DMDM hydantoin (and) methyl paraben	McIntyre
Parcel MCX		Givaudan
Paricin 9	Fatty acid ester	CasChem
Parsol HS	UV-B sunscreen	Givaudan
Parsol MCX	Octyl methoxycinnamate	Givaudan
Parsol 1789	UV-A sunscreen	Givaudan
Parsol 5000	UV-B sunscreen	Givaudan

RAW MATERIALS	CHEMICAL DESCRIPTION	SOURCE
Pationic ISL	Sodium isostearyl lactylate	Rita
Pationic ISL/85	Sodium isostearyl lactylate@85%	Rita
Pationic SSL	Sodium stearoyl lactylate	Rita
Patlac IL	Isostearyl lactate	Rita
Patlac LA (44%)	Lactic acid	Rita
PC 9043	Silicone emulsion	ICI
PC 9044	Silicone emulsion	ICI
PC 9045	Silicone emulsion	ICI
PC 9046	Silicone emulsion	ICI
PC 9047	Silicone emulsion	ICI
PC 9048	Silicone emulsion	ICI
PCV 1454	Fragrance	
PEG-120 Methyl Glucose Dioleate		Rita
Pemulen TR-1	Polymeric emulsifier	Goodrich
Pemulen TR-2	Polymeric emulsifier	Goodrich
Permulgin 3220	Custom blended wax	Koster
Penreco Mineral Oil #9		Penreco
Peptein 2000x1	Hydrolyzed collagen	Hormel
Perfecta USP	Petrolatum	Witco
Perfume Dinalaya 129.051		
Perfume #44575		Fritsche
Perlglanzmittel GM 4055		Zschimmer
Perlglanzmittel GM 4175		Zschimmer
Permethyl 104A	Aliphatic hydrocarbon	Permethyl

RAW MATERIALS	CHEMICAL DESCRIPTION	SOURCE
Pert Type #189-724	Fragrance	
Petrolatum Alba		Witco
Perlglanzmittel GM4175		Zschimmer
Phenonip		Nipa
Phosfetal 201K		Zschimmer
Phospholipid EFA	Biomimetic phospholipid	Mona
Phospholipid PTC	Biomimetic phospholipid	Mona
Phospholipid SV	Biomimetic phospholipid	Mona
Phosphoteric PTC		Mona
Phosphoteric QL-38	Trisodium lauroampho PG acetate phosphate chloride	Mona

Pigment paste white: Nr. 93975
Pigment paste yellow: Nr. 75577
Pigment paste red: Nr. 68775
Pigment paste black: Nr. 78375

Pluronic F-127	Polyoxyalkylene glycol block polymer	BASF
PNC 400	Sodium carbomer	3V
Polawax	Emulsifying wax NF	Croda
Polawax A-31	Emulsifying wax NF	Croda
Polyaldo 10-1-S	Polyglyceryl-10 stearate	Lonza
Polyaldo 10-6-0	Polyglyceryl-10 hexaoleate	Lonza
Polymer JR 400		Union Carb
Polyox WSR	Water soluble resin	Union Carb
Polyquta-400	Polyquaternium-10	Rita
Polyquta 3000	Polyquaternium-10	Rita

RAW MATERIALS	CHEMICAL DESCRIPTION	SOURCE
Polysorbate 60	Polyoxyethylene (20) sorbitan monostearate	ICI
Polysynlane	Hydrogenated polyisobutene	Polyesther
Prisorine 3505	Glycerine	Unichema
Prisorine 3515	Glycerine	Unichema
Prisorine 9083	Glycerine	Unichema
Procetyl AWS	PPG-5 Ceteth 20	Croda
Prodew 100	Sorbitol & Sodium Lactate & Proline & Sodium PCA & Hydrolyzed Collagen	Ajinomoto
Promois W-32	Hydrolyzed collagen	Rita
Promois WK	Hydrolyzed keratin	Rita
Promois WK-HQ	Hydroxypropyl trimonium hydrolyzed keratin	Rita
Promulgen D	Cetearyl Alcohol(and)Ceteareth-20	Amerchol
Promyristyl PM3	PPG-3 myristyl ether	Croda
Pronalan Anticellulite		Centerchem
Propellant A-46	Hydrocarbon propellant	
Propolis Wax		Koster
Protopet 1S	Petrolatum	Witco
Protopet 2A	Petrolatum	Witco
Protox T-25	Ethoxylated amine	Protomeen
Purasal S/SP 60%	Sodium lactate	Purac Amer
Pur. Navy Blue #7110		Whittaker
Pur Oxy Yellow BC & Pur Oxy Black BC	Pigment	

RAW MATERIALS	CHEMICAL DESCRIPTION	SOURCE
Purton CFD		Zschimmer
Purton SFD		Zschimmer
PVP-K-30	PVP	ISP
Pyroter CPI-40		Ajinomoto
Pyroter GPI-25		Ajinomoto
Raluben TL		Raschig
Resyn 26-1314	Hair spray polymer	Nat Starch
Resyn 28-2913	Hair spray polymer	Nat Starch
Resyn 28-2930	Hair spray polymer	Nat Starch
Rewomid S-280		Rewo
Rezal 36G	Al Zr tetrachlorohydrex, gly.	Reheis
Rezal 36GP	Al Zr tetrachlorohydrex, gly. Super ultrafine.	Reheis
Rheolate 5000	Acrylate/Va copolymer	Reheis
Rhodacal A-246/L		Rhone
Rhodapex MA-360	Ether sulfate	Rhone
Rhodapon L-22	Alkyl sulfate	Rhone
Rhodapon LSB	Alkyl sulfate	Rhone
Rhodapon LT-6	Alkyl sulfate	Rhone
Rhodapon SB 8208/S	Alkyl sulfate	Rhone
Rhodapon 21LS	Alkyl sulfate	Rhone
Rhodigel	Xanthan gum	Vanderbilt
Rhodorsil Oils 70041 VO.65		Rhone
Rhodorsil 700 45V2	Cyclomethicone	Rhone

RAW MATERIALS	CHEMICAL DESCRIPTION	SOURCE
Rice Bran Oil Filtered		Koster
Ritabate 20	Polysorbate 20	Rita
Rita-BTAC	Behentrimonium chloride	Rita
Rita CA	Cetyl alcohol	Rita
Ritacet-20	Ceteareth-20	Rita
Rita-Cetearyl 70/30	Cetearyl alcohol	Rita
Ritaceti	Cetyl esters	Rita
Ritacetyl	Acetylated lanolin	Rita
Ritachol	Mineral oil and lanolin alcohol	Rita
Ritachol 1000	Cetearyl alcohol (and) poly-sorbate 60 (and) PEG-150 stearate (and) steareth-20	Rita
Ritachol 2000	R.I.T.A. blend	Rita
Ritachol 5000	R.I.T.A. blend	Rita
Rita-CTAC	Cetrimonium chloride	Rita
Ritaderm	Petrolatum (and) lanolin (and) sodium PCA (and) polysorbate 85	Rita
Rita EGDS	Glycol distearate	Rita
Rita EGMS	Glycol stearate	Rita
Rita GC	Guanidine carbonate	Rita
Rita GMS	Glyceryl stearate	Rita
Ritahydrox	Hydroxylated lanolin	Rita
Ritalan	Lanolin oil	Rita
Ritalan AWS	PPG-12-PEG-65 lanolin oil	Rita
Ritalan C	R.I.T.A. blend	Rita
Ritalastin EL30	Hydrolyzed elastin	Rita

RAW MATERIALS	CHEMICAL DESCRIPTION	SOURCE
Ritaloe 40X	Aloe vera gel	Rita
Ritaloe 200M	Aloe vera gel	Rita
Ritamide C	Cocamide DEA	Rita
Ritapan D	d-Panthenol	Rita
Ritapan DL	dl-Panthenol	Rita
Ritapeg 150DS	PEG-150 distearate	Rita
Ritaphenone 3	Benzophenone-3	Rita
Ritapro 165	R.I.T.A. blend	Rita
Ritapro 200	R.I.T.A. blend	Rita
Rita SA	Stearyl alcohol	Rita
Rita-SBC	Stearalkonium chloride	Rita
Rita-STAC	Steartrimonium chloride	Rita
Ritasynt IP	Glycol stearate and others	Rita
Ritavena-5	Hydrolyzed oat flour	Rita
Ritawax	Lanolin alcohol	Rita
Ritawax ALA	R.I.T.A. blend	Rita
Ritoleth 5	Oleth-5	Rita
Ritoleth-10	Oleth-10	Rita
Ritox 35	Laureth-23	Rita
Robane	Squalane	Robeco
Rosemary Special		Dragoco
Rosswaxes	Waxes	Ross

RAW MATERIALS	CHEMICAL DESCRIPTION	SOURCE
Salon #169-122	Fragrance	
Sandopan DTC Acid	Trideceth-7-carboxylic acid	Sandoz
Sandobet SC	Cocamidopropyl hydroxysultaine	Sandoz
Sandoxylate SX-424	PPG-2-isodeceth-12	Sandoz
Setacin 103 Special		Zschimm-
SF 18 (350) silicone	Dimethicone	Dow Corn
Shea Butter		Koster
Shebu	Shea butter	Rita
Shebu WS	PEG-50 shea butter	Rita
Shellsol T	Petroleum distillates	Shell
Shellsol 71	Petroleum distillates	Shell
Sicomet Black 80 C.I. 77499		BASF
Sicomet Brown 70 C.I. 77491		BASF
Sicomet Brown 75 C.I. 77491		BASF
Sicomet Patentblau 80E 131 Patent Blue		BASF
Silicone 200/100		Dow Corn
Silicone Fluid DC-200 (200 cps)		Dow Corn
Silicone Fluid 225		Dow Corn
Silicone Fluid 344 and 345 (Volatile)		Dow Corn
Silkience Type #169-120 Fragrance		
Siliconyl Beeswax		Koster
Siltex -50 +100 Mesh	Fused silica	Kaopol
Simchin Refined	Jojoba oil	Rita
Sipon ES	Sodium lauryl ether sulfate	Rhone

RAW MATERIALS	CHEMICAL DESCRIPTION	SOURCE
Sipon ES-2	Sodium laureth sulfate	Rhone
Sipon ESY		Rhone
Siponate DDB-40	Alkyl aryl sulfonate	Rhone
Sipon LCP	Sodium lauryl sulfate	Rhone
SL 79-1224	Herbal fragrance	PFW
Softigen 701	Glyceryl ricinoleate	Huls
Softigen 767	PEG-6 caprylic/capric glycerides	Huls
Softisan 100	Hydrogenated coco-glycerides	Huls
Softisan 154	Hydrogenated palm oil	Huls
Softisan 378	Caprylic/capric/stearic triglyceride	Huls
Softisan 601	Glyceryl cocoate (and) hydrogenated cocoanut oil (and) ceteareth-25	Huls
Softisan 645 & 649	Bis-diglyceryl caprylate/caprate/isostearate/hydroxystearate adipate	Huls
Solar Chem O		Caschem
Soltrol 100	C9-11 isoparaffin	Phillips
Solulan PB5	Alkoxylated lanolin derivative	Amerchol
Solulan 16	Laneth-16 & ceteth 16 & oleth-16 & steareth-16	Amerchol
Solulan 25	Laneth-25 & ceteth-25 & oleth-25 & steareth-25	Amerchol
Solulan 98	Polysorbate 80, acetylated lanolin alcohol, cetyl acetate	Amerchol
Soluvit CLR		CLR

RAW MATERIALS	CHEMICAL DESCRIPTION	SOURCE
Sorbo	Sorbitol solution, USP	ICI
Span 60	Emulsifying agent	ICI
Span 65	Emulsifying agent	ICI
Span 80	Emulsifying agent	ICI
Spermwax	Synthetic spermaceti	Robeco
Spicy Lime #4851-AN	Fragrance	IFF
S&P White NF 422P	Beeswax	
S&P #1 Yellow 73	Carnauba wax	
S&P 18	Microcrystalline wax	
S&P 75	Candelilla wax	
Span 40	Sorbitan monopalmitate	ICI
Squeeky Clean #165-047 Fragrance		
Standapol EA-2	Ammonium laureth sulfate	Henkel
Standapol ES-1	Sodium laureth sulfate	Henkel
Standapol ES-2	Sodium laureth-2 sulfate	Henkel
Standapol ES-3	Sodium laureth-3 sulfate	Henkel
Standapol WAQ-LC	Sodium lauryl sulfate	Henkel
Stepanol AM	Ammonium lauryl sulfate	Stepan
Stinging Nettle Special		Dragoco
Super Corona	Luxury USP lanolin	Croda
Supersat AWS-4	PEG-20 hydrogenated lanolin	Rita
Syncrowax ERCL	C18-36 acid glycol ester	Croda
Syncrowax HGLC	C18-36 acid triglyceride	Croda
Silicon Oil AK500		SWS

RAW MATERIALS	CHEMICAL DESCRIPTION	SOURCE
Sipon ES2	Sodium laureth sulfate	Rhone
Sedermasome	Soy lecithin	
SMDI Polyolprepolymer-2 PPG-12		
Softisan 378	Caprylic/capric/stearic triglyceride	Huls
Solulan-75	PEG 75 lanolin	Amerchol
Spectrapearl BLG	Pearlescent pigment	ISP
Spectraveil TG	Ultra-fine ZnO	Tioxide
Spectraveil 90/MOTG	Ultra-fine ZnO(40% solids Disp)	Tioxide
SS4267	Dimethicone and trimethyl-siloxysilicate	GE Silicone
Standamid KD	Cocamide DEA	Henkel
Standamide T	Teals	Henkel
Standapol ES-1&ES-2	Sodium laureth sulfate	Henkel
Standapol ES-3	Sodium laureth sulfate 28%	Henkel
Standapol T	TEA-lauryl sulfate	Henkel
Standapol WAQ&WAQ-SP	Sodium lauryl sulfate	Henkel
Sulfetal KT400		Zschimmer
Sulfetal TC50		Zschimmer
Sunarome OMC		Felton
Super Amide L-9	Coconut fatty acid diethanol-amide	Onyx
Super Hartolan	Lanolin alcohol	Croda
Super Refined Babassu Oil		
Super Refined Wheat Germ Oil		
Super Sterol Ester	C10-30 Cholesterol/Lanesterol Esters	

RAW MATERIALS	CHEMICAL DESCRIPTION	SOURCE
Super White	Petrolatum USP	
Suttocide A	Sodium hydroxymethylglycinate	Sutton
Syncrowax AW1-C	Synthetic wax	Croda
Syncrowax HGL-C	C18-36 acid triglyceride	Croda
Synthetic Beeswax JH-1508		Ross
TABS-D		Union Camp
Tagat S	Solubilizer	Goldschmidt
Takasago	Fragrance	
Talc Supra A	Premium talc, coarse ground	Cyprus
Tamol N	Dispersant	Rohm&Haas
Tapioca Flour		Nat Starch
Tegin M	Glyceryl stearate	Goldschmidt
Tegin T4753	Organic emulsifier	Goldschmidt
Tegobetaine L-7	Cocamidopropyl betaine	Goldschmidt
Teg. "P"		Goldschmidt
Tenox & Tenox 4	Antioxidant	Eastman
Tenox BHA	BHA	Eastman
Tensagex EOC-670		ICI
Tergitol NP-10	Surfactant	Union Carb
Terry AG002	Aloe vera gel	Terry
Ticaxan	Xanthan gum	Tic Gums
Timbuktu O/186901	Fragrance	
Timiron MP-10	Interference pigment	Rona Pearl

RAW MATERIALS	CHEMICAL DESCRIPTION	SOURCE
Timiron Pearl Sheen MP-30	Interference Pigment	Rona
Timiron Super Red	Interference Pigment	Rona
Tioveil AQ/FIN/MOTG	Ultra-fine titanium dioxide	Tioxide
TL-10	PEG-20 sorbitan monolaurate	
Tocopherol		Rita
Tomah AO-728 Special		Tomah
Tritisol	Soluble wheat protein	Croda
Trivent NP-13 & SS-20	Tridecyl neopentanoate	Trivent
Trycol 5963	Ester	Henkel
Tween 20	Polysorbate 20	ICI
Tween 40	Polysorbate 40	ICI
Tween 60	Polysorbate 60	ICI
Tween 61	Polyoxyethylene sorbitan monostearate	ICI
Tween 80	Polysorbate 80	ICI
Tween 81	Polyoxyethylene sorbitan monooleate	ICI
Tween 85	Emulsifying agent	ICI
Unipertan P-24		Lipo
Usnat AO	Anti-dandruff	
Uvinul M-40	Benzophenone-3	BASF
Uvinul MS-40	Benzophenone-4	BASF
Uvinul O-18	Octyl salicylate	BASF

RAW MATERIALS	CHEMICAL DESCRIPTION	SOURCE
Vanclay	Kaolin	Vanderbilt
Van Gel B		Vanderbilt
Vanox PCX	BHT	Vanderbilt
Vanseal CS	Sarcosinate surfactant	Vanderbilt
Vanseal NACS-30	Sodium cocoyl sarcosinate	Vanderbilt
Vansael NALS-30	Sarcosinate surfactant	Vanderbilt
Veegum/F/HV/Pro/ Regular/Ultra	Magnesium aluminum silicate	Vanderbilt
Velsan P8-3	Isopropyl C12-15 Pareth-9- Carboxylate	Sandoz
Veragel Liquid	Aloe vera gel	Dr. Madis
Veragel Liquid 1:1	Aloe vera gel	Dr. Madis
Veragel 200 Powder	Aloe vera	Dr. Madis
Versatyl-42	Hair spray polymer	Nat Starch
Versene NA	Disodium EDTA	Dow
Versene-220	Chelating agent	Dow
Vitamin A Palmitate Type PIMO/BH		
Vitamin E	Tocopherol	
Vitamin E Acetate	Tocopherol acetate	
Volatile Silicone 7207 Cyclomethicone		
Volpo 3	Oleth-3	Croda
Volpo 5	Oleth-5	Croda
Volpo S-2	Steareth-2	Croda
Volpo S-10	Steareth-10	Croda

RAW MATERIALS	CHEMICAL DESCRIPTION	SOURCE
Wachsemulsion 1864		Zschimmer
Wheat Germ Oil		Rita
White Jojoba Oil		Ross
White Protopet	Petrolatum	Petrolite
White Protopet 1S	Petrolatum	Petrolite
Wickenol 163	Dioctyl adipate (and) octyl stearate (and) octyl palmitate	Caschem
Wickenol 386	Aluminum zirconium tetra-chlorohydrex Gly	Caschem
Witconate AOS	Sodium C14-16 olefin sulfonate	Witco
Yellow BC Iron Oxides		Warner
Z-Cote		Sun Smart
Zetesol MS		Zschimmer
Zetesol NL		Zschimmer
Zetesol 856T		Zschimmer
Zetesol 2056		Zschimmer
Zusolat 1005/85		Zschimmer

Section XIV
Suppliers' Addresses

Ajinomoto USA, Inc.
Glenpoint Ctr. W
500 Frank W. Burr Blvd.
Teaneck, NJ 07645
(201)-907-3244

Akzo Chemicals, Inc.
300 S. Riverside Plaza
Chicago, IL 60606
(312)-906-7500

Allied Signal, Inc.
P.O. Box 2332R
Morristown, NJ 07962
(201)-455-2000

Alpine Aromatics Int'l Inc.
51 Ethel Rd. W
Piscataway, NJ 08854
(908)-572-5600

Amerchol Corp.
P.O. Box 4051
136 Talmadge Rd.
Edison, NJ 08818
(908)-248-6000

American Colloid Co.
Hwy 212W
Belle Fourche, SD 57717
(605)-892-2591

American Lecithin Co.
33 Turner Rd.
P.O. Box 1908
Danbury, CT 06813
(203)-790-2700

Amoco Chemical Co.
200 E. Randolph Dr.
Chicago, IL 60601
(312)-856-3200/(800)-621-4567

Angus Chemical Co.
1500 E. Lake Cook Rd.
Buffalo Grove, IL 60089
(708)-215-8600/(800)-323-6209

Aqualon
1313 N. Market St.
Wilmington, DE 19899
(302)-594-5000/(800)-345-8104

Atochem North America
900 Milk St.
Cartaret, NJ 07008
(908)-541-4414

BASF Corp.
100 Cherry Hill Rd.
Parsippany, NJ 07054
(201)-316-3000/(800)-526-1072

Bernel Chemical Co., Inc.
174 Grand Ave.
Englewood, NJ 07631
(201)-569-8934

Brooks Industries, Inc.
70 Tyler Place
South Plainfield, NJ 07080
(908)-561-5200

Buckman Laboratories Int'l Inc.
1256 N. McLean Blvd.
Memphis, TN 38108
(901)-278-0330/(800)-727-2772

Cabot Corp.
Cab-O-Sil Div.
Rte. 36W
Tuscola, IL 61953
(217)-253-3370/(800)-222-6745

Calgon Corp.
P.O. Box 1346
Pittsburgh, PA 15230
(412)-777-8000

CasChem, Inc.
40 Avenue A
Bayonne, NJ 07002
(201)-858-7900/(800)-CAS-CHEM

Centerchem, Inc.
225 High Ridge Rd.
Stamford, CT 06905
(203)-975-9800

Ciba-Geigy Corp.
410 Swing Rd.
Greensboro, NC 27419
(919)-632-7327/(800)-221-0453

CLE Inc.
11220 Grader St.
Dallas, TX 75238
(214)-341-4949/(800)-638-7947

CLR: Dr. K. Richter GmbH
Chemisches Laboratorium
Bennigonstrabe 25,
D-1000 Berlin

Croda, Inc.
7 Century Dr.
Parsippany, NJ 07054
(201)-644-4900

Cyclo Products Inc.
1922 E. 64 St.
Los Angeles, CA 90001
(213)-582-6411

Cyprus Industrial Minerals
P.O. Box 3419
Englewood, CO 80155
(800)-325-0299

Degussa Corp.
65 Challenger Rd.
Ridgefield Park, NJ 07660
(201)-641-6100

Dow Chemical USA
2020 Dow Center
Midland, MI 48674
(800)-258-CHEM

Dow Corning Corp.
Box 0994
Midland, MI 48686
(517)-496-4000

Dragoco, Inc.
10 Gorden Drive
Totowa, NJ 07512
(201)-256-3850

Eastman Chemical Co.
P.O. Box 431
Kingsport, TN 37662
(615)-229-4006/(800)-EASTMAN

ECC America
5775 Peachtree-Dunwoody Rd.
Atlanta, GA 30342
(800)-843-3222

Exxon Chemical Americas
13501 Katy Frwy
Houston, TX 77079
(713)-870-6000/(800)-231-6633

Fanning Corp.
2450 W. Hubbard St.
Chicago, IL 60612
(312)-563-1234

Felton Worldwide
599 Johnson Ave.
Brooklyn, NY 11237

Finetex, Inc.
418 Falmouth Ave.
Elmwood Park, NJ 07407
(201)-797-4686

Florida Food Products Inc.
2231 W. Hwy 44
Eustis, FL 32726
(904)-357-4141

Fritsche Dodge & Olcott Inc.
76 Ninth Ave.
New York, NY 10011
(212)-929-4100/(800)-221-7095

GE Silicones
260 Hudson River Rd.
Waterford, NY 12188
(518)-237-3330/(800)-255-8886

Givaudan-Roure Corp.
100 Delawanna Ave.
Clifton, NJ 07015
(201)-365-8000

Goldschmidt Chemical Corp.
914 E. Randolph Rd.
Hopewell, VA 23860
(804)-541-8658/(800)-445-1809

B.F. Goodrich Co.
9911 Brecksville Rd.
Cleveland, OH 44141
(216)-447-5000/(800)-331-1144

Haarman & Reimer Corp.
60 Diamond Rd.
Springfield, NJ 07091
(201)-912-5707/(800)-432-1559

Hampshire Chemical Co.
55 Hayden Ave.
Lexington, MA 02173

Henkel Corp.
11501 Northlake Dr.
Cincinnati, OH 45299
(513)-530-7300/(800)-543-7370

Hoechst Celanese Corp.
3340 W. Norfolk Rd.
Portsmouth, VA 23703
(804)-483-7530/(800)-526-4960

Hormel
P.O. Box 800
Austin, MN 55912
(507)-437-5676

Huls America, Inc.
80 Centennial Dr.
Piscataway, NJ 08854
(908)-980-6946/(800)-526-0339

Humko Chemical Div.
Witco Corp.
755 Crossover Lane
Memphis, TN 38117
(901)-684-7047

ICI Americas Inc.
Concord Pike & New Murphy Rd.
Wilmington, DE 19897
(302)-575-3034/(800)-822-8215

Ikeda Corp.
New Mexico Bldg. 3-1,
Marunouchi 3-Chome
Chiyoda-Ku, Tokyo 100, Japan
03-3212-8791

Inolex Chemical Co.
Jackson & Swanson Sts.
Philadelphia, PA 19148
(215)-271-0800/(800)-521-9891

ISP: International Specialty Prod
1361 Alps Rd.
Wayne, NJ 07470
(201)-628-3000/(800)-848-7659

IFF: International Flavors &
  Fragrances
521 W. 57 St.
New York, NY 10019
(212)-765-5500

Kaopolite Inc.
2444 Morris Ave.
Union, NJ 07083
(908)-789-0609

Kelco Div.
Merck & Co., Inc.
8355 Aero Drive
San Diego, CA 92123
(619)-292-4900/(800)-535-2656

Koster-Keunen, Inc.
P.O. Box 447
90 Bourne Blvd.
Sayville, NY 11782
(516)-589-0456

Lanaetex Products, Inc.
151 3 Ave.
Elizabeth, NJ 07206
(908)-351-9700

Laurel Industries, Inc.
29525 Chagrin Blvd.
Cleveland, OH 44122
(216)-831-5747/(800)-221-1304

Lipo Chemicals, Inc.
207 19th Ave.
Paterson, NJ 07504
(201)-345-8600

Lonza, Inc.
17-17 Rte. 208
Fair Lawn, NJ 07410
(201)-794-2400/(800)-777-1875

Dr. Madis Labs Inc.
375 Huyler St.
South Hackensack, NJ 07606
(201)-440-5000

McIntyre Group Ltd.
1000 Governors Hwy.
University Park, IL 60466
(708)-534-6200

Mearl Corp.
41 E. 42 St.
New York, NY 10017
(212)-573-8500

Mobay Corp.
Mobay Rd.
Pittsburgh, PA 15205
(412)-777-2000/(800)-662-2927

Mona Industries, Inc.
76 E. 24 St.
P.O. Box 425
Paterson, NJ 07544
(201)-345-8220

Morflex Inc.
2110 High Point Rd.
Greensboro, NC 27403
(919)-292-1781

Morton International, Inc.
150 Andover St.
Danvers, MA 01923
(508)-774-3100

National Starch & Chemical Co.
10 Finderne Ave.
Bridgewater, NJ 08807
(908)-685-5000/(800)-532-1115

Nipa Laboratories, Inc.
104 Hagley Bldg.
Concord Plaza
3411 Silverside Rd.
Wilmington, DE 19810
(302)-478-1522

Novarome Inc.
30 Stewart Pl.
Fairfield, NJ 07004
(201)-575-4550

Noville Essential Oil Co.
1312 Fifth St.
North Bergen, NJ 07047
(201)-867-9080

Penreco
138 Petrolia St.
Karns City, PA 16041
(412)-283-5600/(800)-245-3952

Petrolite Corp.
6910 E. 14 St.
Tulsa, OK 74112
(918)-836-1601

Phillips 66 Co.
376 Phillips Bldg. Annex
Bartlesville, OK 74004
(806)-274-5236/(800)-858-4327

Polyesther Corp.
61 Hill St.
P.O. Drawer 5076
Southampton, NY 11969
(516)-283-4400

PPG Industries
3938 Porett Drive
Gurnee, IL 60031
(708)-244-3410/(800)-CHEM-PPG

Purac America Inc.
111 Barclay Blvd.
Lincolnshire Corporate Center
Lincolnshire, IL 60069
(708)-634-6330

Raschig Corp.
P.O. Box 7656
Richmond, VA 23231
(804)-222-9516

Reheis, Inc.
235 Snyder Ave.
Berkeley Heights, NJ 07922
(908)-464-1500

Rheox, Inc.
Wyckoff Mills Rd.
Hightstown, NJ 08520
(609)-443-2320

RITA Corp.
1725 Kilkenny
Woodstock, IL 60098
(815)-337-2500/(800)-426-7759

Rhone Poulenc Inc.
Prospect Plains Rd.
Cranbury, NJ 08512
(609)-860-4000

Robeco Inc.
99 Park Ave.
New York, NY 10016
(212)-986-6410

Robertet Inc.
125 Bauer Dr.
Oakland, NJ 07436
(201)-337-7100

Roche Chemical Division
Hoffman-LaRoche, Inc.
Nutley, NJ 07110
(201)-235-8077/(800)-526-0189

Rohm & Haas Co.
Independence Mall W
Philadelphia, PA 19105
(215)-592-3000

Rona
EM Industries
5 Skyline Drive
Hawthorne, NY 10532
(914)-592-4660

Sandoz Chemicals Corp.
4000 Monroe Rd.
Charlotte, NC 28205
(704)-331-7234/(800)-631-8077

Shell Chemical Co.
P.O. Box 2463
Houston, TX 77002
(713)-241-6161

Stepan Co.
22 W. Frontage Rd.
Northfield, IL 60093
(708)-446-7500

Sun Chemical Corp.
411 Summit Ave.
Cincinnati, OH 45232
(513)-681-5950/(800)-343-2583

Sun Smart Inc.
P.O. Box 1451
Wainscott, NY 11975
(516)-324-8061

Sutton Laboratories, Inc.
116 Summit Ave.
Cincinnati, OH 45232
(513)-681-5950/(800)-343-2583

Terry Laboratories, Inc.
390 N. Wickham Rd.
P.O. Box 566
Melbourne, FL 32935
(407)-259-1630/(800)-367-2563

TIC Gums Inc.
4609 Richlynn Dr.
Belcamp, MD 21017
(410)-273-7300

Tioxide Specialties Ltd.
Billingham, Cleveland TS23 1PS
United Kingdom
0642-370300

Tomah Products
1012 Terra Drive
Milton, WI 53563
(608)-868-6811

Trivent Chemical Co.
45 Ridge Rd.
P.O. Box 597
South River, NJ 08882
(908)-251-1116

3V Inc.
1500 Harbor Blvd.
Weehawken, NJ 07087
(201)-865-3600

Unichema North America
4650 S. Racine Ave.
Chicago, IL 60609
(312)-376-9000/(800)-833-2864

Union Camp Corp.
1600 Valley Rd.
Wayne, NJ 07470
(201)-628-2680

Union Carbide Chemicals and
 Plastics
39 Old Ridgebury Rd.
Danbury, CT 06817
(203)-794-5300

R.T. Vanderbilt Co. Inc.
30 Winfield St.
P.O. Box 5150
Norwalk, CT 06856
(203)-853-1400

Van Dyk
Main & William Sts.
Belleville, NJ 07109
(201)-450-3264

Vista Chemical Co.
P.O. Box 19029
900 Threadneedle
Houston, TX 77224
(713)-588-3000/(800)-231-3216

Wacker Silicones Corp.
3301 Sutton Rd.
Adrian, MI 49221
(517)-264-8500/(800)-248-0063

Warner Jenkinson Cosmetic Colors
155 Helen St.
South Plainfield, NJ 07080
(800)-543-4524

Welch, Holme & Clark Co.
7 Avenue L
Newark, NJ 07105
(201)-465-1200

Whittaker, Clark & Daniels, Inc.
1000 Coolidge St.
South Plainfield, NJ 07080
(908)-561-6100

Witco Corp. (All Groups)
1)Oleo Surfactants Group
2)Petroleum Specialties Group
3)Polymer Additives Group
520 Madison Ave.
New York, NY 10022
(212)-605-3600

Zschimmer & Schwarz
P.O. Box 2179
D-5420 Lahnstein,
West Germany

COSMETIC AND TOILETRY
FORMULATIONS
Second Edition — Volume 3

by

Ernest W. Flick

This book contains 775 cosmetic and toiletry formulations based on information received from numerous industrial companies and other organizations. This is Volume 3 of the Second Edition of this work; Volume 1 was published in 1989. Volume 2 was published in 1992. There are no duplications in any of these volumes.

The data represent selections from manufacturers' descriptions made at no cost to, nor influence from, the makers or distributors of these materials. Only the most recent formulas have been included. It is believed that all of the trademarked raw materials listed are currently available, which will be of interest to readers concerned with raw material discontinuances.

The cosmetic and toiletry market is projected to increase to close to $2 billion by 1995, thus making the information in the book particularly interesting to anyone considering new products or process variations.

Each formulation in the book is identified by a description of end use. The formulations include the following as available, in the manufacturer's own words: a listing of each raw material contained; the percent by weight of each raw material; suggested formulation procedure; and the formula source, which is the company or organization that supplied the formula.

The formulations in the book are divided into thirteen categories as shown below. In addition, a valuable section on **Trade-Named Raw Materials** is included, which lists trade-names, a brief chemical description, and the supplier's name. The final section contains **Suppliers' Addresses** and will no doubt be a useful tool to the reader.

Section titles are listed below. Parenthetic numbers indicate the number of formulations per topic.

  I. **Antiperspirants and Deodorants (28)**

 II. **Baby Products (21)**

III. **Bath and Shower Products (57)**

 IV. **Beauty Aids (119)**

  V. **Creams (79)**

 VI. **Hair Care Products (118)**

VII. **Insect Repellents (24)**

VIII. **Lotions (62)**

 IX. **Shampoos (120)**

  X. **Shaving Products (4)**

 XI. **Soaps and Hand Cleaners (57)**

XII. **Sun Care Products (68)**

XIII. **Miscellaneous (18)**

XIV. **Trade-Named Raw Materials**

 XV. **Suppliers' Addresses**

ISBN 0-8155-1367-4 (1995)   6" x 9"   464 pages

Other Noyes Publications

COSMETIC AND TOILETRY FORMULATIONS
Second Edition — Volume 2

by

Ernest W. Flick

More than 1900 cosmetic and toiletry formulations are detailed in this volume, based on information received from numerous industrial companies and other organizations. This is Volume 2 of the Second Edition of this popular work, Volume 1 having been published in 1989. No formulations have been repeated.

The data represent selections from manufacturers' descriptions made at no cost to, nor influence from, the makers or distributors of these materials. Only the most recent formulas have been included. It is believed that all of the trademarked raw materials listed are currently available, which will be of interest to readers concerned with raw material discontinuances.

The 1989 market for cosmetic and toiletry raw materials was $1.6 billion. That market is projected to increase to about $1.8 billion by 1994, thus making the information in the book particularly interesting to anyone considering new products or process variations.

Each formulation in the book is identified by a description of end use. The formulations include the following as available, in the manufacturer's own words: a listing of each raw material contained; the percent by weight of each raw material; suggested formulation procedure; and the formula source, which is the company or organization that supplied the formula.

The formulations in the book are divided into fifteen categories as shown below. In addition, a valuable section on **Trade-Named Raw Materials** is included, which lists trade-names, a brief chemical description, and the supplier's name. The final section contains **Suppliers' Addresses** and will no doubt be a useful tool to the reader.

Section titles are listed below. Parenthetic numbers indicate the number of formulations per topic.

1. **Antiperspirants and Deodorants (53)**
2. **Baby Products (52)**
3. **Bath and Shower Products (136)**
4. **Beauty Aids (205)**
5. **Creams (315)**
6. **Fragrances and Perfumes (7)**
7. **Hair Care Products (302)**
8. **Lipsticks (45)**
9. **Lotions (164)**
10. **Shampoos (341)**
11. **Shaving Products (39)**
12. **Soaps (35)**
13. **Sun Care Products (160)**
14. **Toothpastes (32)**
15. **Miscellaneous (83)**
16. **Trade-Named and Other Raw Materials Descriptions**
17. **Suppliers' Addresses**

ISBN 0-8155-1306-2 (1992)   6" x 9"   987 pages

Printed and bound by CPI Group (UK) Ltd, Croydon, CR0 4YY

03/10/2024

01040433-0016